CW01667683

WARSHIP
VOLUME IV

Edited by John Roberts

Conway Maritime Press

Naval Institute Press

WARSHIP Volume IV

Managing Editor Robert Gardiner
Editor John Roberts
Art Editor Mark Stevens

© Conway Maritime Press 1980

All articles published in WARSHIP are strictly copyright and may not be reproduced or stored in a retrieval system without the written consent of the publisher

Published in the UK by
Conway Maritime Press Limited
2 Nelson Road
Greenwich
London SE10 9JB

Published and distributed in the United States of America and Canada by
the Naval Institute Press
Annapolis
Maryland 21402

Library of Congress Catalog Card No 78-55455
UK ISBN 0 85177 205 6
USA ISBN 0-87021-979-0

Manufactured in the United Kingdom

Contents

Editorial (No 13) 1
The Development of Radar in the Royal Navy (1935–45) Part 1 by ALASTAIR MITCHELL 2
HMS ILLUSTRIOUS: The First Armoured Aircraft Carrier by ALAN PAYNE, MINA 15
US Flight-Deck Cruiser Designs by NORMAN FRIEDMAN 24
Centrespread Drawing: Flight-Deck Cruiser Design of 1931 by JOHN ROBERTS 36
Technical Topics: The Preparation of Staff Requirments in the Royal Navy by ANTONY PRESTON 38
The Design and Construction of the Battleship DREADNOUGHT by D K BROWN 39
Warship Pictorial: AUDACE and ARDITO 53
WICHER and BURZA: Big Ships of a Small Navy by PRZEMYSLAW BUDZBON 59
Who was Responsible for the Faults in the TEGETTHOFF class? by RENÉ GREGER 69

Editorial (No 14) 73
German Human Torpedoes and Midget Submarines by ERWIN F SIECHE 74
The Magnificent 7: THE ROYAL SOVEREIGN Class of 1889 Part 1 by R A BURT 86
Historic Hydrofoils of the Royal Navy by D K BROWN 103
Centrespread Drawing: SVERIGE by JOHN ROBERTS 108
German Destroyer Minelaying Operations off the English Coast (1940–41) by PIERRE HERVIEUX 110
The Development of Radar in the Royal Navy (1935–45) Part 2 by ALASTAIR MITCHELL 117
USS WORCESTER by NORMAN FRIEDMAN 135
Book Reviews 143

Editorial (No 15) 145
The Design of HMS INFLEXIBLE by D K BROWN 146
The German Type VII Submarine by ERWIN F SIECHE 153
The JASKOLKA Class minesweepers by MAREK TWARDOWSKI 167
Centrespread drawing: HMS GROVE by N PIPER 180
The ROYAL SOVEREIGN Class of 1889 Part 2 by R A BURT 182
The Origin of the J Class Destroyers by ALAN PAYNE, MINA 192
Amphibious Fire Support by NORMAN FRIEDMAN 199
Warship Pictorial: USS TEXAS 206
A's and A's 213

Editorial (No 16) 217
The 5th Destroyer Flotilla, Autumn 1940 by DONALD L KINDELL 218
USS GALVESTON: The First Talos Guided Missile Cruiser by COMMANDER T BLADES USN (Ret) 226
Amphibious Fire Support: Post War Development by NORMAN FRIEDMAN 234
Technical Topics: Where the Power Goes by D K BROWN 244
The S Class Destroyers 1918–1945 by COMMANDER H StA MALLESON RN (Ret) 251
Centrespread Drawing: HMS SARDONYX, 1919 252
The Original Whalecatcher Escorts by ANTONY PRESTON 264
The CAVOUR and DUILIO Class Battleships by GIORGIO GIORGERINI 267
Warship Pictorial: ML2840 and BRAVE BORDERER 280
Book Reviews 286

The *Jupiter* and the 'V' class destroyer *Vortigern* seen from the *Kelvin* in the autumn of 1940. The *Jupiter* has a camouflage scheme painted up while *Jackal* has a false bow wave (also carried by *Jersey, Kipling* and possibly others of the class) and *Vortigern* is in plain grey. *Jupiter* carries a DF aerial on her masthead while that in *Jackal* has been replaced by the aerials of radar Type 286.
IWM

Editorial

Of the 142 iron and steel battleships which served in the Royal Navy between 1861 and 1960 only one survives. By chance she happens to be the very first of this long line – the broadside ironclad *Warrior* which was launched in December 1860 at the yard of the Thames Iron Works in Blackwall. She was not only the Navy's first iron-built battleship but the world's first 'sea-going, iron, armoured warship'. The precise definition is necessary as she was not the first ironclad, there having been several previous coastal batteries of this type, nor the first sea-going ironclad, that honour belongs to the wooden-built French battleship *La Gloire,* and she was certainly not the first iron warship. Nevertheless, the combination of these features made her a vessel of major importance to warship development in general and to the Royal Navy in particular.

The *Warrior* commissioned in August 1861 for service with the Channel Fleet, with which she remained for the following 11 years, with the exception of a 2 year period for refit and rearmament. After a second major refit, during 1872-75, she became the coastguard ship at Portland and then, in 1881, the RNR drillship on the Clyde. Her active service ended in 1884 when she was placed in reserve, although she still appeared in the 'Navy List' as an armoured cruiser! In 1904 she was attached to *Vernon* as a torpedo depot ship and finally, in 1928 she was transferred to South Wales for use as a jetty by the Naval oil fuel depot at Milford Haven.

In mid 1979 the MoD(N) announced that the hulk of the *Warrior* (officially named *C77* since 1945) was to be handed over to the Maritime Trust for preservation and early in September she arrived at Hartlepool for restoration under the direction of Locomotion Enterprises Ltd. The ship was regularly docked while in service at Milford Haven and her hull is consequently in very good condition. However, most of her original equipment and fittings have gone including most of the rig, bulkwarks, figurehead and, of course, her armament. The work of reconstruction is expected to occupy 5 years, which includes 6 months to clear the vessel of junk and modern additions, but it is intended to open the vessel to the public as soon as it is convenient. When completed it is hoped to provide her with a permament berth at Portsmouth near to HMS *Victory.* Needless to say the project will prove costly – between £4 and £8 million – and would not be possible without substantial public support. Any contributions should be sent to the Maritime Trust (marked 'Warrior Fund') at 16, Ebury St, London, SW1W 0LH.

John Roberts

The *Warrior's* sister, *Black Prince,* as completed

P A Vicary

THE DEVELOPMENT IN THE ROYAL NAVY (1935-45)

THE PREWAR PERIOD

From the early days of radio, observers had noted that metal objects moving close to a transmitter reflected the waves, causing fluttering or distortion of the received signal. Several put forward theories that detection by radio was therefore possible but it was not until the mid 1930s that radio engineering advanced sufficiently to allow the development of a practical system. In the UK interest, initially, was focused on detection of aircraft and following R Watson Watt's first demonstration of the possibilities in February 1935 a development group was set up, initially at Orfordness[1], to design a warning system for the air defence of Britain and to investigate other radar applications.

In the inter-war period all radio experimental and engineering work for the Royal Navy was carried out by the SS (Admiralty Signal School)[2] at Portsmouth. Staff from the School visited Orfordness in July 1935 to acquaint themselves with the new developments and following their visit the Director Signals Department (DSD) directed that work should begin as soon as possible on shipborne radar. Admiralty approval was given in October and it was decided that development would be undertaken at the Signal School rather than staff joining the RAF/Army group at Orfordness, although liaison with that group was to be maintained. Strong links also existed between the Orfordness group and the power-valve section of the SS.

The initial tentative requirements for radar location were; against aircraft: warning of approach at 60 miles range, precise location at 10 miles range, against ships: warning at 10 miles range, precise location at 5 miles range.

The basic problem facing the designers in meeting these requirements was choosing the optimum operating frequency. Radar works on the principle of sending out short pulses of radio energy and measuring the time taken for each pulse to reach the target and return. As the amount of energy reflected is very small, high powers are required from the transmitter to achieve a reasonable range. In 1935 this requirement confined the operating frequencies to the lower end of the spectrum, 30-100 megahertz (MHz), if existing power-output valves were employed. For pure range measurement this did not matter but, if accurate bearing information was desired, the low frequencies created problems in shipborne aerial design. Basically the rule that the higher the frequency the smaller the aerial for a given beamwidth applied, so for shipborne radar, where topweight was of importance, high-frequency working was an attractive proposition. The early development of radar, and its improvement in all round performance, was therefore directly related to the continued improvement in power-valve performance and design.

EARLY DEVELOPMENT

In 1935, the above meant that any practical design had to be a compromise between range and directivity and the choice of frequency made on the basis of the intended function of the sets. Work started in 1936 on several frequencies but most of the effort was concentrated on producing a long-range air-warning set on 75MHz. By the end of the year this set, numbered Type 79X, was installed in the Signal School's tender, the minesweeper *Saltburn,* for trials. For the initial trial in December 1936 a simple aerial system strung between the two masts was used and with this an aircraft flying at 5000ft was tracked from 17 miles, the ship being stationary at the time. Work immediately went ahead for a second trial using rotatable aerials. The system chosen consisted of two parallel dipoles fitted on each mast, one to transmit, the other to receive; rotation was by hand. The trial took place in July 1937 but the results were disappointing, a range of only 8 miles being recorded. Work continued for some time longer on 75MHz but eventually, in March 1938, it was decided to drop the frequency to 43MHz to gain the advantage of the higher-powered valves available at the lower frequency.

[1]Later named TRE (Telecommunications Research Establishment). Moved several times, finally to Malvern. Now Royal Signals and Radar Est (RSRE)
[2]Later named ASE (Admiralty Signal Establishment). Moved to Haslemere 1941 then to Portsdown. Now ASWE (Admiralty Surface Weapons Establishment).

1 The Hunt class minesweeper *Saltburn* in 1939. She was the first RN ship to be fitted with radar.

NMM

2 A view of the midship portion of the battleship *Rodney* in 1939 showing, at the rear of the AA control platform above the bridge, the mast supporting the transmitting aerial of her experimental Type 79 radar set.

NMM

OF RADAR

PART 1 by Alastair Mitchell

Alastair Mitchell served in the Royal Navy during the 1950s during which time he was extensively employed in radar maintenance. He is now working in the commercial electronics industry.

1

2

At the time of the frequency change the whole radar programme was reviewed and on 25 March the order of priority in development was laid down. First was the Type 79, with a hoped for range of 50 miles at 5000ft. Second was a low-angle gunnery set with 1° accuracy at 20,000yds and third was short-range ranging on aircraft up to 5 miles with sufficient accuracy for AA fire control. The latter two requirements were of course the progenitors of the wartime Type 282-285 series although the AA requirement had to wait until the centimetric Type 275 for complete fulfilment.

The frequency change for the Type 79 produced a viable system and in May 1938 the Admiralty directed that two Type 79Y sets were to be fitted in ships of the Home Fleet for trials. The cruiser *Sheffield* was the first to be equipped and in October she reported detection ranges of 53 miles at 10,000ft and 30 miles at 5000ft with a power output of 15–20 kilowatts. The battleship *Rodney* was fitted with the second set by January 1939.

With the success of the Home Fleet trials design work on an improved set, Type 79Z with an output of 70 kilowatts, was begun in June 1939 and installed on the AA cruiser *Curlew* by September. Orders were also placed for thirty more to be built. This final version of the Type 79 was therefore the culmination of nearly four years work and was the only 'production' surveillance radar at sea. It operated on a frequency of 43MHz, ie 7½ metres wavelength, and the transmitter output consisted of pulses with a length variable between 8 and 30 microseconds transmitted 50 times a second. Separate aerials for transmitting and receiving were used, each approximately 11ft high with a 14½ft sweep diameter, consisting of two parallel dipoles with reflectors. Power was fed to the transmitting

1 The *Curlew* in 1937, shortly after conversion to an AA cruiser. She was the Navy's third major warship to be fitted with radar, being fitted with a Type 79Z air-warning set in 1939

Wright and Logan

2 An amusing contrast of early warning system is provided in this photograph of the foremast of the cruiser *Sheffield* in 1940. In the lookout position, visible just above the DCT, a seaman searches the horizon with a pair of binoculars, over his head, at the top of the mast, the receiving aerials of one of the first radar sets to go to sea – Type 79Y

CPL W/2/008

1

2

aerial by twin open-wire feeders via a flexible final section that allowed hand rotation through 400°. The beamwidth was roughly 75° but by using DF methods (ie locating the bearing where the return signal was strongest) a bearing accuracy of 5° was obtained. Bearing resolution with such a wide beamwidth was of course very poor. The receiver was fairly conventional and the echoes were displayed on an 'A' scan with the scale calibrated in miles.

Although the main effort prewar had been directed towards the Type 79, the gunnery requirements had not been ignored by the Signal School. In 1937 a second development group worked on a 1300MHz set using a split-anode magnetron (not to be confused with the resonant-cavity magnetron of 1940), and driven aerials consisting of 6ft diameter parabolic dishes. Ranges of up to 3 miles were recorded but this advanced work appears to have been dropped and, after the March 1938 restatement of priorities, work started on the design of a 600MHz gunnery system. By 1939, using newly developed GEC valves with outputs of 20 kilowatts and recently developed coaxial signal cables, a successful trial was carried out on board the destroyer *Sardonyx.* This design work carried on into the war and led to the Type 282–285 series introduced during 1940–41.

Summing up the RN's efforts prewar, it can hardly be said that spectacular results had been

1

2

1 A very clear view of the Type 282 'Yagi' aerial array mounted above one of the pompom directors of the battleship *Prince of Wales* in 1941. The curved plates, at the backs of the aerials, are the reflectors and the seven short rods mounted in front of them are the director dipoles which served to reduce the beamwidth. The eighth dipole on each aerial, which actually carried the radar signal, is hidden by the reflector. The support bars, and their triangular stays, for the dipoles were constructed of insulating material. Until 1942-43, when T/R switching was introduced, one aerial was used for transmitting and the other for receiving

IWM

2 An onboard photograph of the cruiser *Ajax* taken at New York, Navy yard, on 16 October 1943. Viewed from the port side this shows the back of the bridge and the port HACS Mk III director with the aerials of Type 285 mounted on its roof

USN

achieved but at least a warning set that gave good service throughout the war had been produced and a start made on solving the gunnery radar problems. The only other nations with radar at sea in 1939 were Germany and the United States. In Germany development had started in 1933 and by 1937 the Navy had a gunnery set (Seetakt) at sea for trials. With an output of 7 kilowatts on 375MHz giving a range of 10 miles, it was far advanced on anything in the UK, but after 1937 development slowed down and by 1939 only two ships, including the armoured ship *Graf Spee*, were fitted. An air-surveillance radar on 125MHz (Freya) was also developed by the Navy but was not fitted on ships.

In the United States work had started in 1934 and there too an experimental set was tested at sea in 1937 on the destroyer *Leary*. In December 1938 a development of this set, Type XAF, was fitted on the battleship *New York* for trials. This set worked on 200MHz with a power output of about 15 kilowatts feeding a 17ft rotatable 'mattress' array. Range was similar to the Type 79 but the beamwidth was much narrower at 14° giving a bearing accuracy of 3° and reasonable resolution. The US use of a single aerial for transmission and reception was technically a big advantage and represented a major advance on the then current British practice. From this is can be seen that the two countries were leading Britain technically but the RN was moving faster in getting the sets into operational service, a situation that paralleled that on land.

THE EARLY WAR YEARS – SURVEILLANCE RADAR

With so few ships fitted, the initial impact of radar on fleet operations was virtually nil. However, the new First Lord, Winston Churchill, had been most impressed by a demonstration of a coastal gunnery set and actively pressed for faster development of radar for gunnery at sea. As an initial step the RN acquired an Army AA gunnery set, the GL1, and fitted it in the AA cruiser *Carlisle* late in 1939 for trials, giving it the number Type 280X.

The GL1 had been conceived as an addition to the Army's optical fire control system and it was intended to give warning of the approach of aircraft, a rough bearing for optical alignment and then accurate ranging. For the latter task it was equipped with a precision potentiometer for off-setting the zero and moving the 'A' scan trace along until the target was in the middle of an expanded trace. An accuracy of plus and minus 25yds up to 24,000yds was claimed. The set operated over a frequency band of 54–84MHz with a pulse length of 3 microseconds, 1000 times a second. The Army aerial system consisted of one dipole and a reflector for ranging and two dipoles for bearing information but the Navy modified this.

Trials were conducted off Malta early in 1940 but no results are quoted in official documents. Three more Type 280s were acquired and fitted in the auxiliary AA ships *Alynbank, Springbank* and *Ariguani* but the set was not generally adopted for naval use. Like many hybrids it was probably inferior to the separate, purpose built, surveillance and gunnery sets. The ranging system was however adopted and incorporated into the Type 79, the set then becoming Type 279, and the planned long-range surveillance set Type 281.

The design specification of the Type 281, issued in the autumn of 1939, reflected the desire for a multi-purpose set. Intended

2

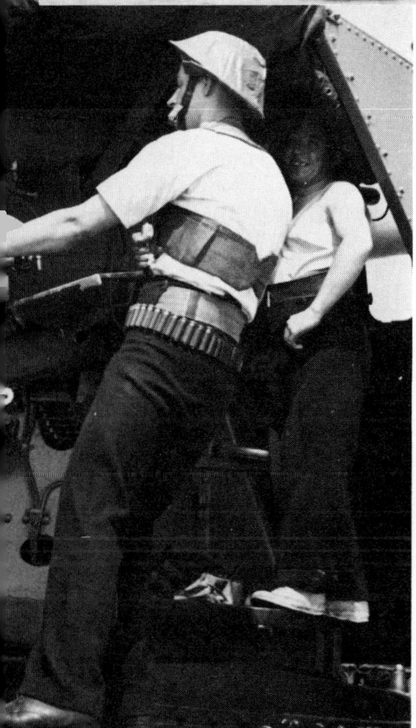

1

1 The foremost 4.7in gun mounting of the destroyer *Whitshed* in 1941. In the background, above the lookout position on the foremast, is the aerial array for Type 286

CPL W/13/002

2 On 17 March 1941 the destroyer *Vanoc,* seen here early in her career, located the submarine *U 100* with her Type 286 radar and initiated the first U-boat kill attributable to radar

Gieves Ltd.

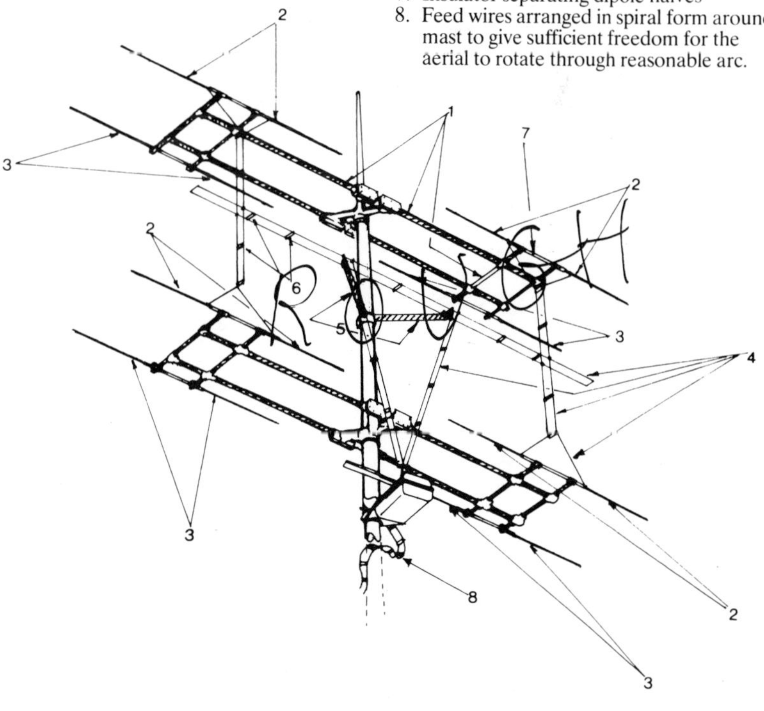

The radar aerial for Type 281 was virtually a doubled version of the Type 279 aerial with the two large horizontal 'H' aerials replaced by four of smaller dimensions.

1. Supporting framework
2. Transmitting dipoles
3. Reflector dipoles
4. Open-wire feed carrying radar signals to transmitting dipoles
5. Support arms for feed wires
6. Insulators between feed wires
7. Insulator separating dipole halves
8. Feed wires arranged in spiral form around mast to give sufficient freedom for the aerial to rotate through reasonable arc.

primarily as a long-range surveillance radar with improved surface range it was also to incorporate facilities for long-range gunnery ranging. In addition to fitting the GL1 ranging unit this involved using two different pulse lengths on the transmitter and most of all providing an aerial system to

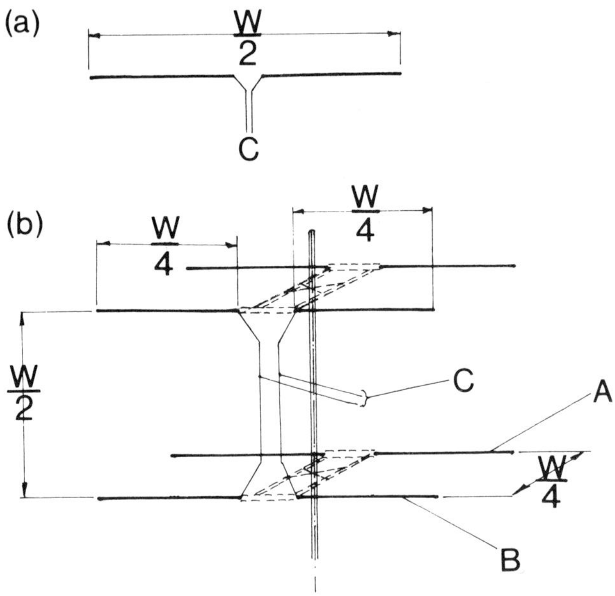

Metric radar aerial dimensions
(a) Radar dipoles were normally half-wave, ie the total length (excluding the insulated gap between the two poles) was W/2, where W= wavelength, and each half was W/4 in length. In the case of Type 279, which operated on 7.5 metres, each dipole was 3.75 meters long and each half of the dipole 1.875 meters long. **(b)** In parallel dipoles, of the type used in 279, and 281, the top and bottom dipoles were a half-wave apart and the transmitting and reflector dipoles were a quarter-wave apart.

A Reflector dipole
B Transmitting dipole
C Twin open-wire feed.

give a bearing accuracy of plus and minus $\frac{1}{2}$°. To achieve this high accuracy, beam-switching of the receiver aerial was adopted. This system involved switching between two dipoles and required a symmetrical aerial array. To keep the new aerial roughly the same size as that of the Type 79 the frequency of the set was doubled to 90MHz. The resulting aerial system consisted of separate transmitting and receiving aerials each of four dipoles with reflectors supported on a rectangular frame. The transmitting aerial had a beamwidth of approx 35° and a higher gain than the Type 79. Rotation was by hand and the aerials had to be stopped to use the beam-switching facility.

The increase in frequency was made possible by the advances made in 1939 in the design of high-power-output valves. By early 1940 powers of up to 1 megawatt were being achieved on 90MHz and, with the valve design stabilised in January 1940, development of the Type 281 went ahead. First fitted in the cruiser *Dido* in September 1940, the Type 281 operated in the 86–94MHz band with a transmitted power of 350 kilowatts, on a 15 microsecond pulse, for long-range warning and a 1 megawatt, 2–3 microsecond pulse, for gunnery ranging. The pulse repetition rate, at 50 pps (pulses per second) was the same as the Type 79. The results in *Dido* were good, with ranges of 60–110 miles being obtained on aircraft and up to 12 miles on surface ships. Performance against low-flying aircraft was better than that of Type 279 but not very satisfactory. The basic reason for this variable performance by low-frequency radars on aircraft is the phenomenon known as 'lobing'. To avoid large reflections from the sea causing 'clutter', radar aerials were normally horizontally polarised. When they radiated the signals took two paths, one in the direct line of sight, the other bouncing off the surface of the sea. When the two signals combined in space, the phase of the reflected signal's varied in relation to the direct one at different angles of elevation, adding together at some and completely cancelling at others. The number and elevation angles of the resultant 'lobes' depended on the operating frequency and the height of the aerial above sea level. For long-range work only the low angles, up to about 3–4° , were of interest so the most important 'lobe' was the bottom one. Equally important were the 'nulls' between the 'lobes' and of course the no-signal area below the bottom 'lobe'. Type 279 and 281 aerials were normally mounted 100–150ft above sea level giving bottom 'lobe' angles in height of approximately 2.6 and 1.3° respectively. For Type 281, this gave an optimum detection height of 10,000ft at 100 miles. The one advantage of the 'lobe' structure was that by careful calibration over a period of time the elevation of the incoming aircraft could be determined by noting the distance at detection and then when the echo faded as it passed through the 'nulls'. Until the introduction of centimetric height-finders this was the only method of determining height.

A second Type 281 was fitted on the battleship *Prince of Wales* in January 1941 and with it, ranging to 22,000yds with an accuracy of plus and minus 25yds was possible. Production deliveries began in February 1941 and the set was widely fitted, a total of 59 being made.

SMALL SHIP RADAR

During 1940 the question arose of fitting some form of radar on ships smaller than cruisers, mainly to detect surfaced U-boats. The problem was one of size and, with the frequencies then obtainable,

A midships view of the destroyer *Cambrian*, while engaged in transferring a patient at sea. On the platform, at the centre of the picture, is a twin 'Hazemeyer' Bofors mounting carring radar Type 282, the aerials of which are mounted on the framework above the guns. A mechanical drive maintained the aerials in-line with the gun barrels at all angles of elevation. Mounted on the arm, extending forward from the mast at the extreme right of the photograph, can be seen the 'egg-timer' aerial of the Type 253 interrogator for Type 291

IWM

putting an aerial high enough to get a reasonable surface range. To meet immediate requirements the Navy turned to the RAF and acquired several airborne ASV Mk II sets. Designed for fitting on Coastal Command aircraft, this set initially worked on 176MHz, later increased to 200MHz, and employed a system of fixed 'Yagi' aerials which scanned an area below and ahead of the aircraft.

For naval use the fixed 'Yagis' were clamped to the foremast with the transmitted aerial pointing forward and the two receiving 'Yagis' angled outwards. As in the aircraft, the receiving aerials were beam-switched to give an indication of bearing and the ship had to be manoeuvred until equal echoes were received showing the target to be dead ahead. Basically the set did not perform well in service due to the inexperience of the crews and the short surface-range obtainable with the aerial at a comparatively low level. In the first U-boat 'kill' attributable to Type 286 the destroyer *Vanoc* detected a submarine at 1 mile range; it was the first time she had detected anything! To partially overcome the problem, fleet destroyers had a modified aerial fitted at the masthead consisting of a rectangular frame supporting three sets of twin parallel 'Yagis', the centre set pointing forwards and the outer pairs again angled outwards for beam-switching. Type 286 was, however, only intended as a stopgap and the ASE began development of a naval version using a new, 100 kilowatt peak, transmitter which was designed to be small enough for fitting in submarines and fleet minesweepers. This ultimately formed the basis of the Type 290/291 small ship's air-warning set.

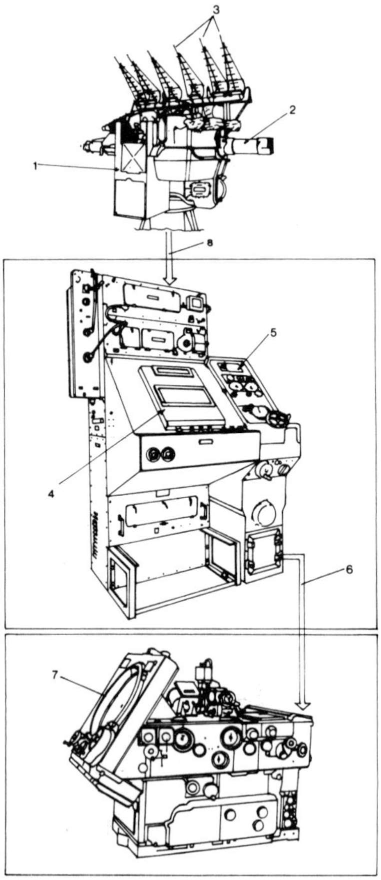

Radar Type 285 as applied to the High Angle Control System (HACS) Mk III
1. HACS director
2. Range finder
3. 'Yagi' aerials
4. Ranging display (panel L12) of radar set
5. Range transmission unit (RTU)
6. Range information transmitted to HACS in high angle control position (HACP)
7. HACS Mk III – a mechanical computor which processed the fire control information for AA weapons.
8. Radar signals to radar office

NB. In its later development, two ranging displays (Panels L24 and L22) together with a spotting tube were fitted in the HACP and the ranging panel in the radar set was removed.

THE INTRODUCTION OF GUNNERY RADAR

The basic difference between surveillance and gunnery radar is the degree of accuracy in range and bearing required. The main criteria for a warning set were long range and good all-round cover and, to achieve this, high powers, contained in long pulses, and rotatable aerials were used. Gunnery radar was not required to range beyond the limits of the ship's armament but an accuracy of a few yards was sought. This meant lower powers but shorter, sharp rising, pulses to give good accuracy and resolution. Initially, range was all that was wanted as the existing optical methods of gunnery control were good on bearing but poor on range.

As mentioned previously, development of a 600MHz system had started in March 1938 and by 1939 transmitter powers of 15–20 kilowatts were achieved. At the end of 1939 the Admiralty stated a requirement for a radar rangefinder for control of close-range AA armaments and it was agreed that the 600MHz system was sufficiently advanced to meet this need. An aerial system consisting of two single 'Yagis' mounted on a pompom director was used, the set being numbered Type 282. Shore trials were carried out in February 1940 and detection at ranges of 5000yds obtained, but the type was not fitted in ships until 1941.

Late in 1939 investigations had begun into using aerial arrays consisting of a parabolic reflector excited by a line of dipoles. These proved successful and a prototype array consisting of an 11ft x $2\frac{1}{2}$ft parabolic cylinder, or trough, with 12 dipoles was fitted to the main armament DCT of the battleship *Nelson* in June 1940. Detection ranges of 30,000yds were obtained but this, unfortunately, proved to be due to abnormal atmospheric conditions and 24,000yds was a more realistic maximum. The system, however, went into production as the Type 284 low-angle gunnery set, the first production model being fitted in the battleship *King George V*. The aerial for this set was increased to 24 dipoles making it 21ft long which was the same length as the director's optical rangefinder. Two aerials were fitted, one for transmitting and the other receiving. Variations to the fit, depending on the type of ship, were $12\frac{1}{2}$ft troughs or a mixture of $12\frac{1}{2}$ft and 21ft troughs. Trials in the *King George V* at the end of 1940 gave 20,000yds detection range on a cruiser and 12,000yds on a destroyer. The beamwidth of the 21ft array was 8° and by finding the bearing of maximum signal an accuracy of $\frac{1}{2}$° was obtained.

The accurate ranging unit of the GL1 was not available in sufficient quantity for Type 284 so the

The low-angle gunnery radar system Type 284 in its later form with displays in the transmitting station (TS)
1. Director control tower (DCT)
2. Radar aerial
3. Ranging display (Panel L12) – normally removed when display panels fitted in TS.
4. Radar signals to displays in TS
5. Ranging display (panel L24)
6. Spotting tube
7. Training tube
8. Radar signals to radar office

NB. In addition to the equipment shown above the TS was fitted with a second ranging display (Panel L18 – part of the Type 273QR) for use as a stand-by or with the after gunnery radar set.

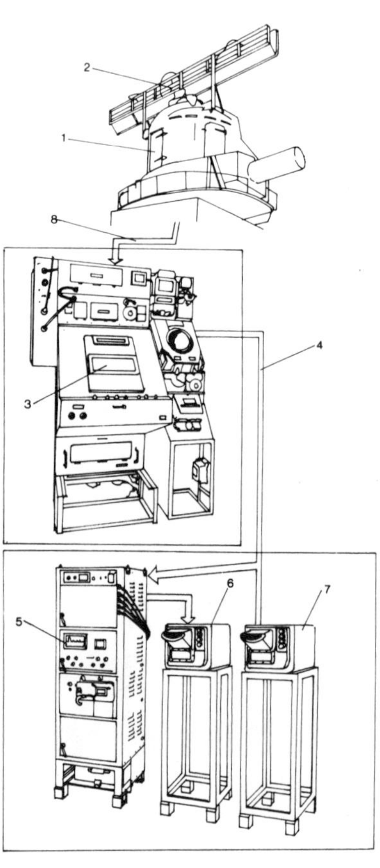

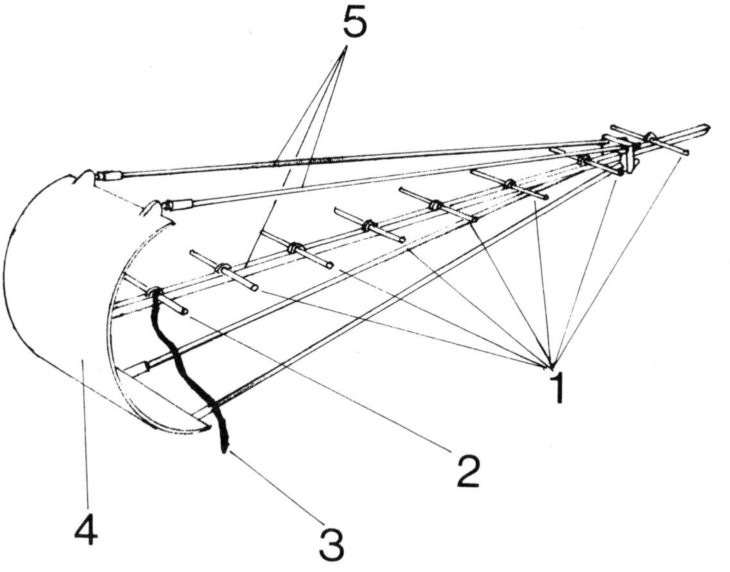

The 'Yagi' aerial array was named after its Japanese inventor but was more popularly known as the 'fishbone' aerial. It was employed for radar Types 282, 283 and 285.
1. 'Passive' director dipoles
2. 'Active' transmitting dipole
3. Feed to active dipole
4. Parabolic reflector
5. Non-metallic supports

ranging equipment developed for the Type 282 was used, giving an accuracy of plus and minus 120yds up to 24,000yds. Normally, for surface action, range-rates (i.e. the rate of change of range) were relatively slow so the operator read off the range by comparison with a calibration trace. It was then passed verbally to a second operator who set it on a 'drum counter' which transmitted the information to the fire-control table in the 'transmitting station' where a range plot was kept.

The third set under development in 1940 was designed to give range information for high-angle gunnery and was basically an extension of Type 282. The same equipment was used but the aerial system was enlarged to six 'Yagi' or 'fishbone' arrays, three for transmitting, three receiving (on some directors there was only room for five so the transmitting side was reduced to two). Numbered Type 285, the set was fitted on the 'Hunt' class destroyer *Southdown* in November 1940 for trials during which detection ranges of up to 18,000yds were recorded on aircraft and 13,000yds on surface ships. Up to 15,000yds range, the accuracy was plus and minus 150yds. Beamwidth was appoximately 18° in the horizontal plane and 43° in the vertical.

Finally the basic Type 282 appeared in 1941 as a rangefinder for pompoms. Intended to combat dive-bombing attacks, it was designed with a simple ranging-unit that was easy to use under stress. It had a range of 5000yds, an accuracy of plus and minus 50yds and the whole range was displayed on a 12in tube. The operator set the range on a dial and transmitted it to the director by a rate-aided range transmission unit which moved a cursor across the face of the tube. With a beamwidth of nearly 40° in both planes, tracking of the aircraft was done optically.

GUNNERY RADAR IMPROVEMENTS

After being in service for a few months it was obvious that the three gunnery systems were in need of improvement. The Type 284 was the most satisfactory, mainly because surface gunnery was the least demanding, but it's range was less than that of a battleship's main armament. In the *Bismarck* action of May 1941 the set had not particularly distinguished itself, although four of the capital ships involved were fitted with it. With Types 282 and 285 the situation was even worse. Type 282 was unreliable due to the high voltage required for the large-tube display, the mechanical cursor caused errors and the wide beamwidth created the situation whereby the range operator had no way of knowing if he was ranging on the target at which the director was pointing if several aircraft were around. The problem with Type 285 was mainly in the interface between the radar and the guns. To achieve a hit did not just require range but a solution to the problem of predicting where the aircraft would be when the shell arrived. The Navy's mechanical computers could not assimilate the range information fast enough, nor set the time fuses fast enough, to produce a correct solution so little improvement in AA fire resulted from fitting radar.

All the above problems, coupled with a demand for a blind-fire capability which involved much higher bearing accuracies, led to a series of modifications during 1941–42. With Type 284, increases in the performance of the output valves led to the transmitted peak-power increasing from 25 to 150 kilowatts, the pulse length at the same time being reduced to 1 microsecond. This increased the range to 30,000yds and reduced the range resolution to 164yds. Beamswitching was applied to the 21ft aerial giving an accuracy of plus and minus 5 minutes of arc, unless there were two targets close together at maximum range. Finally common-aerial working was introduced which eliminated one aerial but, in the case of Type 284, did not materially affect performance. The resulting set was numbered Type 284M and reached its peak in the action between the *Duke of York* and *Scharnhorst* in December 1943.

In the case of Type 285 the initial solution was to introduce barrage-fire. Under this system it was arranged that the guns fired automatically at a preset range between 1000 and 5000yds. Although it was a technical success it was not liked in the fleet as fire had to be withheld until the target was close and then there was only one chance to destroy it. There were still too many variables for the liking of those at the receiving end if it did not work!

Eventually both Type 282 and 285 benefited from the increases in transmitter power and the application of beam-switching, although the latter was only applied in the horizontal plane. In their case additional accuracy, in bearing, was obtained as common-aerial working allowed all the 'Yagi' arrays to be used for transmission and reception.

Ranging accuracies were also improved by the development of a phase-shifter network that allowed 1000yd sections of the range to be displayed on the 'A' scan instead of the full range. A handwheel moved a calibration pip from zero along the scale until the target echo was reached; the range was then read off a dial. An improved accuracy of 4yds in 40,000 was claimed. Some cyclic errors were present in the phase-shifter – these were smoothed out when range rates were measured by hand-following, as in the Type 284, but caused problems when attempts were made to apply auto-follow to the Type 282.

The application of the various modifications produced, in the case of Type 284 and 285, reasonable systems for use in conjunction with optical methods but still insufficient for complete blind-fire capability. In addition, Type 285 still could not follow in elevation and was inherently incapable of doing so without a change of director type. In the case of Type 282, which was widely fitted to pompom directors, beam-switching gave a bearing accuracy of 1½° and it was much improved although remaining essentially a rangefinder. In 1942 the ASE developed an improved tachometric predictor involving the use of Type 282 on a Hazemeyer 40mm Bofors mounting but work stopped in 1943 in favour of the 3 centimetre Type 262. The mounting was fitted on several destroyers during 1944–45 but did not prove successful, being difficult to maintain and, like Type 285, having no radar elevation information.

In 1943 Type 283, with a similar aerial system to Type 282, was introduced to control barrage-fire of the main armament but it contained nothing novel. The whole series remained in service until the end of the war but was gradually supplanted by the centimetric radars.

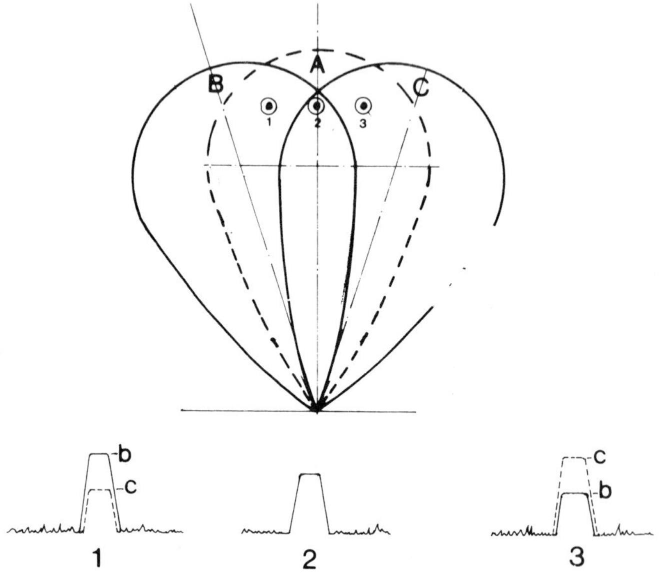

Beam-switching
In the upper, plan, view the transmitting aerial directs maximum power ahead, forming lobe A, while the receiving aerials are switched alternately producing lobes B and C. The results produced on the 'A' scan by the three targets (numbered 1, 2 and 3) are shown below the lobe diagram. Target 1 produces a stronger signal on the B lobe than the C lobe giving correspondingly high and low pips (b and c). When the target is dead ahead the return signals are equal so target 2 produces a single (combined) pip. Target 3 produces the reverse of target 1.

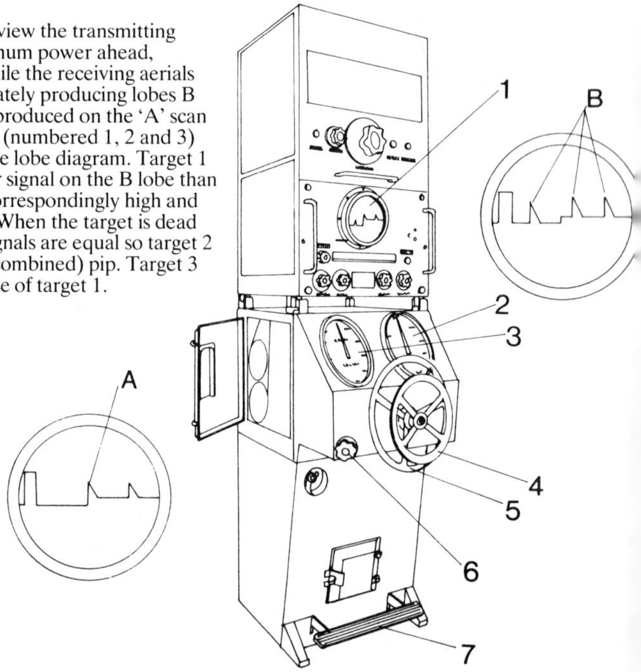

Auto-barrage unit with ranging display (panel L22) mounted above it. Inset are typical 'A' scan displays – on the right a normal trace showing several target echoes (B) none of which are aligned for range (the set-range being the step in the centre of the trace), on the left a normal trace with a target (A) aligned for range.

1. 'A'-scan ranging tube
2. Range dial
3. Rate dial
4. Fine seting range and rate handwheel
5. Course setting range handwheel
6. Rate setting knob
7. Firing pedal

HMS ILLUSTRIOUS

THE FIRST ARMOURED AIRCRAFT CARRIER

by Alan Payne MINA

The *Illustrious* in the Mediterranean in August 1940 accompanied by her immediate predecessor, the *Ark Royal*

CPL W/13/001

In response to the growing menace of Nazi Germany, the 1936 shipbuilding programme provided for the construction of two aircraft carriers, as well as two battleships and seven cruisers – the largest British shipbuilding programme since the First World War. The aircraft carrier design section of the Admiralty was, at the time, swamped with work on the detailed building drawings for the *Ark Royal* of the 1934 Programme. Under these circumstances, it was decided that the new carriers would have to be virtually repeat *Ark Royal's,* with the displacement increased by 1000 tons to bring them up to the new Treaty displacement of 23,000 tons, which would probably involve some minor modifications and additional armour.

The latter would would have been the most feasible way of approaching the design and would have been carried though but for the fact that the Controller of the Navy, and Third Sea Lord, happened to be Sir Reginald Henderson who was determined not to build repeat *Ark Royals* and demanded fully armoured carriers. The idea was that the new 'armoured' carriers should be capable of surviving heavy air attacks, from land based bombers, in the confined waters of the North Sea, Mediterranean and other areas. The design of such a vessel would normally have taken about two years to complete but the implacable Admiral Henderson forced the design through in less than three months and then obtained the approval of the Board of Admiralty for its construction.

It so happened that Admial Henderson was by far the most experienced carrier officer in the Navy – after commanding the carrier *Furious* – he had been appointed the first 'Rear Admiral, Aircraft Carriers' in 1931. In 1936 there were no staff requirements for an armoured carrier, probably because nobody considered it possible to design one on a displacement of 23,000 tons. Nonetheless Henderson was determined to go ahead and never seems to have considered the design might be impossible to obtain within the 23,000 ton limit.

DESIGN

Fortunately the head of the aircraft carrier design section at the Admiralty had considerable experience of carrier design and operation. His name is W A D Forbes and, in a letter to the author, he described the unorthodox design of the first armoured aircraft carrier:

'Reggie hated paperwork and just could not wait for Staff Requirements to be prepared and it was his decision to go ahead without them, which led to the great saving in time in placing orders for the ship. Sir Arthur Johns (then Director of Naval Construction)

Alan Payne lives in Australia and has spent much of his time researching the design history of British warships. Much of his information comes from the records of his father, a constructor with the RCNC from before the First World War, and correspondence with many of the constructors and naval officers that have been concerned with the design of the Royal Navy's warships.

The *Ark Royal* in April 1939; note the high freeboard compared with that of the *Illustrious* and her sisters

Wright and Logan

was sick and absent from the Admiralty (he never returned), his deputy Mr Fred Bryant was fully occupied in running the department, which was exceedingly busy, so Reggie just sent for me and told me he wanted a fully armoured carrier with a flight deck proof against 500lb bombs and armour on the sides of the hangar equal to that used for contemporary cruisers. Most of the rest he left to me.

'For about two months he sent for me each Friday afternoon and after I had told him how I was getting on, he would stride up and down that great room of his at the Admiralty and talk and talk, and if he stopped for breath I would put in a word and off he would go again. Nobody else was present at these talks.

'Several alternative designs had to be worked out before I could get a fully armoured carrier of 23,000 tons and one of the main secrets was that the 3in armour plate which formed the flight deck was used both for protection and longitudinal strength. No backing was used under the armour. The 3in armour, worked structurally with riveted and rabbeted laps and butts, was worked on the flight deck and 4½in on the hangar sides.

'Immediately Reggie was satisfied, an official "sketch design" was submitted to the Board by Mr Bryant and Reggie bulldozed it through and got Board approval in a month, an unprecedented speed. Shortly after this approval of the "sketch design" was given Mr, afterwards Sir, Stanley Goodall was appointed Director of Naval Construction and so the 'Boulding drawings and specification' were prepared under his supervision and he signed them in due course before they were submitted for Board stamp.

'Sir Stanley had not had much previous experience of carriers. It so happened that I was uniquely equipped to undertake the job. I was familiar with all existing carriers and had witnessed many hundreds of landings of all types of aircraft, and moreover I had the recent experience of being the constructor in charge of the *Ark Royal* design. There were few people who had the necessary experience of carriers to do the job; your father was one of the few.' In fairness to Mr Forbes I think he is being over generous, and that the only constructor who could have done such a first class job in the time was the one man with many years experience of carriers and that man was Mr Forbes.

CONSTRUCTION

In November 1926 the Admiralty invited tenders for the construction of the two new carriers, *Illustrious* and *Victorious.* The lowest tender was from Vickers Armstrong, who built both ships, at Barrow and Newcastle-upon-Tyne respectively. The two carriers were ordered on 13 April 1937, the cost per ship was to be £2,395,000 and the time allowed for construction 36 months. *Illustrious* completed on the 20 April 1940, and *Victorious* did not complete until May 1941 due to various delays. Two more carriers of the same class were ordered under the 1937 programme – the *Formidable,* built by Harland and

Wolff, actually completed in November 1940, well head of *Victorious*. The fourth carrier, *Indomitable,* was also late in completing but this was due to her being altered while under construction to accommodate a second hangar deck to carry more aircraft. Lady Henderson launched HMS *Illustrious* on the 5 April 1939 at Barrow but, tragically, Sir Reginald Henderson had been forced to relinquish his office in March due to ill health and died on 2 May. For nearly five years as Controller of the Navy, he had been reponsible for an unprecedented activity in naval rearmament. After years of starvation, the Navy expanded at a higher rate than ever before and it was essential that there should be a continuity, as well as great skill and experience, in its guidance. This Admiral Henderson provided with great distinction at the cost of his life.

Illustrious left Barrow on the 24 April 1940 and ran her acceptance trials in Liverpool Bay on 24 May. When completed her standard displacement proved to be 23,207 tons, only 207 tons above the design figure. In view of the wartime additions made to the ship this is an incredible figure and shows that Mr Forbes always kept something up his sleeve.

DESIGN FEATURES

Because *Illustrious* was armoured she had only one hangar deck and could therefore carry only 36 aircraft, half the number that the *Ark Royal* was designed for, although the latter only carried 60 in service, and with this reduced number of aircraft, she needed only two lifts compared with three in *Ark Royal.* It was only by limiting the design to a single hangar deck that the designer was able to obtain satisfactory stability in the armoured carrier. In the standard condition the metacentric height was 5.35ft and this increased to 8.26ft in the deep condition, which was 28,619 tons. The design of an aircraft carrier presents the naval constructor with some of the most difficult problems encountered in warship design. It also presents great difficulties for the designer of the machinery. It is a particular requirement of the carrier that the machinery be capable of rapid changes of power compared with other warships. Also, she has to be capable of steaming at high speeds for a large proportion of the time and must be highly manoeuvrable. Partly for this latter reason it was decided that, as in the *Ark Royal,* the new design should adopt triple shaft machinery. Twin shafts would have meant far too much power per shaft, although the main reason was

The *Illustrious* in July 1946 shortly after the completion of an extensive postwar refit during which the close-range armament was strengthened, the flight deck lengthened and modified, aircraft fuel stowage increased and the catapult altered to launch heavier aircraft.
MoD

The *Indomitable,* fourth unit of the *Illustrious* class, seen here in 1944, was modified while under construction to incorporate a half-length lower hangar in order to increase the aircraft complement.

USN

Midship section at the engine room of *Illustrious* showing armour thickness.

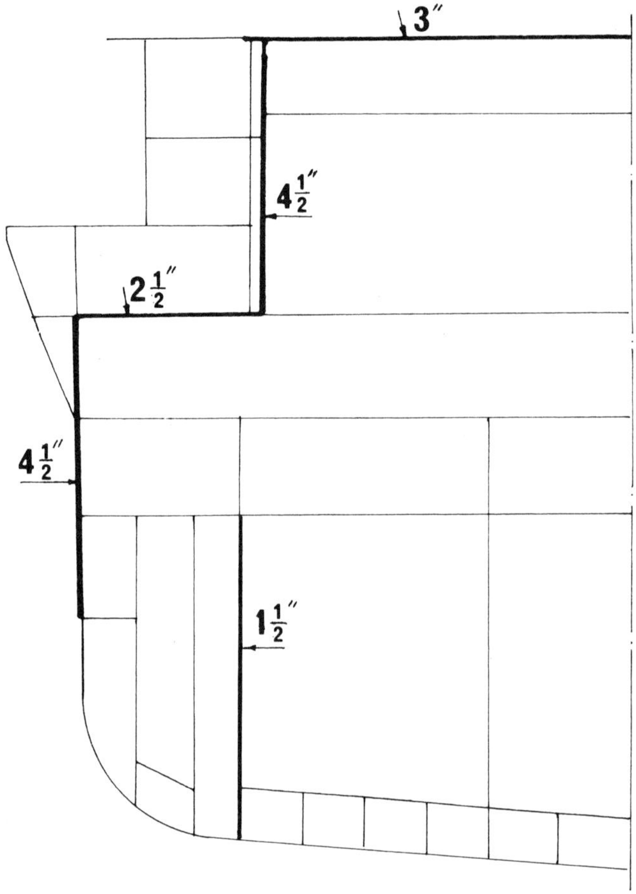

that the triple shaft arrangement was more economical in weight and the ship had to be built to Treaty limits. Another advantage was that the centre screw immediately forward of the rudder improved manoeuvrability. Sir Stanley Goodall had been rather surprised with the excellent speed of 31¾kts obtained by *Ark Royal* on trials with 102,000 shp and did not expect that the *Illustrious* would do so well on 111,000 shp and she was in fact slightly slower. Unfortunately no speed trials were carried out in *Illustrious* at the time of her completion.

It will be noted (see Table) that the *Ark Royal* was a slightly longer ship but with a foot less beam than *Illustrious* and much more freeboard, due to her having two hangar decks, in fact the difference in freeboards is almost exactly equal to the hangar deck height of 16ft. By the end of the war the deep displacement of *Illustrious* had increased by 3000 tons to 31,630 tons and the resultant reduced freeboard accentuated the tendency of this class to ship water over the flight deck when steaming at speed into a head sea. The design really required more freeboard and this would have been provided if Treaty limitation had not prevented it. The Admiralty after a while decided that the price paid for the armour was not worth the severely reduced number of aircraft carried, and that future British carriers should carry the maximum number of aircraft possible, the necessary reduction in the thickness of the armour being accepted. *Indomitable,* the third of the class, was therefore modified while under construction to accommodate a double hangar at the after end. This was achieved by reducing the thickness of the hangar armour to 1½in. The last two carriers of the class, designed in 1938, were fitted with double hangars and reduced hangar protection, these were the *Implacable* and *Indefatigable.*

THE BOYD COMMITTEE

The actual weight of armour in the *Illustrious* was 4,941 tons, just double the weight of the machinery and 21 per cent of the standard displacement. Some experts considered that a modified *Ark Royal* with more armour might have given a better design. The

An aerial view of *Illustrious* during the Coronation Review at Spithead in 1953

Real Photographs

other side of the picture is given by Admiral Sir Denis Boyd, the first captain of *Illustrious*. 'The history of the *Illustrious* design, as far as I am concerned, was that I formed a committee of the brightest naval officers at the Admiralty to study the Navy of the future', wrote Admiral Boyd, 'we were in a fury because we thought we were building the Navy of the past instead of the Navy of the future. Our chief target was the out of date battleships, the first of which was the *Prince of Wales* [sic]. Our idea was that the carrier was the prime fighting ship. In all this the constructors were most sympathetic and helpful. But then we came up against a snag. Who suggested an armoured carrier I cannot remember, what I am certain is that we took advice on the menace against which the carrier had to be protected. The RAF assured us that a 500lb bomb was the largest we would be liable to meet in the next ten years. This was in 1935. So a carrier was designed to keep out a 500lb bomb. As no one ever threw such a small bomb at us, we were well off the mark. All the rest is wisdom after the event.'

It would appear that the Boyd Committee was an informal one, which kept no records and made no submissions to the Board of Admiralty. But the Boyd Committee were right in one main issue. At the outbreak of war in 1939 Britain had nine battleships building or authorised but only six carriers, two of which were not completed until 1944. If those six

2

1

1 The *Illustrious* in Scottish waters in the late 1940s

Ralston

2 The last two ships of the *Illustrious* group, *Implacable* (pictured here in May 1949) and *Indefatigable,* were fitted with full-length lower hangars which increased their aircraft stowage to 72, compared with 36 in *Illustrious*

Wright & Logan

carriers or even more had been completed by the end of 1941, the war at sea might have taken a very different direction.

ACTION DAMAGE

In January 1941 *Illustrious* survived some of the heaviest damage inflicted on any Allied warship at sea. It has generally been assumed that it was the armoured flight deck that saved the ship but Admiral Boyd did not entirely agree. 'The only bomb that hit the armour went right through it and exploded on the floor of the hangar', he wrote. 'Now, had there been no armour, that bomb would probably have exploded in the engine room and this could well have been disastrous. All the other bombs struck outside the area which had been armoured. Three in the after lift well, one right forward, which was touched-off by striking something solid. It exploded outside the ship and blew about two hundred holes in our side creating quite a fire.

'As usual there is no easy answer to your question. The design of the *Illustrious* shook the world as the words "an armoured carrier" had a magic of their own, but as you see it was not quite true. When she was in the design stage I asked if I could be considered for commanding her. That I did is my lasting pride.'

Admiral Boyd was clearly disappointed with the design of the armoured carrier when subjected to very heavy bombs; they were in fact double the weight the RAF had advised as the maximum to be expected back in 1935. 'If large bombs had to be kept out it was obvious that more armour would be required', Admiral Boyd wrote. 'This could not be put high up without making for instability and therefore the armour must be on the floor of the hangar, the flight deck having say 1½in just to touch off the fuze of the bomb'.

The final and supreme test of the armoured carriers did not come until April and May 1945 when three were hit by Japanese Kamikaze attacks. All three were back in action in a few hours. Admiral Sir Philip Vian commanded the carriers of the British Pacific Fleet and his comments sum up the merits of the armoured carrier. 'The armoured flight deck, which was a feature of British fleet carriers, paid a dividend on this occasion. In spite of the direct hit, *Indefatigable* was able to operate aircraft again within a few hours. American carriers similarly struck were invariably forced to return to a fully equipped Navy Yard for repairs. This led to discussion as to the relative virtues of British and American designs. The British carriers, with their armoured deck, high safety precautions and closed hangars, could stow fewer aircraft than their American counterparts. On the other hand events were to show that had our carriers not had armoured decks, Task Force 57 would have been reduced to negligible proportions.' Admiral Henderson's conception of the armoured carrier had indeed paid dividends.

Particulars of Illustrious and Ark Royal

	Illustrious	*Ark Royal*
Length (oa):	743ft 6in	800ft
Length (wl):	710ft	*721ft 6in*
Beam (wl):	95ft 9in	94ft 9in
Standard displacement:	23,000 tons	22,000 tons
Freeboard to flight deck:	43ft 4in	59ft 11in
SHP:	111,000	102,000
Speed (deep):	30½kts	30¾kts
Oil Fuel:	4850 tons	4400 tons
Armament:	16–4.5in HA/LA, 6–eight barrelled pompoms	16–4.5in HA/LA, 4–eight barrelled pompoms

US Flight-Deck Cruiser Designs

by Norman Friedman

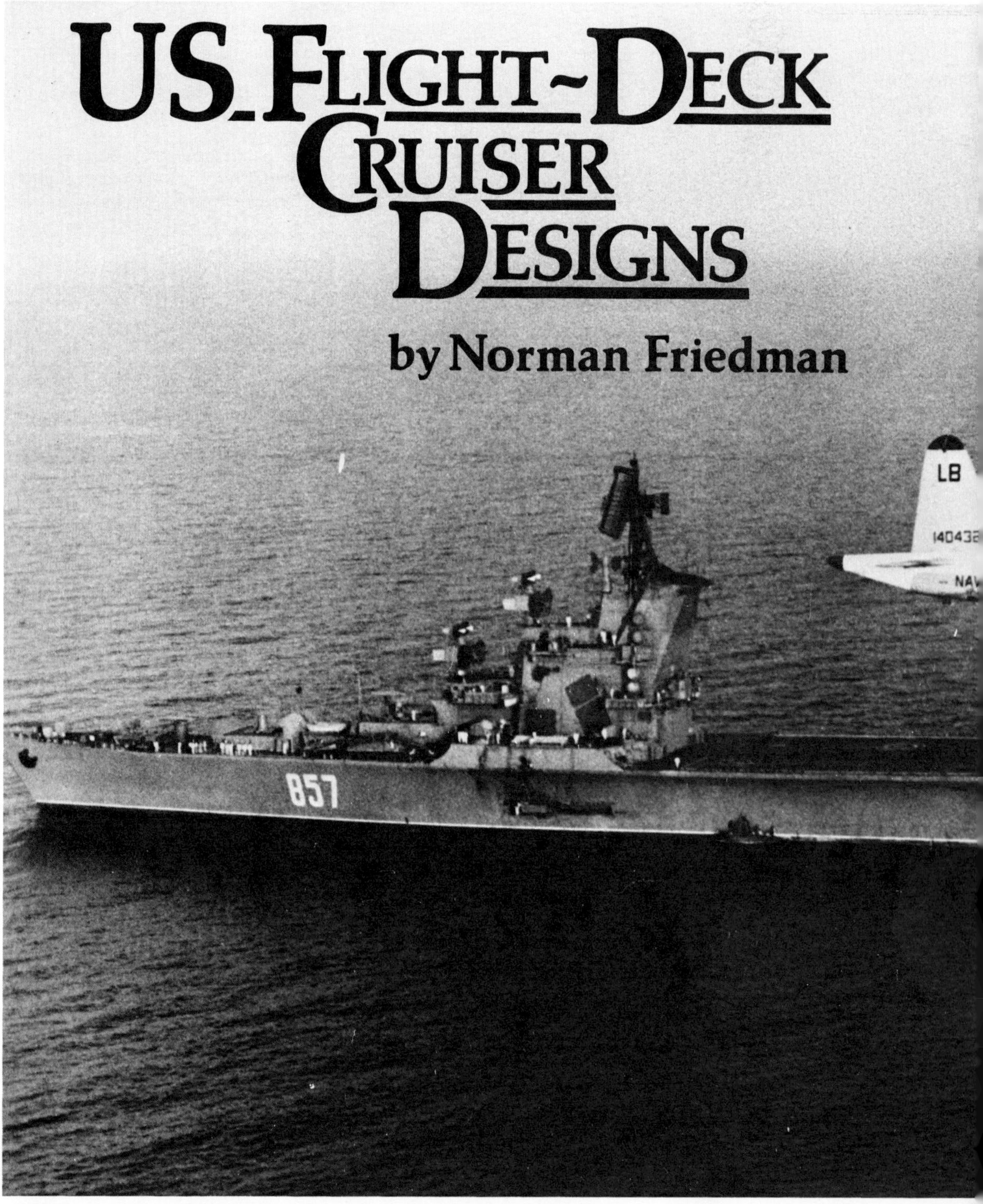

Norman Friedman is a physicist employed by the Hudson Institute in New York. He is the author of several monographs and books including *USS Yorktown* (in the 'Ships Data' series), *Battleship Design and Development* and *Modern Warship Design and Development* for Conway Maritime. He is currently completing the third in this series, on aircraft carrier design.

The Soviet helicopter carrier *Moskva* off the coast of Turkey in the Mediterranean on 5 November 1968. The aircraft is a P2 Neptune of the US 6th Fleet

United Press

Treaties, like other legislation, often engender strange and wonderful creations, such as the German 'pocket battleship', the 'treaty' cruiser and the US *Erie* class gunboats. Another such, which came close to fruition but did not quite materialise, was the US flight-deck cruiser, a hybrid cruiser-carrier invented in the course of negotiations for the 1930 London Treaty. It was very nearly built as cruiser No 39 but the idea seemed to disappear when a conventional cruiser, *Quincy,* was built instead. It was revived in 1938-40 but finally fell victim to a combination of increased carrier construction and progress in aircraft design which made the hybrid's short flight-deck impractical.

The US concept was by no means equivalent to that represented by the Japanese hydrid cruisers and the converted battleships *Ise* and *Hyuga.* In these ships, the flat deck aft was intended merely to simplify the handling of aircraft, which were launched by catapult and recovered, if at all, from the water. All the US designs, on the other hand, called for a flying-off/landing deck which was to replace much of the superstructure and at least one turret of a conventional cruiser. The closest any foreign navy came to this concept was HMS *Vindictive* which, before her large refit in the mid-1920s, was more carrier than cruiser. However, she had been converted to a carrier at a time when carriers were badly needed, and indeed was most closely analogous to the later US *Independence* class. She may have inspired a pair of proposals presented by the Preliminary Design Section of the Bureau of Construction and Repair (C and R) in the course of the study which eventually led to the *Pensacola* class: they called for retention of two or three 8in gun turrets and the erection of a long flight deck amidships, with 250ft for landing-on, 164ft for taking-off, and an elevator and funnels in between. In the three turret scheme, one of the three was mounted under the flight deck, the other two at the ends; in the alternative scheme one of the two turrets was a quadruple. These designs did not originate in the Bureau of Aeronautics and do not appear to have gained the support of that Bureau; the General Board selected a far more conventional design.

In 1925 the US Navy had only one carrier in operation, the experimental *Langley:* by 1930 she had been joined by the immense *Lexington* and *Saratoga,* and a fourth carrier, the *Ranger,* was well advanced. The General Board, and indeed the Service as a whole, was convinced of the potential of naval aviation and in fact the US Navy led the world in what seemed to be the most effective aerial anti-ship weapon, the dive bomber. However, there was one great problem: at the Washington Conference the United States had agreed to a limit on total carrier tonnage which would soon curtail new carrier construction. At the London Conference of 1930 similar limits were placed on cruisers and the question soon arose of whether cruisers might carry flight decks.

The latter appears to have been the idea of Rear Admiral W A Moffett, Chief of the Bureau of

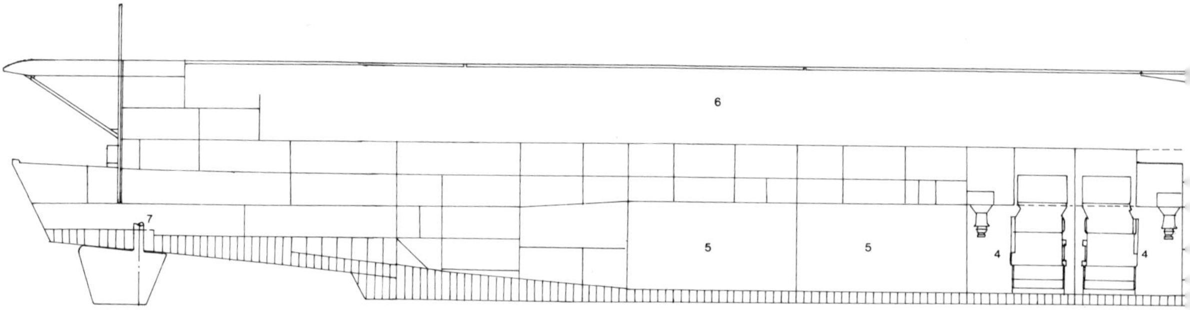

A

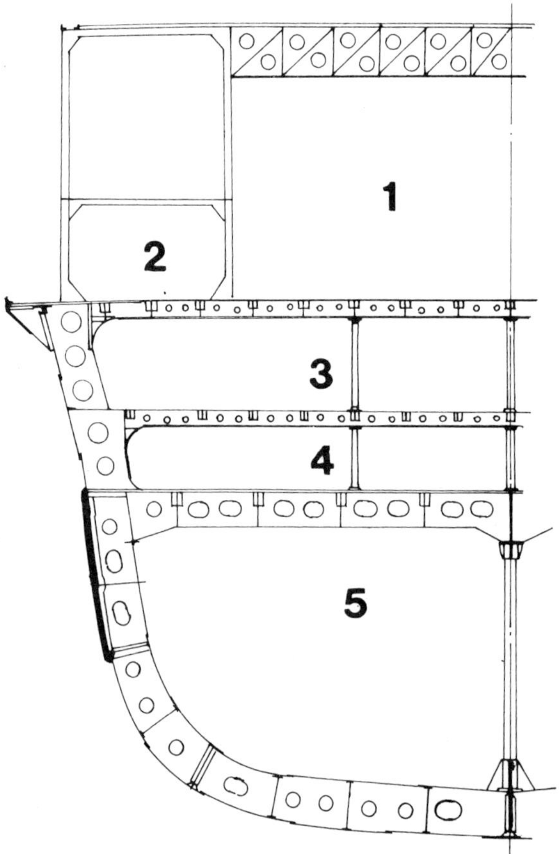

B

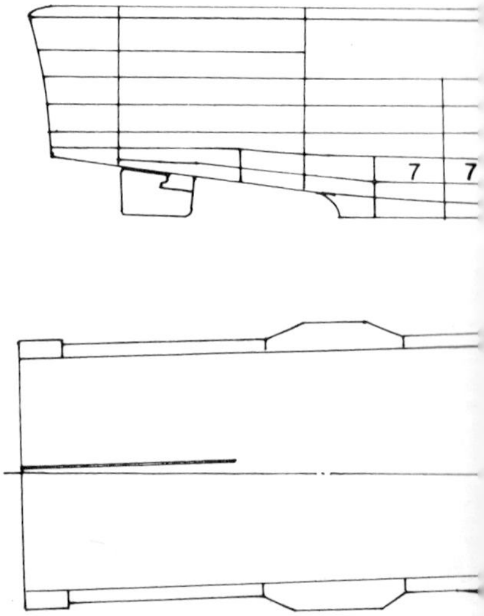

C

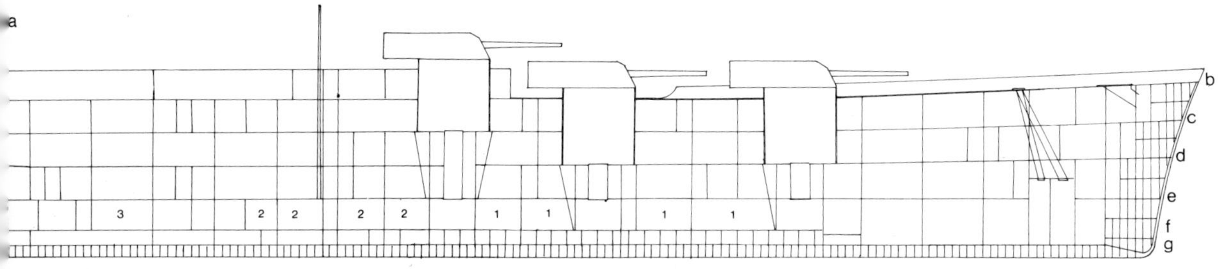

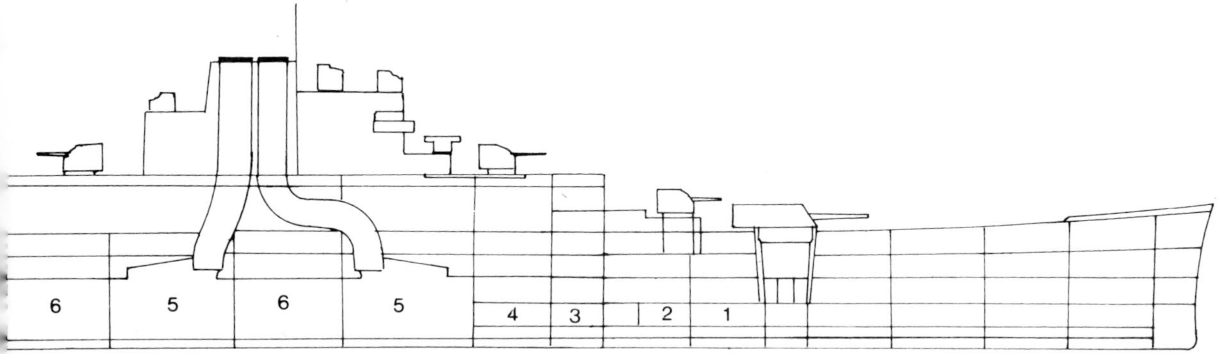

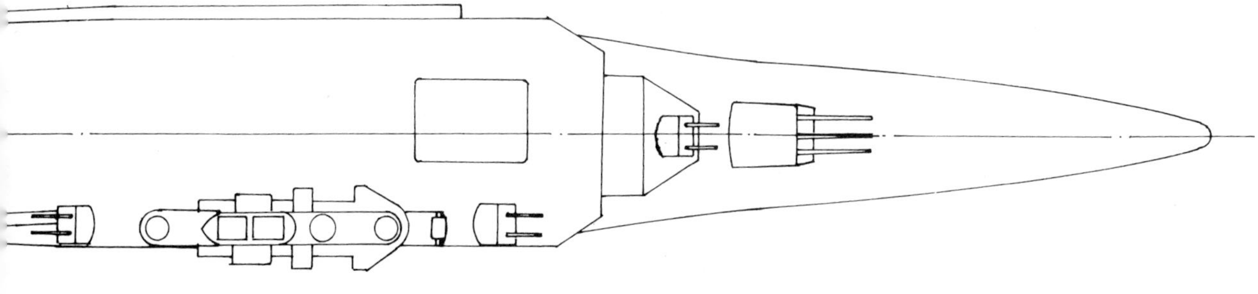

A. Longitudinal section of the C and R flight-deck cruiser design of 1931.
1. 6in magazines
2. Gasoline tanks
3. 5in magazines
4. Boiler room
6. Hangar
7. Steering gear compartment
a. Flight deck
b. Upper deck
c. main deck
d. 2nd deck
e. 1st platform
f. 2nd platform
g. Hold

B. Midships section of the C and R flight-deck cruiser design of 1931.
1. Hangar
2. Crews wash room
3. Crews mess
4. Aircraft stores
5. Engine room

C. The proposed flight-deck cruiser CF1 of December 1939.
1. 8in magazine
2. 5in magazine
3. Bomb room
4. Plotting room
5. Boiler room
6. Engine room
7. Gasoline stowage
8. Hangar

Aeronatics, who, in March 1930, argued that 'I do not think there can be any question whatever that she would be properly charged to cruiser tonnage, because an aircraft carrier under the Washington Treaty is defined as ". . . a vessel of war with a displacement in excess of 10,000 tons, standard, designed for the specific and exclusive purpose of carrying aircraft . . ." The proposed ship is not designed for the specific and exclusive purpose of carrying aircraft and if she carried nine 6in guns and a battery of anti-aircraft guns certainly her primary purpose is guns and not aircraft. However, it makes it all the more important that the United States insist that all combatant vessels between 2000 and 10,000 tons, and not exempt, be charged to cruiser tonnage, or that all combatant vessels between 2000 and 10,000 tons, and not exempt, will be charged to either the cruiser or the aircraft carrier tonnage, at the discretion of the nation concerned.'

The US delegation to the Conference included a strong design team which was able to test some of the design implications of proposed qualitative limits; it produced sketches of flight-deck cruisers and even of flight-deck battlecruisers. In the argument over 8in versus 6in cruisers this team had shown that, on 10,000 tons, a cruiser with four triple 6in mountings could be as well protected as the then standard *New Orleans* class. It appeared, too, that a flight-deck cruiser could be built to mount three triple 6in, to make 31 knots (90,000 shp and to accommodate at least 20 aircraft, with a clear run of 234ft on the flight-deck. Moffett saw such a ship as perhaps 'a connecting link between the present light cruiser fitted for the accommodation and handling of from four to eight planes and some future type of vessel in which the handling of aircraft may be considered as important from the standpoints of offence and defence as the main battery guns.

PRELIMINARY DESIGNS

At first the idea was quite popular; for example, the Senate Naval Committee decided that the next cruiser should be a flight-deck cruiser, unless the Navy should decide otherwise – as it ultimately did. Preliminary Design actually exceeded Moffett's hopes, providing 24 aircraft (which it assumed would be equally split between fighters and dive bombers), three 6in gun turrets, six 5in AA guns, a 332ft flight deck (including elevator at forward end) and a speed of 32.5 knots with 80,000 shp (this reduction in power was a consequence of a considerable growth in length, from 560 to 637ft on the waterline). All of the aircraft could be spotted on deck leaving 130ft which would suffice for them to fly off with a 30kt wind over deck, although the dive bombers would have to sacrifice some fuel. However, landing-on would require that only twelve occupy the forward end of the flight deck.

There were, to be sure, sacrifices as compared to the pure cruiser. The hull could be protected only against 6in shellfire (immune zone, at a target angle of 60 degrees, was 10 to 20,000yds) and even then was described as somewhat spotty. Worse, in the original design gasoline tanks were between the outer shell and the magazines; in the subsequent General Board Hearing the Bureau of Ordnance argued for a location much further inboard, far better protected. This redesign cost Aeronautics the hinged smoke stacks that it wanted but Construction and Repair felt that they would be impractical in any case – as turned out to be the case with the *Ranger,* then under construction. All of the 6in guns had to be mounted forward, when Moffett had wanted one, or even two, turrets aft; in the 1931 design there were, at best, only two 5in/25 guns on the fantail, which could not be counted as AA in view of the limits imposed on their elevation by the overhang of the flight deck; these were in addition to six 5in/25 AA alongside the superstructure forward.

One point brought out at a General Board Hearing in July was that if, in fact, the flight-deck cruiser was able to use her own aircraft for self-defence she might well dispense with the 5in AA guns; the design was so tight that any major weight saving was very welcome, and there were problems associated with the location of the 5in director. Lighter anti-aircraft weapons would, however, remain important as a counter to dive bombing. Captain (later Vice Admiral) R K Turner of the Bureau of Aeronautics pointed out that in fact carrier-borne fighters would not be able to function as interceptors; they would not be able to climb to counter high level bombers in the absence of early warning; such warning might be available in a fleet with pickets spread out well away from fleet centre but not in the case of a solitary cruiser, for which the fighters would function mainly as escorts for the cruiser-borne dive bombers. Turner therefore wanted to retain the heavy 5in guns, which would be essential to break up level bombing attacks. It was even suggested that improvements in bomb-sights might make level bombing of ships quite practical; after all, only the US Navy was even interested in dive bombing.

WAR COLLEGE ANALYSIS

In the 1930s the Naval War College functioned as the Navy's 'think tank', testing new weapons and tactics in standard Pacific (Orange) war scenarios; it gamed the proposed flight-deck cruiser in a variety of typical engagements. Necessary assumptions inlcuded that the cruiser could operate its aircraft in any sea state in which a fleet carrier could and that a 500 lb bomb hit from a dive bomber would be equivalent to a 14in shell hit (terms in which the War College's tables of ship survivability were calibrated). The first assumption, at least, rather favoured the cruiser, which in practice would probably have rolled badly – as did the *Ranger,* which was about the same size but had not been completed at the time of the study. Flight-deck cruisers with 8in protection and 12 aircraft, and 6in protection and 24 aircraft were considered in combat against 8in cruisers protected against 6in (eg, *Northampton*) and 8in (eg, *New Orleans*) fire. Furthermore, it was assumed that AA fire would reduce the effect of dive bombing by about 30 per cent – an arbitrary figure that the War College agreed might well be incorrect. One interesting feature of the ship-on-ship engagements was that the flight-deck cruiser had to defend herself against the catapult aircraft of the opposing cruisers; it was tacitly assumed that battle would be joined after the two ships made visual contact, without aircraft scouting. Both gun cruisers were sunk in these battles. One question was whether a rough sea would interfere more with 8in gunfire or with aircraft operation. The pure cruisers won only under conditions of low visibility, as might be expected.

The War College reported that 'When the CLV (flight-deck cruiser) has planes in the air or has freedom of movement sufficient to launch planes, a bombing attack by the CLV's planes will decide the action. The exact value of such an air attack is dependent upon the problematical relative values of dive-bombing versus modern anti-aircraft fire but, from the best information obtainable, it is believed that the damage done would be sufficient to cripple the CL hopelessly at least before the CLV would be seriously damaged. If the CLV has its planes on deck warmed up and ready to launch, weather permitting they can be launched before the CL can inflict sufficient damage to prevent it, at 24,000 yards range or over, under any condition of relative bearing, or at ranges down to 18,000yds if the CLV is to windward or to leeward in a light breeze. If the CLV is to leeward at 18,000yds in a strong breeze, such that the ship must turn, rearrange its planes and launch them over the stern, a chance shell hit might prevent the launching but, according to War College rules, the planes would be launched even in this case before sufficient damage is received to prevent it. If the CLV is brought under effective fire at about 14,000yds or less when planes are still on deck, they probably could not be launched. The presence of planes in the hangars of a CLV and gasoline containers on deck and in the hangars probably increases the fire hazard as compared to a CL, but under the present College rules this cannot be evaluated.'

A second series of tests pitted CLs and CLVs against a destroyer attack on the battle line; during a day engagement the CLV was decidedly superior, thanks to the rapid fire on its 6in guns; the same held at night. As for the more general question of screening, the War College observed that, 'Screening against aircraft has assumed such importance that means for its efficient accomplishment must be devised. These means are still in the experimental stage but obviously they involve the use of aircraft projected into screening areas or along screening lines. The use of a considerable number of aircraft for this purpose at once presents for solution the problem of the most practicable method(s) of carrying aircraft in the fleet . . . A limited number of planes have been placed on the cruisers – the screening vessels themselves. It has been found that the limited number of such planes in cruisers must be supported or supplemented in the practice of screening by others from the fleet carrier(s). The

Originally classified as a through-deck cruiser the *Invincible,* seen here fitting out, is now referred to as an anti-submarine command cruiser – a more accurate definition. She differs in concept from the vessels discussed in this article in that her aircraft are her only offensive armament, which places her in the aircraft carrier rather than the cruiser/carrier category.

Vickers Ltd

employment of fleet air-force aircraft for other than what may be considered their legitimate function has led, among other reasons, to the concept of the CLV type of cruiser with its own flying-off deck to accommodate a considerable number of its own planes, rather than the limited number hitherto provided.

However, it was clear that the CLV would present a larger target than most cruisers, due to its hangar and flight deck, and that it would probably lose its carrier feature early in a gun engagement. The War College argued that even then it would retain its cruiser features, whereas a carrier detached for fleet air screening would have no value at all after the destruction of its flight deck. In distant screening, the CLV might be able to overcome even the new German 'pocket battleship', which in 1931 was creating a considerable sensation. The same could not be said of any conventional cruiser. CLV advantages were even more evident in such traditional cruiser functions

as scouting, convoy escort against surface raiders (which might be detected and dealt with well out of range of their guns) and commerce raiding.

Although the General Board was not entirely enthusiastic, characteristics were written and a design (which accompanies this article) drawn. However, it was never translated into a contract, due to the Depression which caused the suspension of US Naval construction. When construction resumed, the new programme included two 20,000 ton carriers and a specialised flight-deck cruiser was far less attractive.

Even so, work on flight-deck cruiser designs continued through 1932. It appears to have been suspended in 1933, as the General Board and C and R were then occupied with the 6in cruiser design which became the *Brooklyn* Class. However, the question was revived early in 1934 when the War Plans Division of the Office of the Chief of Naval Operations observed that a flight-deck cruiser, which it called a CF, would be handicapped by the fact that its carrier capability might well be destroyed by shellfire but that it would still be quite useful in the distant scouting and screening operations characteristic of a Pacific War, and also in trade warfare in a Red (ie, versus Great Britain – at that time the standard Atlantic war) war. At least one should be built – 'and it should conform to the spirit of the London Treaty in every aspect – which I do not consider the present design to do.' This meant that the 1931 design devoted more than half its length to air operations; it was not a carrier only in that it was not exclusively devoted to aircraft. The CF should, rather, devote a limited length to its aircraft.

C and R proposed a new design based on the *Brooklyn* but with the two after turrets eliminated in favour of a flight deck; two of the forward turrets were moved to port to balance the weight of an island superstructure; the flight deck was also moved to port for balance. The result was not far from that of 1931 and the General Board was sufficiently unexcited for the project to be halted soon after it had been begun. However, by May C and R had an alternative approach: it eliminated No 3 turret of the *Brooklyn* and installed a 200ft landing-only deck with a catapult for launching at its forward end. This solved the problem of astern fire but greatly reduced the appeal of the ship as a whole, since there was now very little aviation capability left and aircraft handling would be difficult at best. Even the President got into the act; he suggested a portable after extension for the flight deck, to be fitted over or around the after guns during flight operations only. This soon proved impractical and by mid June the project had died. A Bureau of Aeronautics suggestion at this time that all cruisers incorporate at least limited flight decks was turned down by the General Board and no more was heard of the CF for about four years.

FLIGHT-DECK CRUISERS RECONSIDERED

Then, in November 1938, the subject was raised by the Director of War Plans. It had just been decided to continue to build 20,000 ton carriers rather than the 10,000 ton type favoured by some, largely on the basis of economies of scale. War Plans pointed out, however that:

'In considering our future carrier needs, the number of planes *carried* by a single carrier is not the ultimate desideratum, which should be the amount of *useful plane employment* in a given projected sphere. The War Plans Division believes that it is necessary – *first* to consider what needs or functions new carriers must provide, *next* decide the necessary characteristics to accomplish these functions *and then* determine the size which will best, and most economically, embody these characteristics. Global and ship limitations in tonnage may bring about an incorrect approach to the problem resulting in ships too small to provide the characteristics required or ships too large to accomplish their functions . . .

'Most of our thoughts concerning the addition of ships to the Navy have been guided by the consideration of adding to the power of a concentrated fleet. We have given thought largely to the ability to *establish command of the sea* rather than to the *exercise* of that command. By the exercise of command is meant the interruption of enemy trade routes, and sea communications, while protecting our own trade routes and sea communications (including air routes overseas).

'In the World War England assumed command of the sea immediately at the beginning of the war. But even with her outstanding strategic position and her outstanding Navy, including units designed (for that time) to *exercise* command of the sea, she encountered difficulties. A large number of warships were required for many weeks to bring to an end the *Emden* raids. Similar difficulties with other raiders were encountered from time to time. Submarine warfare forced the construction of thousands of special ships and the development of special methods of protection and anti-submarine warfare, in order to *exercise* the benefits of the command of the sea – unforeseeable then but a definite problem for us now. The complementary factor, in the exercise of sea command, the denial of use of sea routes by Germany, was comparatively simple for Great Britain. For us, the denial of the use of sea routes by any prospective enemy would be a major problem . . .

'Whatever the situation, the problem of *exercising* command of the sea will be a difficult one. Our superior strength may force, and will probably cause, the other fellow to spread out. We must be ready to spread out to get him . . . When we consider the problem of running down *Emdens,* we immediately recognize the advantage, or necessity, of covering an extended area with a widespread mobile based air scouting force supported by gun power adequate to the need.

'The current (US) conception of a carrier has centred largely on one capable of carrying 72 planes, with gun power, speed and protection modified to fit a varying tonnage that has been somewhat artificially selected to fit into a limiting global tonnage . . . The 'large' carrier . . . concentrates on *air striking power.*

'On a scouting mission there are a number of ways to spread the planes out from the carrier . . . It is obvious that if only two planes are used at a time, one on each side, the maximum scouting effectiveness of the plane will be obtained. If two planes are used on each side, a small amount of duplication scouting occurs in obtaining the scouting distance. With an increase in the number of planes on each side this duplication increases and the added area of search of each individual plane is reduced.

'If it is agreed that the average carrier scouting plane can fly 480 miles on an average flight, and if high visibility is assumed, plane radius of visibility of 10 miles, then with 12 planes on each side (in the sweep method) the outer plane will be able merely to proceed to the outer lane 240 miles and return to the carrier. For a single carrier the area that can be swept with reasonable effectiveness does not increase as the number of planes . .

'Generally speaking the area that can be patrolled effectively or searched by a carrier with 72 planes exceeds very little that which can be searched by a carrier with 36 planes. If it were found necessary to patrol an area 500 miles by 1280 miles, it could be done better with three small carriers carrying a reduced number of planes than with two large carriers. If large carriers only were available, three would be used for the task just the same.

'For scouting we need numbers o carriers in order to spread them out we need to reduce their vulnerability (as regards the numbe of planes put out by a damaged flight deck or by underwater damage); we need some increased gun power in order to reduce their dependence on close and constant cruiser protection. We can reduce the number of planes per carrier and increase the efficiency of search operations for a given number of planes, by making the primary function of these planes scouting. They should be able to carry bombs

The Italian helicopter cruiser *Vittorio Veneto* leaving Cagliari harbour on 20 February 1978

Dr S Cioglia

FLIGHT-DECK CRUISERS – COMPARATIVE DATA

Scheme	Moffett	C & R	Scheme II	CF–1	CF–2
Date	5/30	6/31	5/34	12/39	2/40
Standard displacement (tons)	10,000	10,000	10,000	12,000	12,200
Trial displacement (tons)	–	11,580	11,607	14,250	14,600
Length (wl)	560ft	637ft	600ft	640ft	640ft
Beam (wl)	66ft	62.5ft	62.5ft	67ft	67ft
Draught	–	19.4ft	19.7ft	21.7ft	22ft
Speed	31kts	32.5kts	32.5kts	33kts	33kts
SHP	90,000	80,000	–	100,000	100,000
Endurance at 15kts	–	10,000nm	–	10,000nm	10,000nm
Armament Main:	9x6in	9x6in	12x6in	3x8in	6x6in
Secondary:	8x5in	8x5in	8x5in	8x5in	4x5in, 16x1.1in
AAMG:	–	–	–	8x0.5in	8x0.5in
Aircraft:	20	24	12	24–36	36
Armour Belt:	4in–2.5in	4.5in	4in	5.6in	4in
Deck:	2.25in–2in	1.4in	1.6in	2.25in**	2.5in
Turret faces:	6in	5in	5.25in	8in	6.5in
Flight deck length	234ft*	332ft	200ft	420ft	390ft

*clear of elevator
**Has, in addition, 1in main deck

and have sufficient manoeuvrability and gun equipment to act in the capacity of fighters, but the primary requisite is scouting. 18 to 24 of such planes will provide a very efficient unit for this purpose.

'A carrier constructed for the general purpose of operating with cruisers in *exercising control of the seas* will fill other functions or needs also . . .'

War Plans envisaged it as a ship of roughly 10,000 tons, with a flight deck long enough for landings (about 400ft) and steady enough to permit flight operations under most sea conditions; planes would be launched by flight-deck catapult and there would be 18 to 24 (preferably 24) of them, of a special scout bomber type. They would be housed in a hangar proof against shellfire. The 'scout carrier' would be armed with eight of the new 6in DP guns, which the Bureau of Ordnance was then promoting as the main battery of the next *(Cleveland)* light cruiser, mounted in twin gunhouses. 'Gun power should be sufficient so that two such carriers will be at least equal to one 10,000 ton light cruiser in gun power, and sufficient in fighting power to handle converted merchantmen. Anti-aircraft protection to be provided by dual-purpose 6in guns and 1.1in and other machine guns.'

War Plans also considered its scouting carrier an ideal ASW weapon, as well as an extremely valuable fleet scout. The Division did not want it to be built out of a limited carrier or cruiser tonnage but rather out of some fresh authorisation, as the new scout would fill deficiencies in fleet capability previously not addressed.

The result was clearly a flight-deck cruiser, not very different from proposals made in 1934, although its rationale was very different – partly because in 1936 the total tonnage (global) restrictions on various categories of warships had been abandoned, although the authorisation Acts under which new US ships were built still carried such limits on total Navy tonnage; hence the call for a new category.

NEW DESIGNS

In November 1938 there was considerable interest within the Navy in new types of warships; clearly a large naval war was in prospect and with it the end of all Treaty limits. Thus, at about the same time, there was an ongoing study of super-cruisers which ultimately resulted in the building of the *Alaskas*. In October 1939 the General Board passed along tentative Characteristics for a 'Flying-Deck Cruiser' to the technical Bureaus; they called for a displacement of about 12,000 tons, a flight deck long enough to carry a complement of 18 scout bombers (dive bombers) spotted for take-off, and with two flush-deck catapults at its forward end, a speed of 35 knots, 70,000 gallons of gasoline and a battery of twelve 5in/38 guns in twin mountings, four quadruple 1.1in mountings and a pair of triple torpedo-tubes. Side armour would protect magazines, machinery, gasoline and steering gear against 6in cruiser fire at 10,000yds range (60 degree target angle). These characteristics were only rough, and the following January they were amended to show a triple 6in turret at each end, with two twin 5in/38 mountings on the flight deck, one forward and one aft; the torpedo tubes were eliminated, the air group increased to 36 scout bombers, and the flight deck length set at 420ft, with only a single catapult. Although reliance (tacit) on catapults in the October requirements seems unremarkable, at this time catapulting was a relative rarity in US carrier operations: indeed, it would not become common until after the escort carriers, which had flight decks about the size of the one contemplated for the cruiser, found that they could not operate effectively *without* their catapults. Thus the January shift, with an enlarged air group, shows a desire to be able to fly off most or all of the air group under its own power, with the catapult as a backup against insufficient wind over the deck. In fact, however, the Bureau of Aeronautics would observe a month later that a typical dive bomber, the SB2U-2, required a run of at least 337ft in a 25 knot wind; moreover, 'the run listed is an optimistic value obtained under favourable conditions. For average carrier operation, it is usual to reserve from 35 to 50 per cent additional deck length for safe launching.' Maximum ship speed was reduced in these later characteristics to 33 knots, and bomb stowage increased with the increased air group.

C & R produced two sketch designs, CF-1 (one triple 8in/55) and CF-2 (two triple 6in); CF-1 had eight 5in/38, no 1.1in guns, and eight 0.50 calibre machine guns; the use of only a single turret made possible the inclusion of a 420ft flight deck on her 640ft (roughly *Baltimore* class) hull. CF-2 showed only that the tentative characteristics could not quite be realised on about 12,000 tons; her flight deck, for example, was only 390ft long, and even then corners had to be cut fore and aft, with a hinged flap forward for use when the forward turret was not in use. The Bureau of Aeronautics was less than pleased: CF-2 could launch a scout bomber, but only if no other aircraft were on deck. Elevator operation during flight operations would impose a delay of 2.5 to 3 minutes between launches (1.25 min between catapult launches, for which up to 28 planes could be spotted on the flight deck). Similarly, each plane would have to be struck below as it landed, to present a clear landing deck; and upon take-off, the 5in gun mounts and island would present an unusual hazard extending all the way along the take-off run. The Bureau felt, indeed, that it would take unusual piloting skill to avoid these obstacles on either take-off or landing. Matters would be far simpler for a spotting plane but such an aircraft would be far less effective than the scout bomber and indeed would largely negate the value of the flight-deck cruiser. Admiral Towers, the Chief of the Bureau of Aeronautics, concluded his gloomy assessment: 'No promising rearrangements are apparent for improving the carrier features of the design without seriously curtailing existing cruiser characteristics.'

By this time the flight-deck cruiser had already lost its last battle. In the Autumn of 1939, with a large US cruiser programme in prospect, the General Board polled major commanders and some senior officers as to its composition: alternatives were CA1, a *Baltimore* class 8in gun cruiser; CA2, which became the *Alaska* class 12in gun cruiser; the *Cleveland* class 6in gun cruiser; CLD, a kind of super *Atlanta;* and the flight-deck cruiser, CF. Of eight organisations polled, only War Plans supported its child, the CF; the next least loved, the CLD, received three 'yesses'. Everyone but War Plans saw the possibility of so many real carriers that there would be no need for hybrids. Thus the flight-deck cruiser moved from a means of increasing the number of aircraft at sea by cutting Treaty corners (1930-34) to a means of exercising sea control (1939-40) to the limbo of ships so compromised as to be ineffective for any task. Superficially, it seems akin to the Soviet *Kiev* and to other ships armed with a combination of VTOL aircraft and surface-ship weapons, such as (potentially) the Italian *Vittorio Veneto* group. Moreover, every once in a while such ships are perceived as preferable to a big carrier whose only weapon is the air group she carries.

MODERN COMPARISONS

These analogies are not quite perfect, but they are interesting. As in the War College's studies of 1931, it is necessary to ask just what the aircraft add to those surface ship weapons. The air capability of a ship like the *Kiev* is both limited and extremely expensive in terms of ship steel and (unlike 1931) electronics, eg for bad-weather landing and for aircraft navigation. The *Moskva* design demonstrates that far less is needed if only ASW helicopters are to be carried: *Kiev* adds both surface-to-surface missiles (SSN-12) *and* VSTOL attack aircraft, where a naive person might suspect that one *or* the other would suffice, at least for surface attack. The planes add a very limited air-to-air capability; it appears, for example, that they have no intercept radars. One wonders whether Admiral Gorshkov carried some Soviet equivalent of the 1931 War College study when he sold his carrier-cruiser programme to the Politburo, presumably in about 1965.

There has, of course, been considerable speculation. Those VTOL fighters would not be too effective against modern jet attack-aircraft but, on the other hand, they would be extremely effective against slow ASW patrol aircraft. The SSN-12 can really rearrange the average Western destroyer or frigate; in effect the *Kiev* is a means of achieving sea control over a limited area of sea and the air above it (not to mention the sea below, given her helicopters and other ASW weapons) – as long as opposition is limited to Western surface warships, submarines, and ASW aircraft – a condition which may well obtain in the strategic Greenland-Iceland-UK Gap in a future war between the Soviets and NATO. The concept is not too far from that espoused by War Plans in 1938, although the details are of course very different, and the motivation, to secure the holding area for Soviet ballistic missile submarines north of the Gap, is by no means analogous to any of the tasks envisaged for the US Navy before Second World War. After all, the US and Soviet Navies have quite different tasks even now.

There is one final irony. The flight-deck cruiser originated in the US Navy as a means of extending an already well established carrier aviation organisation. The Soviet Navy, although it considered carrier construction at least before the Second World War and probably after the war as well, by the 1960s claimed publicly that carriers were obsolete. The *Kiev* may just possibly be the forerunner of true carriers, in which case she will have performed just the opposite of the role claimed, originally, for the flight-deck cruiser.

As for the scout carrier idea, although it never found much favour in the US Navy, it is not too far from some British ideas, in which small carriers (without much gun armament) were actually to have replaced cruisers after the Second World War. There is also some affinity between the scout carrier idea and the ASW carrier group concept of the 1950s and 1960s, and in some ways the current Through-Deck Cruiser is not too far from this idea. Perhaps the moral is that technology, but not basic strategy (the latter the province of War Plans), changes over the years.

The Russian aircraft carrier *Minsk,* like the British *Invincible,* is intended primarily as an AS vessel. This photograph was taken by an RAF Nimrod in the Eatern Mediterranean, south of Crete

MoD

THE US FLIGHT-DECK CRUISER DESIGN of 1931

Drawn by John Roberts

The accompanying drawing by John Roberts is based on the official general arrangement plans prepared by the US Bureau of Construction and Repair for the flight-deck cruiser proposal of 1931 – described in this issue by Norman Friedman. Particulars of the vessel were as follows:

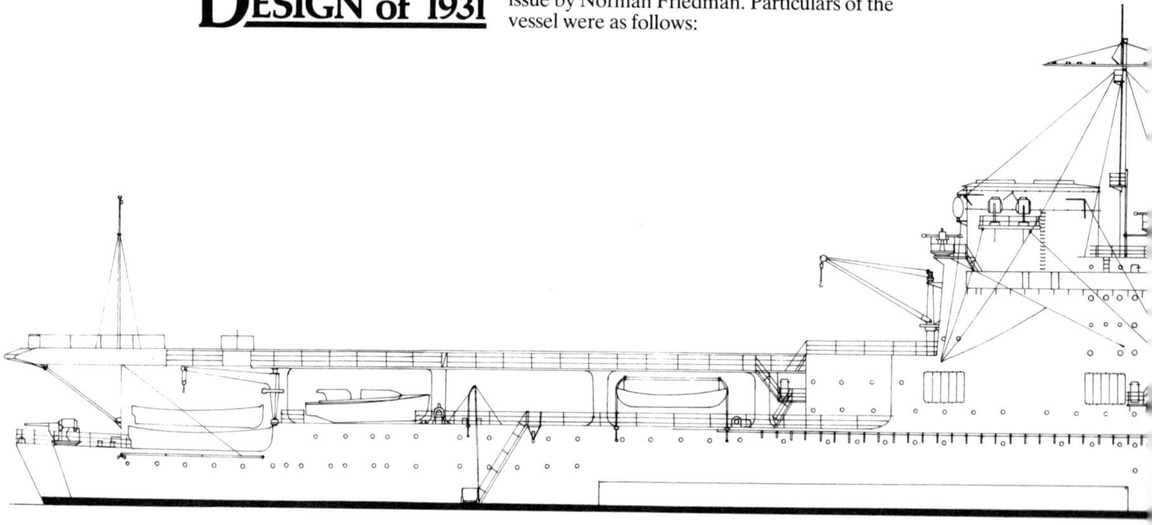

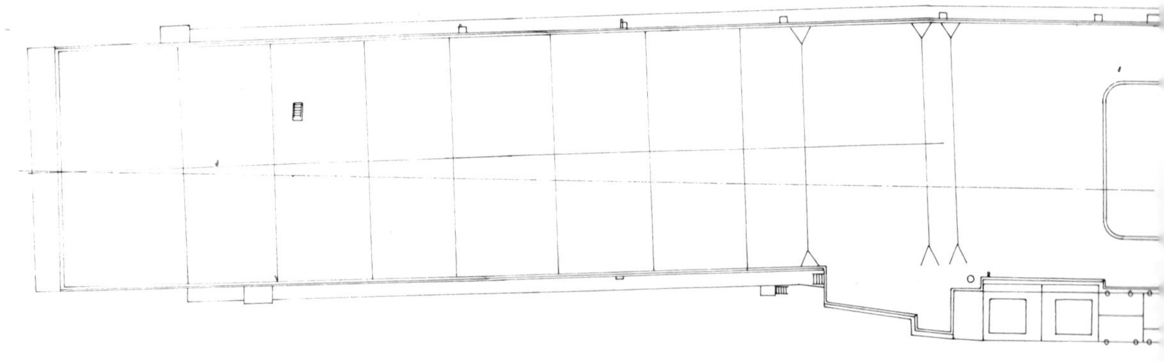

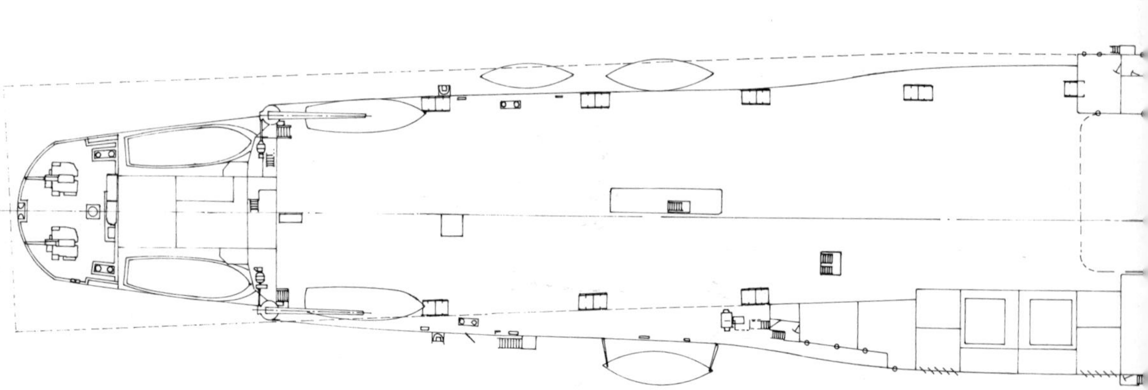

Displacement: 10,000 tons standard
Length: 627ft (wl) 654ft 7½in (oa)
Beam: 62ft 6in(wl)
Draught: 19ft 5in
Armament: 9-6in (3x3), 8-5in (8x1), 24 aircraft
Armour: 4.5in belt, 1.4in 2nd deck, 5in turret faces
Machinery: 4 shaft, geared steam turbines, 80,000 shp = 32.5kts, 6 boilers
Boats: 2-40ft motor launches, 2-36ft motor launches, 2-35ft motor boats, 2-30ft whaleboats, 1-26ft motor whaleboat
Searchlights: 4-36in

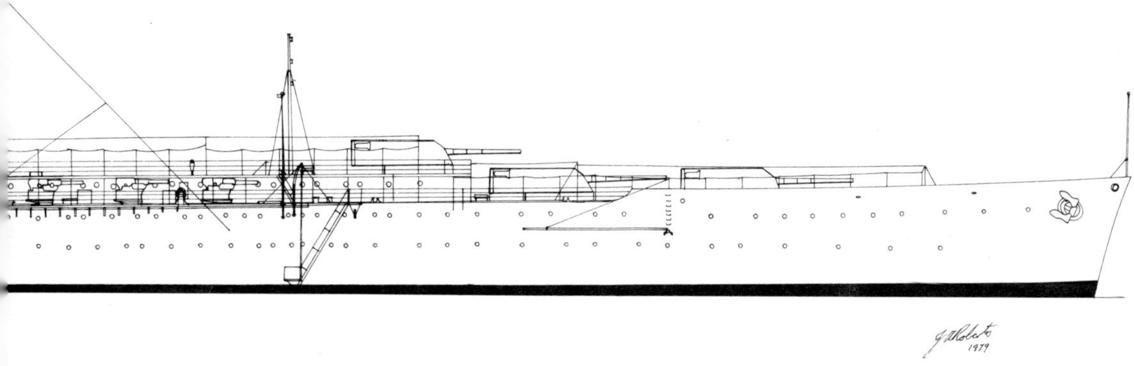

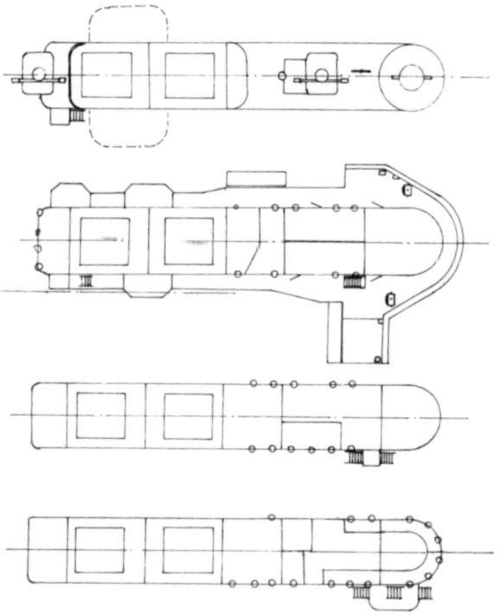

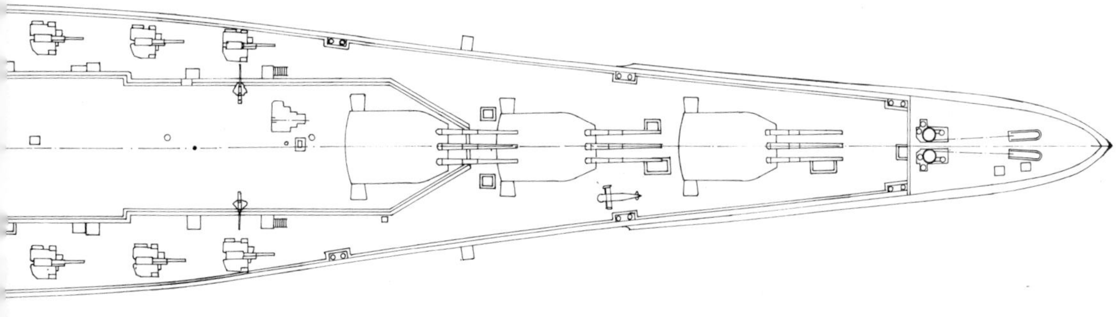

THE PREPARATION OF STAFF REQUIREMENTS IN THE ROYAL NAVY by Antony Preston

It is often said that for every bad warship designed there are ten bad Staff Requirements. In the light of such severe criticism it is opportune to discuss how a Staff Requirement is generated in the Royal Navy today.

The Royal Navy's principles for manning and complementing are laid down in two documents, the Naval Staff Code (NAVSTARCODE) and 'Complementing Instructions for HM Ships' (BR 4017). The NAVSTARCODE covers the whole process, within the Ministry of Defence, for raising Naval Staff Targets, first in outline and then clearly listing the options (which are still left open). They are then progressed into Naval Staff Requirements with sufficient detail to lead to a specification for building a ship. The Naval Staff Target looks at the following points:

1. Function – a description of the ship and its purpose.
2. Concept of operation – the method of operating, employment with other forces and an estimation of the number of ships required.
3. Major characteristics – speed, endurance, armament etc, and the order of complement, likely cost and the timetable to bring the class into service.

When the Naval Staff Requirement comes to be drawn up, further questions are asked:

1. The estimated number of equipments to be fitted.
2. Numbers of maintenance personnel by rank and rating.
3. Numbers of operating personnel by rank and rating.
4. The number of man-years required for specialised training.
5. The number of maintenance and instructional staff which will be required.
6. The total manpower bill.

The Naval Staff Requirement will then be considered for each section, ie, general, armament, navigation and radar, communications, domestic, medical and dental, education and training, and even air conditioning arrangements. These decisions are considered in the light of the number of people requiring support and the volume involved.

The Complementing Instructions ask the following questions:

1. What personnel are needed to meet the normal tasks?
2. What personnel are needed for operating weapons and associated systems when the ship is at Action Stations?
3. What is the shore support policy, the refit and upkeep cycle, usage and availability? The total maintenance task must be stated and what proportion of it is to be undertaken by the ship's company.
4. What requirements must normally be met simultaneously?
5. What number of men is borne purely for communal cleaning and ship husbandry?

Clearly the staff directorates dealing with weaponry strive for the best environment for their missile systems, gun mountings, helicopters etc, but many of these systems are interdependent. Thus, a decision to install a helicopter leads in turn to the provision of a hangar (necessary to increase availability) and then to a sophisticated surveillance radar with the ability to provide helicopter-control (necessary to allow the best use to be made of the helicopter). The use of the helicopter will in turn depend on the main role of the ship (is she intended mainly for ASW, surface strike or merely surveillance?). In former days it was said that a warship was like a rookery, in which human beings found whatever perch they could, after all the weaponry had been packed in, but in today's navy it is not so easy. The RN's Director-General (DG) Ships issues guidelines in the form of a specification defining the areas to be allowed for each rating, as well as allowances for his 'support services' in the form of washing and eating facilities, etc. There are also Naval Magazine and Explosive Regulations (NMERs) and shock standards to be allowed for.

The criticisms which arise from such a rigid framework inevitably come, for the most part, from shipbuilders, anxious to tailor existing designs for export, but also from serving naval personnel, impressed both by British-built export designs and by Soviet warships. It is often forgotten that British (and for that matter American) warships are designed in the light of an enormous amount of practical experience gained not only in the Second World War but in Korea and off Vietnam. Some critics have, for example, pooh-poohed DG Ships prejudice against all-aluminium superstructures for warships, but the disaster to the USS *Belknap* in November 1975, when the entire superstructure melted during a fire, and a later fire in HMS *Amazon,* when aluminium ladderways melted, show that it can be wise to hold back. Another problem faced by the official designer is that he is often designing from someone else's drawing board, in the sense that he does not know how big the new surveillance radar will be, or how much topweight he must allow for a new fire-control system, and so on. It is sometimes said that the new *Broadsword* class frigates took 14 years to design, but what actually happened is that the Naval Staff Requirement changed radically during this period, to meet changes in the threat. Thus a ship envisaged as a replacement for the *Leander* class frigate, and meant to serve primarily as an escort, changed into a deep-water nuclear submarine-hunter, with a much more powerful sonar set, two helicopters in place of one, and a new missile defence against surface-to-surface missiles. In the intervening period habitability standards were also greatly improved. Small wonder that the size doubled and that the design process was protracted.

THE DESIGN AND CONSTRUCTION OF THE BATTLESHIP DREADNOUGHT

by D K Brown

David Brown is a naval architect who combines employment in modern warship design with a strong personal interest in the history of warships, particularly during the nineteenth century. He is currently working on a detailed history of the Royal Corps of Naval Constructors.

2

3

4

The *Dreadnought* was the logical outcome of many years of steady development in ship and engine design, hastened by rapid developments in gunnery and fired by the imagination and drive of Admiral Sir John Fisher, then First Sea Lord. Like all 'Jackie' Fisher's great projects the *Dreadnought* showed brilliant thinking, mixed with muddled detail and personal favouritism. There are many threads in the skein of her development and amongst these were Fisher's relationships with two leading members of the Royal Corps of Naval Constructors (RCNC). Fisher was Captain of the *Inflexible* when the young Phillip Watts came on board to try the anti-rolling tanks which he had designed. This friendship endured and contributed to the choice of Watts as successor to Sir William White as Director of Naval Construction (DNC) in 1902.

As C in C, Mediterranean Fleet in 1900, Fisher became friendly with the Chief Constructor at Malta, W H Gard, and discussed ideas for battleships with him. Fisher and Gard came together again at Portsmouth in 1902 when the former became the C in C. Fisher had also been discussing his big ship with Sir Andrew Noble of Armstrongs who proposed Design A with an armament of sixteen 10in guns. Captain Bacon's arguments for the 12in gun demolished this design and Gard was asked to produce Design B with eight 12in guns.

Other people's ideas hd been moving in the same direction. Deadman and Narbeth, the

This series of photographs was taken at Portsmouth Dockyard while the *Dreadnought* was under construction and illustrate extremely well the rapidity with which the ship was built. They are reproduced with the permission of the General Manager, HM Dockyard Portsmouth.

1 Monday 2 October 1905 – the day the *Dreadnought* was laid down!

2 Wednesday 4 October 1906 – the middle-deck beams and inner-bottom plates are starting to be fitted

3 Saturday 7 October 1905 – the majority of the middle-deck beams are in position

4 Saturday 14 October 1905 – most of the middle-deck protective plating ($1\frac{3}{4}$in flat, $2\frac{3}{4}$in slope) is in place

5 The main-deck beams on 25 November 1905

6 30 December 1905 – the hull almost complete

7 10 February 1906 – the day of launch, just over 4 months after laying down

8 2 June 1906 – the superstructure begins to take shape

9 11 August 1906 – main armament and foremast in place

constructors who designed the *King Edward VII* class, had proposed a heavier armament for these ships, particularly the later vessels of the class, and had then pressed for an all 12in gun armament in the *Lord Nelson* of 1904 (Ref 1). This line of thought was reinforced in an article by the Italian naval architect Cuniberti, in *Jane's Fighting Ships* for 1903. Cuniberti advocated the fast, all big gun ship and, though his article had wide popular appeal, it probably had little influence on naval or constructor thinking since they had already reach the same conclusion. In the following year the US Navy Board prepared the designs for the USS *Michigan* and *South Carolina* which were armed with eight 12in guns.

5

7

8

When Fisher was appointed as First Sea Lord in 1904 he brought Gard to the Admiralty as Assistant Director of Naval Construction (ADNC) and, by Admiralty Letter of 22 December, he set up the Committee on Designs. Though this committee was to discuss battlecruisers, destroyers and submarines as well as battleships, it is only the latter which will be discussed here.

THE COMMITTEE ON DESIGNS

The Committee had a brilliant membership, selected from Fisher's protégés, consisiting of: Rear Admiral Prince Louis of Battenberg (Director of Naval Intelligence), Engineer Rear Admiral Sir John Durston (Engineer in Chief), Rear Admiral A L Winslow (CO Torpedo and Submarine Flotillas), Captain H B Jackson (Controller of the Navy), Captain J R Jellicoe (Director of Naval Ordnance), Captain C E Madden (NA to Controller), Captain R H S Bacon (NA to First Sea Lord), Sir Phillip Watts RCNC (Director of Naval Construction, Lord Kelvin, Professor J H Biles (Glasgow University), Sir John Thornycroft, Alexander Cracie (Fairfield), R Edmund Froude (Superintendent, Admiralty Experiment Works), W H Gard RCNC (ADNC), Commander W Henderson (Secretary), E H Mitchell RCNC (Assistant Constructor and Assistant Secretary).

The formal terms of reference made it clear that the committee was an advisory body and that 'It must be clearly understood that it is no part of the function or purpose of the Committee to relieve the Director of Naval Construction of his official responsibility'. They were to consider a battleship with

12in guns and anti-torpedo-boat armament with no intermediate guns. Speed was to be 21kts and armour 'adequate'.

The first meeting was held on 3 January 1905 and, after reading the terms of reference, the President, Admiral Sir John Fisher, also read a preliminary statement which was adopted. He pointed out that previous Design Committees had largely been concerned with stability and safety of ships but that development in Naval Architecture had now eliminated the need for any more such studies. The committee was to bring officers who would have the future handling of the contemplated types of ship into communication with the scientific, constructive and engineering experts both in and outside the Admiralty. It was for the Naval element to propose the tactical and fighting requirements and for the civilians to state the limits within which the requirements could be fulfilled. Fisher said that 'The two governing conditions which have definitely been fixed on are *Guns* and *Speed,* associated with suitable protection. Both theory and the actual experience of war dictate a uniform armament of the largest guns, combined with a speed exceeding that of the enemy, so as to be able to force an action', and above all, 'The battleship is the embodiment of concentration of gun-power'. Another typical Fisher phrase was 'The Board of Admiralty have decided that the gun is the 12in gun, and the speed is 21kts'. Some light QF guns were also needed to deal with torpedo craft.

The Committee's report makes it clear that the reason for an all big gun armament was concentration of force, to bring an overwhelming weight of heavy guns on a part of the enemy's battle line. The compact line of *Dreadnoughts* would be quicker in manoeuvre than the long, straggling line of older ships. Other reasons given for uniformity of armement were: a) simplicity of fire control, b) interchangeability of all parts, c) simplicity of ammunition supply and a reduction in the total number of rounds to be carried.

9

A wartime view of *Dreadnought.*
CPL

It is clear that neither Fisher nor his committee were up to date with the thinking of Percy Scott and his gunnery school. Indeed Fisher was still writing 'the results of all long-range shooting have gone to prove that if we wish to make good shooting at 6000yds and above, the guns must be fired slowly and deliberately and the shots marked, preferably one gun at a time. Hence the use of a larger number of guns with large bursting charges is overwhelming', Scott, of course, saw the need for 'spotting' the fall of large salvoes, 'shorts' and 'overs' bracketing the correct range. Fisher was only concerned with the devastating effect of large shells with their bursting charges increasing as the cube of the diameter.

The effect of these large charges was later noted by Captain Pakenham, the RN observer at the Battle of Tsushima – 'when 12in guns are firing, shots from 10in guns pass unnoticed, while, for all the respect they instil, 8in or 6in guns might just as well be pea shooters, and the 12pdr simply does not count'. As a result of their experience in that war it was believed that both the Russians and the Japanese were going to all 12in gun ships in their new construction.

The committee concluded that damage to the enemy depended on three factors. Firstly, the shell must hit and this was assisted by a low trajectory (ie high velocity) and accurate spotting. Secondly, the damage depended on the remaining kinetic energy in the projectile, and

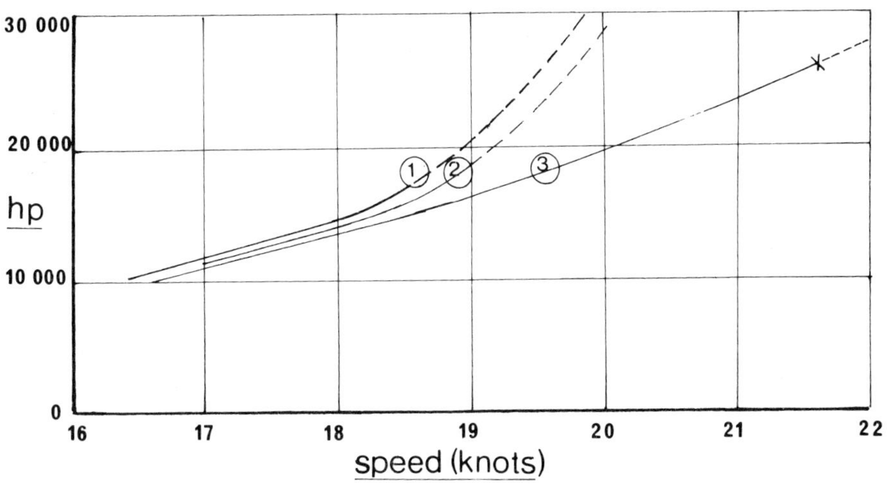

Comparative horsepower curves for *Lord Nelson* (1), *King Edward VII* (2) and *Dreadnought* (3)

on its bursting charge, and finally, damage depended on the rapidity of loading, the number of guns carried and on the effective arcs of these guns. The first two of these factors point clearly to large guns and, while the third appears to indicate a need for larger numbers of smaller guns, this is contradicted by the difficulty of spotting fall of shot.

SKETCH DESIGNS

With this background, the committee started work by considering Designs E and F, prepared mainly in accordance with the ideas of Admiral Wilson, C in C, Channel Fleet. The Director of Naval Ordnance (DNO) did not have information on blast effects with him but it was 'sufficiently clear' that the lower turrets would be untenable with upper turrets firing over them. In fact, it was decided that the acceptable arcs would be only 30° either side of the broadside. This view was incorrect, as demonstrated in the US Ships *Michigan* and *South Carolina* and indeed by later British ships. The committee also thought that the grouping of superfiring turrets in E and F presented too good a target and hence were undesirable. Before adjourning, the committee directed the DNC to prepare G by noon the next day. This has a peculiar arrangement with two turrets side-by-side fore and aft, each with a single turret firing over them.

The next day, the DNO brought along curves showing blast of 30lbs/sq ft, thought to be acceptable by personnel. Design C was also found unacceptable and further consideration was given to design D. This was based on the *Lord Nelson* with a single twin 12in mounting forward and aft but with two twin 12in turrets either side of the superstructure in place of the 9.2in turrets of the earlier ship. The DNC was asked to modify D in various respects and to prepare wooden models.

The Committee met again on 12 January to look at the modified designs, in which the forward turret had been placed on a raised forecastle so that it could be fought in any weather even when steaming at high speed. In D1 the wing turrets were close together, with their magazines below, while in D2 they were widely spaced with a boiler room between them. Blast curves were taken to show that the turrets in D1 were too close together and even in D2 there was interference. A single, centreline turret in place of the after pair of wing turrets was then studied and found to have advantages so this was the layout chosen for design H, the final layout. Apparently sound reasoning had led the gunnery experts to an awkward arrangement, soon to be abandoned in later ships, when slight changes to the sighting hoods of the turrets had been made.

The next day, design H was accepted, despite the objections of the DNC that the internal layout and boat stowage would be difficult. Battle was then joined on the choice of machinery. Sir John Durston favoured turbines and was strongly supported by Phillip Watts. However, the executive officers were worried about the loss of astern power which would result from the adoption of the small, high revving, propellers which were necessary with ungeared steam turbines. Evidence by the Hon Charles Parsons on the 17th and later by Froude in support of the turbine led to prolonged discussion. The climax was described many years later by Stanley Goodall – 'One who had been a secretary of the Commiteee (probably E H Mitchell) was present at some ship trials that I attended and one night after dinner I got him to talk. I was particularly anxious to know what part Sir Phillip Watts played in the discussions. I was told that PW said very little but when he did speak it generally resulted in a quick decision. For example, much time had passed on consideration of the type of machinery to be adopted and the old Push-and-Pull had nearly won when the Chairman turned to Sir Phillip and said "You have expressed no opinion. What do you think?". The reply was in one short sentence. "If you fit reciprocating engines these ships will be out of date in five years". The turbine won.

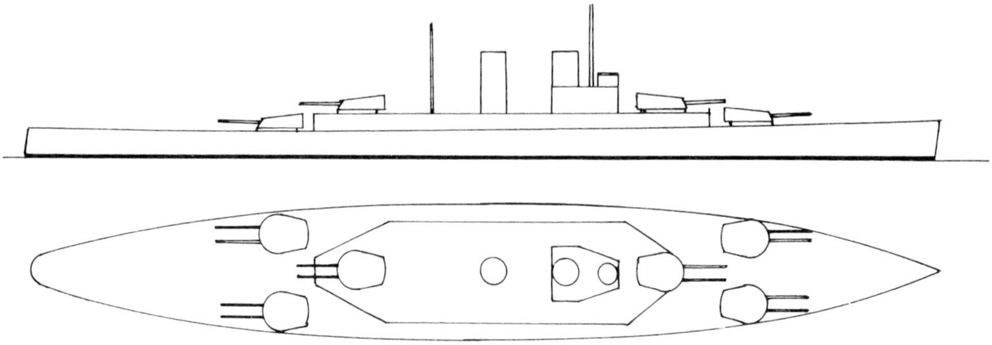

Battleship design D (19,000 tons)

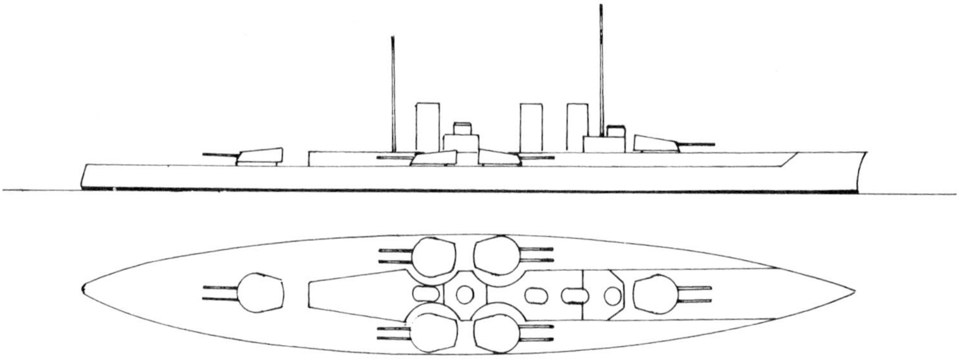

Battleship design D1 (18,500 tons)

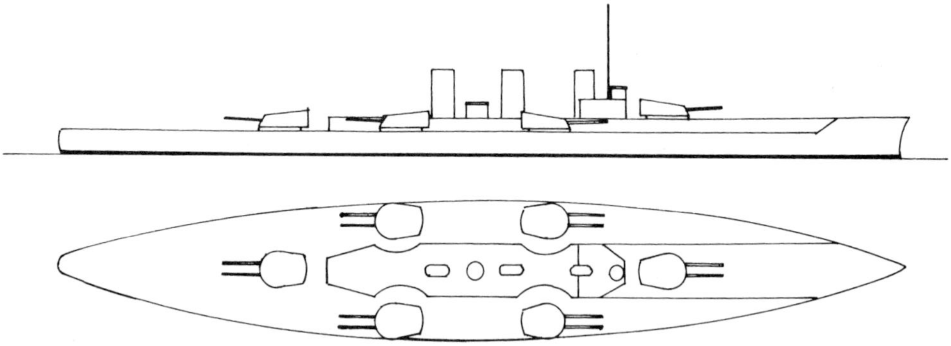

Battleship design D2 (18,500 tons)

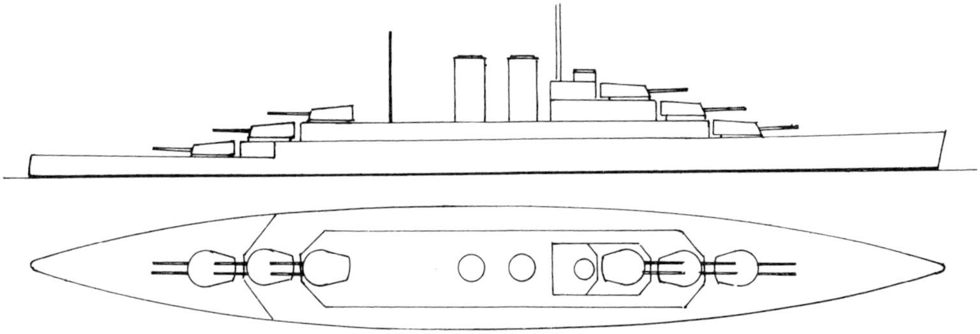

Battleship design E (21,000 tons)

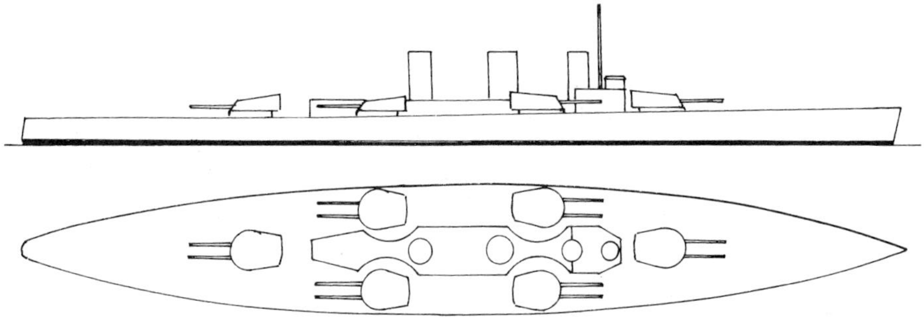

Battleship design G (21,000 tons)

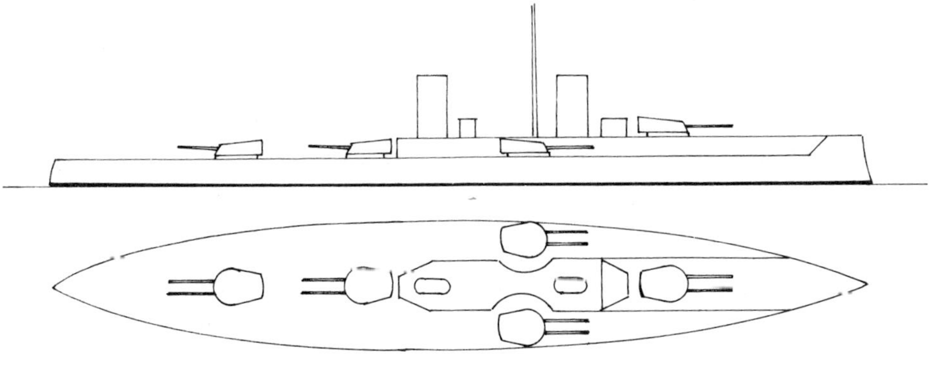

Battleship design H, from which *Dreadnought* was developed, (17,850 tons)

All these designs carried 12-12in guns in twin turrets except H with 10-12in. The design speed in all was 21kts but only D1, D2 and H were turbine powered the remainder having reciprocating machinery.

HMS *Dreadnought* as completed in 1906
NMM

The committee then considered some minor points and quickly decided that, in such a big ship, the officer accommodation should be forward, near the bridge, instead of in the conventional space aft. A sub-committee of Captains Madden, Jellicoe and Jackson then considered boat handling arrangements. It was not an easy problem but the solution they chose, with the fore tripod mast and its spotting top just behind the forward funnel, showed once again that the significance of long-range gunnery had not been appreciated. The position of the foretop was recognised from the start as the worst feature of *Dreadnought* and was corrected in following ships. It is remarkable that a few years later the same error was made in the original layout of *Lion* with even more disastrous consequences.

This Committee agreed its report on 22 February 1904, seven weeks after their first meeting, an incredibly short time for such a variety of naval decisions.

A design group is usually set up for modern designs with representatives from much the same

A close-up of the forward superstructure of *Dreadnought* in 1906

NMM

departmental interests. While the 1905 committee was staffed at Director level, the modern group is mainly at Constructor/Commander level. The higher grade committee was able to make more rapid decisions but there is some evidence, as discussed above, that many of these decisions were incorrect. It seems likely that the lack of professional engineering opinion on the weapons side may have been to blame.

DETAILED DESIGN

The concept of the fast, all big gun battleship having been agreed, the detail design could commence. When the completed ship was shown to the world there was general astonishment over what had been achieved. 'Critics and writers of all sorts vied with each other in endeavouring to discover what had been sacrificed to carry this immense armament at the great speed of 21½ knots. It did not, however, appear to have occurred to any of them that the skill of the Naval Architect and the Marine Engineer had a good deal to do with it'. (Ref 1, reproduced by courtesy of the RINA).

The Marine Engineer must take the greatest share of the credit. The turbine, invented by the Hon Charles Parsons, and encouraged by the E in C (Admiral Durston) and by Phillip Watts had arrived. The introduction of turbines showed a direct saving in weight (for the same power) of about 300 tons (see Table 1), but the heavier machinery would have required a larger hull and the overall saving due to turbines must have been about 1000 tons. The progress of turbines, through experimental destroyers and cruisers, was now complete.

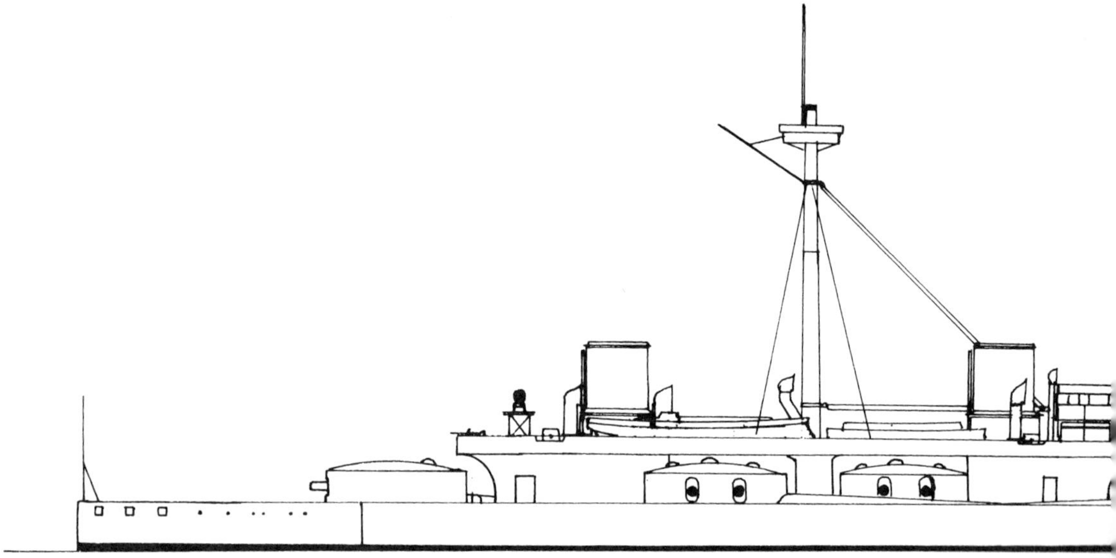

TABLE 1: COMPARATIVE PARTICULARS OF THE MACHINERY OF LORD NELSON AND DREADNOUGHT

	Lord Nelson	*Dreadnought*
Machinery Type	Reciprocating	Turbine
hp	16,750	23,000
Designed maximum speed	18kts	21kts
Machinery weight	1660 tons	1990 tons
Cost	£215,000	£320,000

TABLE 2: FORM OF DREADNOUGHT COMPARED WITH EARLIER BATTLESHIPS

	King Edward VII	*Lord Nelson*	*Dreadnought*
Speed	18.5kts	18kts	21kts
Length (oa)	457ft	443ft	526ft
V/ L	.86	.86	.92
Block coefficient	.645	.666	.591
Prismatic coefficient	.68	.695	.606
M*	5.5	5.3	6.1

$$^*M = \frac{\text{Length} \quad 1}{(\text{Volume}) \; 3}$$

An artists impression of one of the earliest proposals for an 'all big gun armament' based on the following description which is reproduced by courtesy of the RINA.

Many years later, Watts was to say 'I brought away with me (after the trials of *Inflexible* – about 1881) an outline design which had been evolved during the cruise and which was described as a combination of the *Devastation* and *Inflexible* designs. The armament was an *all big gun* armament of four pairs of 16in, 80 tons, guns, mounted in turrets, all placed on the upper deck; one pair at each extremity of the vessel on the middle line, as in *Devastation,* and one pair placed on each broadside as in the *Inflexible* . . . The scheme involved a displacement of 16,000 tons and on this account was considered inadvisable by Sir Nathaniel Barnaby'.

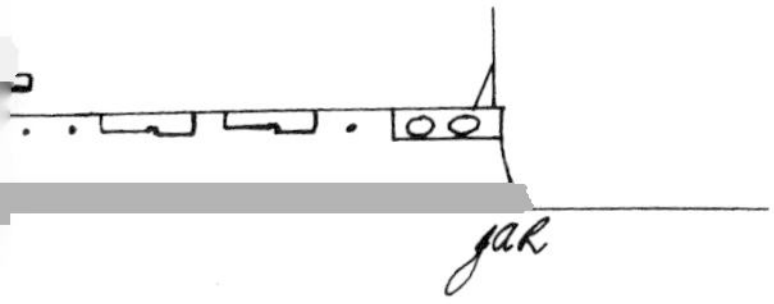

The naval architect was then able to contribute by ensuring that the power was used efficiently. Twenty one knots on a length of 526ft is an awkward speed ($V/\sqrt{L} = 0.92$) as it lies on what is called the prismatic hump of the resistance curve where a one point change in the prismatic coefficient* can cause a 9% change in resistance. The older, 18kt battleships were in a less sensitive area and were able to use a much fuller form, into which it was easier to fit magazines etc, without undue penalty. There was nothing wrong with the forms of the earlier ships for 18kts but the extra speed of *Dreadnought* needed special consideration (See Table 2).

Narbeth, now promoted to constructor at about 40 years of age, took special care over the lines using mathematical curves to ensure fairness. The curve of areas was a sine curve in the fore body, merging into an ellipse aft, with the CB (centre of buoyancy) well aft. The fore body was faired round waterlines which were portions of sine curves while the after body was built up round elliptical buttocks.

Model tests of the Admiralty Experimental Works (AEW) showed that it was a good form, lining up well with Froude's methodical series data and much better than in earlier battleships. Two modifications were made to the model and then two slightly different forms tested, the final version incorporating twin rudders (giving her a smaller turning circle than most earlier ships) and having a slightly narrower beam.

The hull structure and fittings continued the steady reduction in weight and detailed improvements of recent years. The hull weight was the same as that of the *Majestic* class (1894) which were some 3000 tons smaller overall. The improved 12in gun mountings, introduced in the *King Edward VII,* also contributed to weight reduction. A slightly more dubious saving was omission of the upper strake of 8in armour fitted in the *Lord Nelson.* At deep load, 21,845 tons, with 2900 tons of coal (900 tons nominal) on board the draught rose to over 31ft and the 11in belt was completely submerged. The protection of the waterline then depended on about 4ft of 8in armour, with unprotected sides above. This was an unfortunate failing of all the early Dreadnoughts. It is believed that this was due to political pressure to keep the displacement down and against the wishes of Watts.

It was intended to build the ship in record time and 'In order to facilitate this the hull structure was designed so as to secure the utmost simplicity in every way while retaining the exceptional rigidity required in the decks and bulkheads to successfully withstand the shock of firing such a large number of

The *Dreadnought's* Q and Y turrets viewed from amidships during the ship's trials

NMM

$$\text{*Prismatic Coefficient} = \frac{\text{Immersed Volume}}{\text{Length x Midship section area}}$$

heavy guns. The variety of steel sections used was reduced to a minimum; and a great effort was also made to adopt plates, of the several thicknesses required, in standard sizes, so as to avoid the usual labour and waste of time in cutting and sorting plates. Considerable quantities of plates, etc, were ordered to standard sizes, and by this means an ample supply of materials was maintained well in advance of the work on the ship' (Ref 1). A fine example of 'Design to Cost'!

The design team through Deadman, Attwood to Narbeth were filled with enthusiasm for the great ship and simple, effective novelties abounded. The unpierced bulkheads and separate pumping and flooding systems introduced in *Lord Nelson* were retained aiding rapid construction as well as reducing ship vulnerability.

BUILDING

The *Dreadnought* was laid down on 2 October 1905, launched on 10 February 1906 and went on trials on 3 October 1906 – though not completed until December 1906. It was a remarkable achievement. Building times from laying down to completion (months), for previous classes were (average: *Majestic* 32, *Canopus* 42, *Formidable* 42, *London* 39, *Duncan* 47, *King Edward VII* 33 (3 ships out of 6, others not known), *Lord Nelson* 39. These figures obscure some very interesting facts. Portsmouth could build more rapidly than any other dockyard and the dockyards were then much faster than private yards and, usually, much more economical in weight. Portsmouth averaged 31 months for ships of the classes listed above with *Majestic* herself completed in 22 months from February 1894 to December 1895.

The announcement of building plans was contained in the First Lord's statement on the Navy Estimates in early 1905. Work started in early May 1905 and by the official date of laying down £12,217 had been spent on labour and £29,078 on materials – about 6000 man weeks of work, at 48 hours a week, with 1100 men being engaged by laying down.

The high material costs show that much had been ordered in advance. The Dockyards were, as usual, being reorganised and the heads of departments were made responsible for ordering their own materials. The reorganisation made the Chief Constructor, Thomas Mitchell, into MCD (Manager, Constructive Department) and the rapid building of *Dreadnought* made him into Sir Thomas on 1 January 1906. Well earned, this was probably the only knighthood bestowed so low down the Corps Seniority list, and even more notable in that he was promoted from the Foreman grade.

Long hours were worked; at a time when the Yard as a whole worked from 7–12am and 1.30–5.00pm, $5\frac{1}{2}$ days a week, the men on *Dreadnought* worked from 6am–6pm, 6 days a week with only half an hour for lunch. This 69 hour week was not popular but, with large numbers of men being discharged*, it was not difficult to persuade the men to accept it. There was a good deal of discussion on the advantages of double shift working but with nearly 3000 of the 8000 men in the yard already working on *Dreadnought,* this was impossible.

The *Dreadnought* was laid down on Monday, 2 October 1905 and, as the accompanying photographs show, by Saturday the main deck beams were in place. The ship was largely built from the mould loft (as opposed to the drawing office) and 'the excellent work of laying-off and preparing the detail drawings was carried out under Mr J R Bond, who afterwards left the service, but whose splendid work contributed in no small degree to the rapid, economical and efficient building of the vessel. (Ref 1)

Sea trials were very satisfactory, the high freeboard forward making her a dry ship and the full section amidships giving good roll damping. Gunnery trials showed the structure well able to withstand the loads imposed by her heavy broadside and the turbine machinery was very reliable and generated much less vibration than the old Push-and-Pull. One would be tempted to call this ship the high point of the RCNC in conception, design and rapid, economical building but the achievement was to be repeated, many years later, with the nuclear submarine named *Dreadnought.*

Ref 1 'Three Steps in Naval Consruction' J H Narbeth Eng, CBE, MVO, MINA, RCNC. Transactions of the Institition of Naval Architects, 1922.

*Redundancy pay was generous by the standards of the day. The *Morning Telegraph* said '. . . every man having served 7 years or more, on discharge gets a week pay for each year, a minumum of £7. But many men get £1.15 a week, and some of them have served 30 and 40 years which means that they will get between £50 and £70 each for doing no work. Do you call this economy?'

AUDACE AND ARDITO

The photographic feature for this issue comes from Dr S Cioglia of Cagliari, and illustrates two of the most important warships of the Italian Navy – the destroyers *Audace* and *Ardito*. These vessels, whose original design was based on the preceeding *Impavido* class, were constructed between 1969 and 1973 at the Riva Trigosa and Castellammare shipyards respectively. They are intended as multi-purpose escort vessels and combine the functions of anti-aircraft and anti-submarine, with a limited anti-ship capability. They are generally regarded as a highly successful type being large, well equipped vessels with a high speed (33-34 knots maximum) and good seakeeping qualities.

The *Ardito* (foreground) and *Audace*, of the Italian Navy's First Division, during an operational visit to Cagliari harbour

1

2

1 Three generations of Italian post-war destroyers – the *Impetuoso* of the 1950s (left), the *Intrepido* of the 1960s (right) and *Audace* of the 1970s (centre) – a graphic illustration of the increase in the destroyers size and power

2 A starboard broadside view of *Ardito.* Note the separation of the superstructure into two blocks – Otomat missile launchers are to be installed in the space amidships

3 The forward superstructure of the *Audace* showing one of her two 127mm (5in) gun mountings. The majority of the aerials on the bridge structure are for ECM gear but the small dish aerial facing forward is for the Orion RTN 10X radar which controls the 127mm guns via the Argo NA10 Mod 0 fire control system

4 A view of the funnels and mast of *Audace* in which all but one of her outfit of radar aerials is visible. From left to right, the two SPG 51B for tracking and missile guidance, the SPS 52(3-D) surveillance radar and below it one of the three Orion RTN 10X fire control radars (the second aerial is hidden on the port side of the funnel and the third is visible above the bridge forward). Above the flag on the foremast is the SPQ 20 navigation radar and above the forward funnel the SPS 68 search radar, which replaced an SPS 12 set in 1978

3

1

2

3

4

1 The *Audace* in mid-1978, note the high-freeboard – a new feature in Italian warships

2 A close-up of the midships portion of *Audace* showing the two port 76mm Oto AA gun mountings

3 The hangar of *Audace.* Carried on the roof is the Mk 13 missile launcher for the Standard RIM 66A surface-to-air missile of which 40 are carried. The structure and crane on the side of the hangar are for one of the ships boats

4 A view of the stern and helicopter platform of *Audace.* The ships motto, above the hangar door, is 'Gloria Audaciae Comes' (Glory is the companion of boldness); the motto of *Ardito* is 'Nihil Obest' (Nothing opposes)

1

2

1 The hangar of *Audace* from astern with the door fully open

2 Two AB212 ASW helicopters parked on the flight deck of *Ardito.* The openings in the transom stern are the exit points for four fixed torpedo tubes which launch 'Kangaroo' wire guided AS torpedoes. The two ships of the class also carry two triple Mk 32 torpedo tube mountings, for the Mk 44 AS torpedo, and an EDO 610 keel sonar

WICHER AND BURZA

BIG SHIPS OF A SMALL NAVY
by Przemyslaw Budzbon

Despite an ambitious naval programme established in 1919 the Polish Navy in the mid-1920s consisted of only six torpedo-boats, four minesweepers, two gunboats and a few auxiliary vessels. However, the stabilisation of Poland's economy, which took place in 1924, created the possibility of additional naval expenditure and the Polish naval authority – Kierownictwo Marynarki Wojennej (Kier Mar Woj) – worked out a 'small programme' for the building of 2 cruisers, 6 destroyers, 12 torpedo-boats and 12 submarines during the following twelve years. To initiate this programme, the Kier Mar Woj asked, in 1925, for the authorisation of 9 submarines (3 minelaying), a floating dock and the necessary shore facilities, as well as buying the old French cruiser *Desaix* for service as a submarine depot ship.

In the middle of 1925 the Polish economy encountered serious troubles and all but the three minelaying submarines fell victim to budget cuts. But, quite unexpectedly, the Kier Mar Woj received from the War Ministry permission to order two destroyers, but with one condition – both had to be ordered from the new French yard of Chantiers Naval Français, at Blainville (CNF). It was hoped that this would assist the Polish Government in negotiations then taking place with the French for a considerable loan, as some members of the French Government were shareholders of CNF.

Przemyslaw Budzbon is a recently graduated Polish naval architect who spends much of his spare time gathering information on naval history and preparing excellent line drawings of warships, particularly those of Poland and Russia. He is a regular contributor to Polish maritime magazines.

DESIGN

Restricted to building destroyers in a French yard, the Kier Mar Woj had to choose between two designs: 'Contre Torpilleurs' of the *Chacal* class or 'Torpilleurs d'Escadre' of the *Bourrasque* class. As the first was large and costly, it was decided to order two *Bourrasques* and negotiations with the CNF started in 1925, when not even one ship of this type had been completed. After the Torpilleurs had been analysed, the following official opinion of the design was presented at a sitting of the Council of Ministers on 16 March 1926; 'It is important to emphasise that the destroyers of the French type have substantial construction and equipment faults so the Kier Mar Woj are unable to accept the design entirely. As the main fault we would point to the stability which, in case of damage

Reconstruction of both *Wicher* and *Burza* was planned for the early 1940s but, as all design drawings for this proposal were lost during the war, this drawing was prepared from available descriptive sources. After reconstruction the armament of each ship would have consisted of 5 – Bofors 120mm (3x1 on A, B and X mountings, 1x2 on Y mounting), 4 – Bofors 40mm AA(2x2), 8? – 13.2mm AA machine guns (4?x2), 6 – 550mm torpedo tubes (2x3) and 2 depth charge chutes.

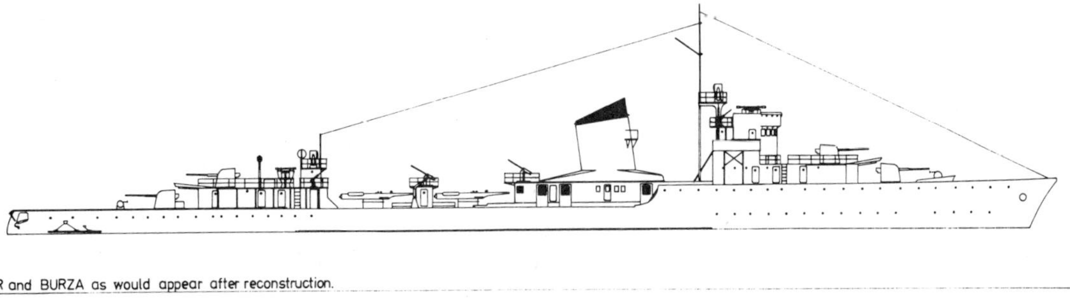

WICHER and BURZA as would appear after reconstruction.

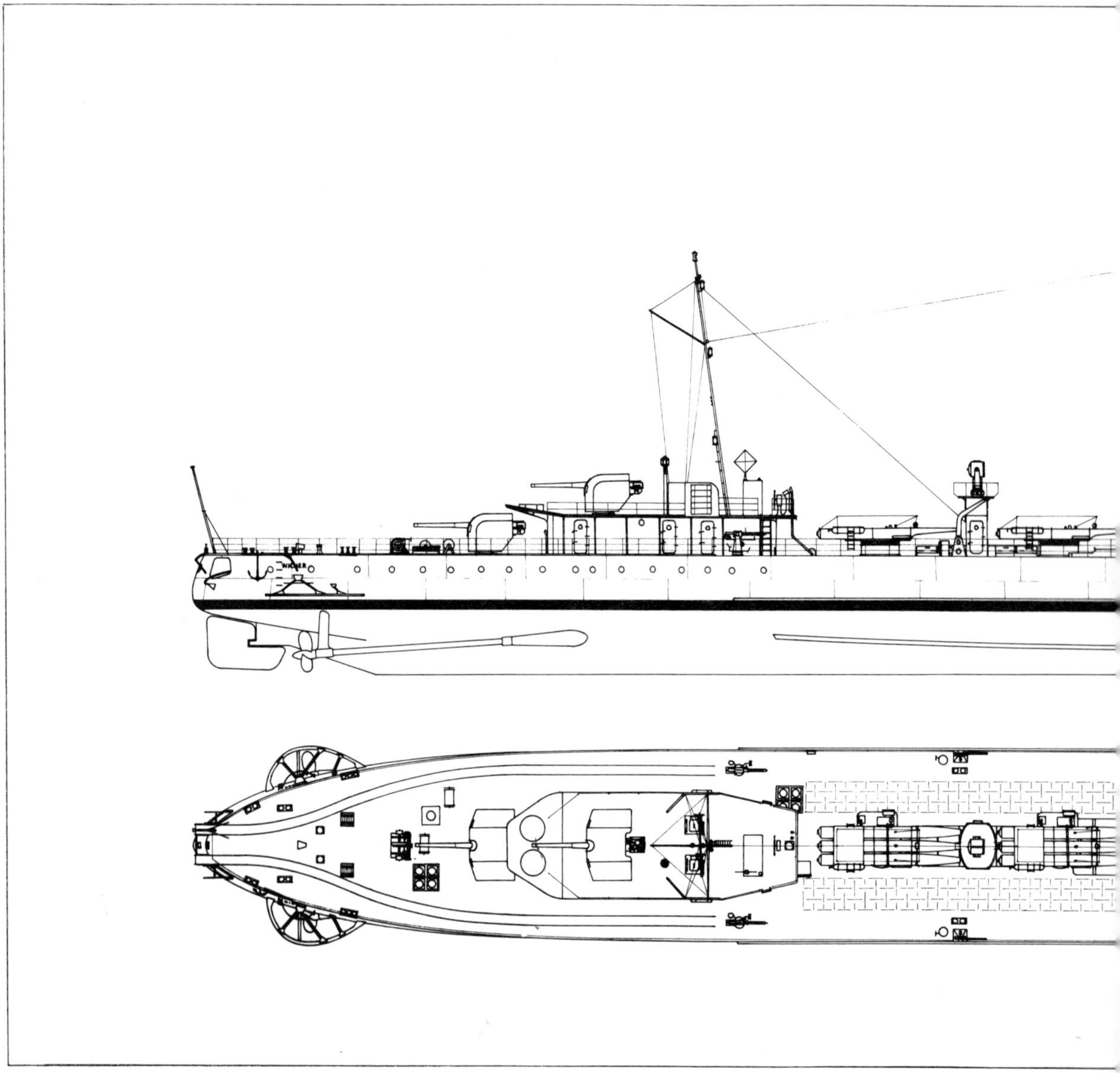

Wicher 1932

and flooding to one of the larger compartments, would be greatly reduced, ie the boat would list and in a heavy sea she would find herself in a critical condition. Furthermore, we note the absence of a double bottom, even in the most vital parts of the ship. In addition, the disposition of the main steam pipes, generators, fire fighting installations in shell rooms, fire pumps, etc, have many shortcomings and if these are maintained in accordance with the original design a single shell would be sufficient to neutralise the whole destroyer.'

These ships were further criticised for: a speed one or two knots slower than foreign contemporaries; slow rate of fire (4 rounds per minute from the 130mm Mod 1919 guns); disposition of crew accommodation in the order of rank and not according to action stations; enormous silhouette with three distinctive funnels.

After the first sea trials of the *Bourrasques* were carried out, all ships of the class were taken in hand, for improvements. These included replacing the 130mm Mod 1919 guns with the lighter and faster firing Model of 1924 (with large gun shields), replacing the 75mm AA gun in Q position with two 37mm AA guns on the main deck, reducing the height of all the funnels by 3ft and adding 20 tons of ballast. However, they were still not considered good sea boats and their maximum speed was reduced to 28–29kts in a seaway.

In consequence of the previously mentioned observations, the design of the two destroyers for the Polish Navy was recast and improved in respect of stability and reliability of the vital systems. Despite these

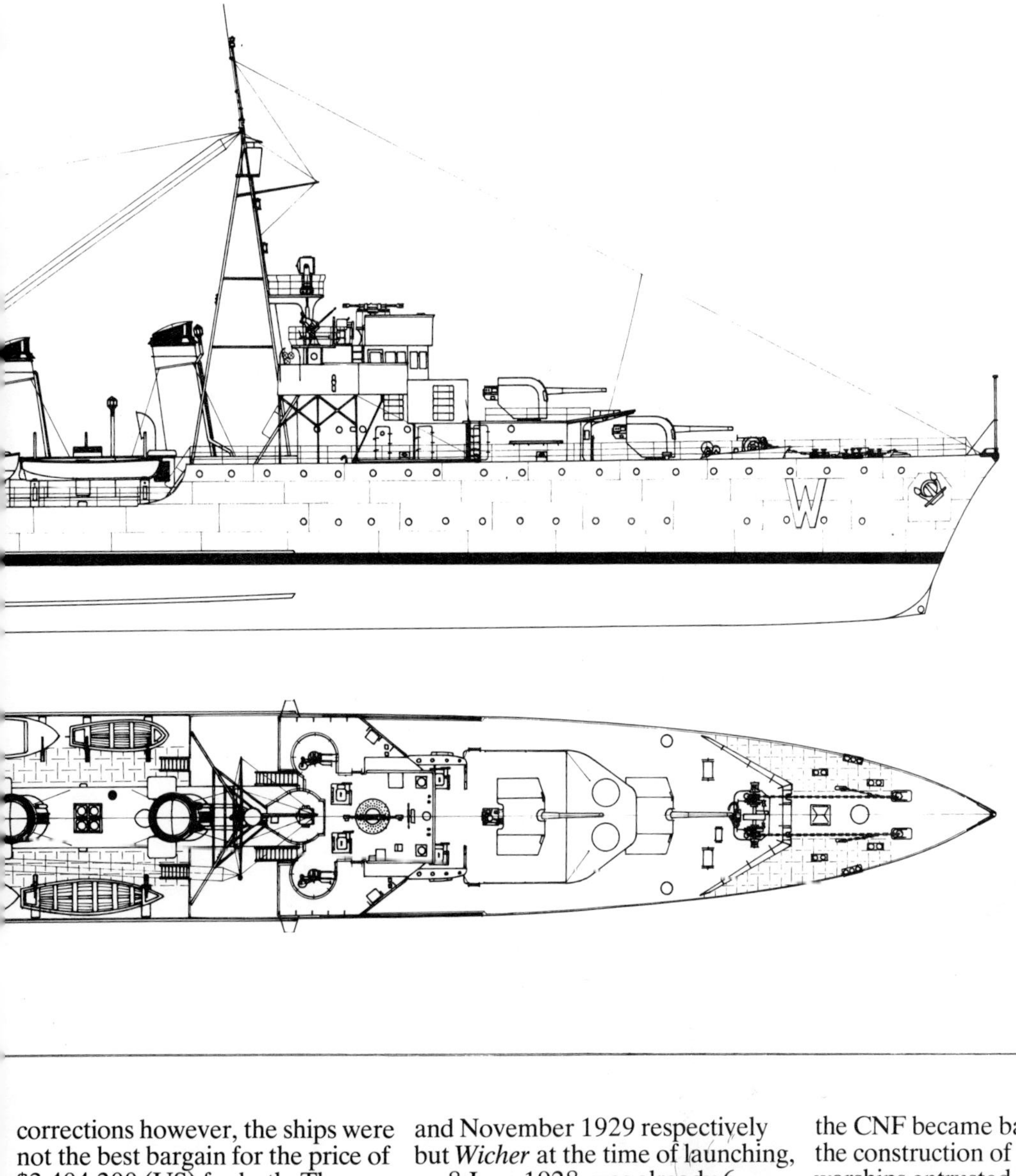

corrections however, the ships were not the best bargain for the price of $2,404,200 (US) for both. The Poles considered that British yards would have been able to produce much more satisfactory ships but the Kier Mar Woj had to place the order with the CNF for political reasons.

PROTRACTED CONSTRUCTION

The first destroyer, *Wicher,* was laid down on 19 February 1927 and the second, *Burza,* on 1 November 1927. They were due to be commissioned by November 1928 and November 1929 respectively but *Wicher* at the time of launching, on 8 June 1928, was already 6 months late – a poor prospect for the remaining five warships ordered by the Polish Government in France. Such protracted building times were notorious in French yards during the 1920s and 30s.

In the middle of 1928 troubles occurred in the construction of *Burza.* Casting defects were found during machining of the casings of the turbines, which caused at least three months delay. As a result of the delays with *Wicher* and *Burza* the CNF became bankrupt but, as the construction of the three Polish warships entrusted to CNF (*Wicher, Burza,* and *Zbik*) had prestige importance to the French Government, the yard was placed under judicial administration. Construction of both destroyers was then resumed. In October 1928 casting defects in a boiler-end intended for *Burza* caused another delay but in the spring of 1929 *Burza* was finally launched although this did not end her chronic series of defects and delays.

On 8 July 1930 *Wicher* entered

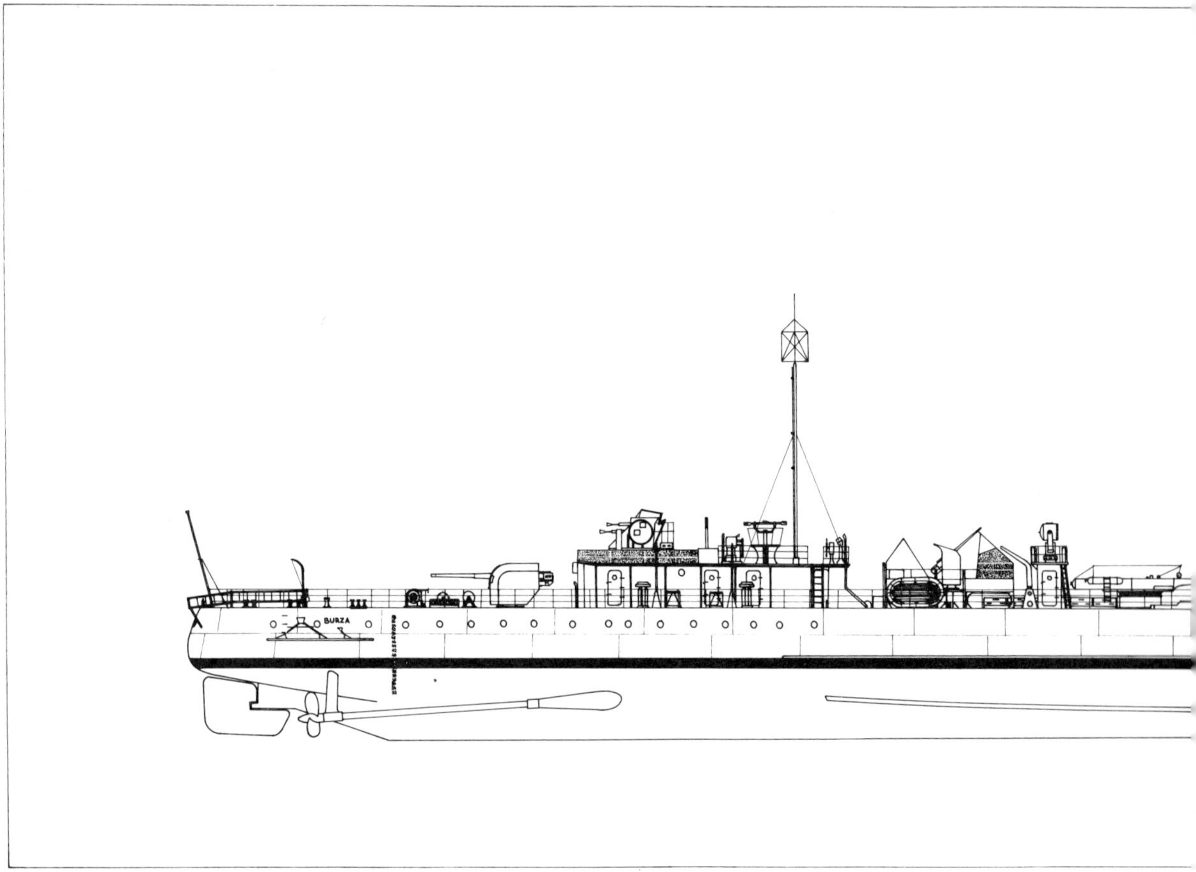

Burza 1943

1

1 The *Wicher* with her original two-masted rig
Museum of Gdynia

2 The *Burza* in the late 1930s, after the removal of her mainmast
TWYP

2

service, 40 months after being laid down, half of which time was the result of delays. A month later *Burza,* after 35 months under construction, was ready for sea but, following the first day of sea trials, deformations were discovered in the boiler drums. After a 6 month dispute, the chief arbiter, Rear Admiral Mark Rundle (RN), ruled that the yard should replace all three boilers. This occupied another year and *Burza* was not commissioned until 10 August 1932, 57 months after being laid down!

IN THE BALTIC

Five days after being commissioned, *Wicher* was welcomed to her home waters, marking a new era for the Polish Navy. The word 'era' is not an exaggeration, sufficient to say that the commissioning of *Wicher* enlarged the total displacement of the Polish Navy by 60%.

On 15 June 1932, *Wicher* welcomed three British destroyers paying a courtesy visit to Gdansk. This demonstration played a significant role in a dispute between Poland and Gdansk, related to a 'port d'attache' problem. After *Burza* was commissioned both destroyers enjoyed a peaceful life, dividing their time between manoeuvres and courtesy visits, and during the following years they called at every major Baltic port. Being the first large, modern Polish warships they attracted the critical eye of the nation, which had just

Burza 1955

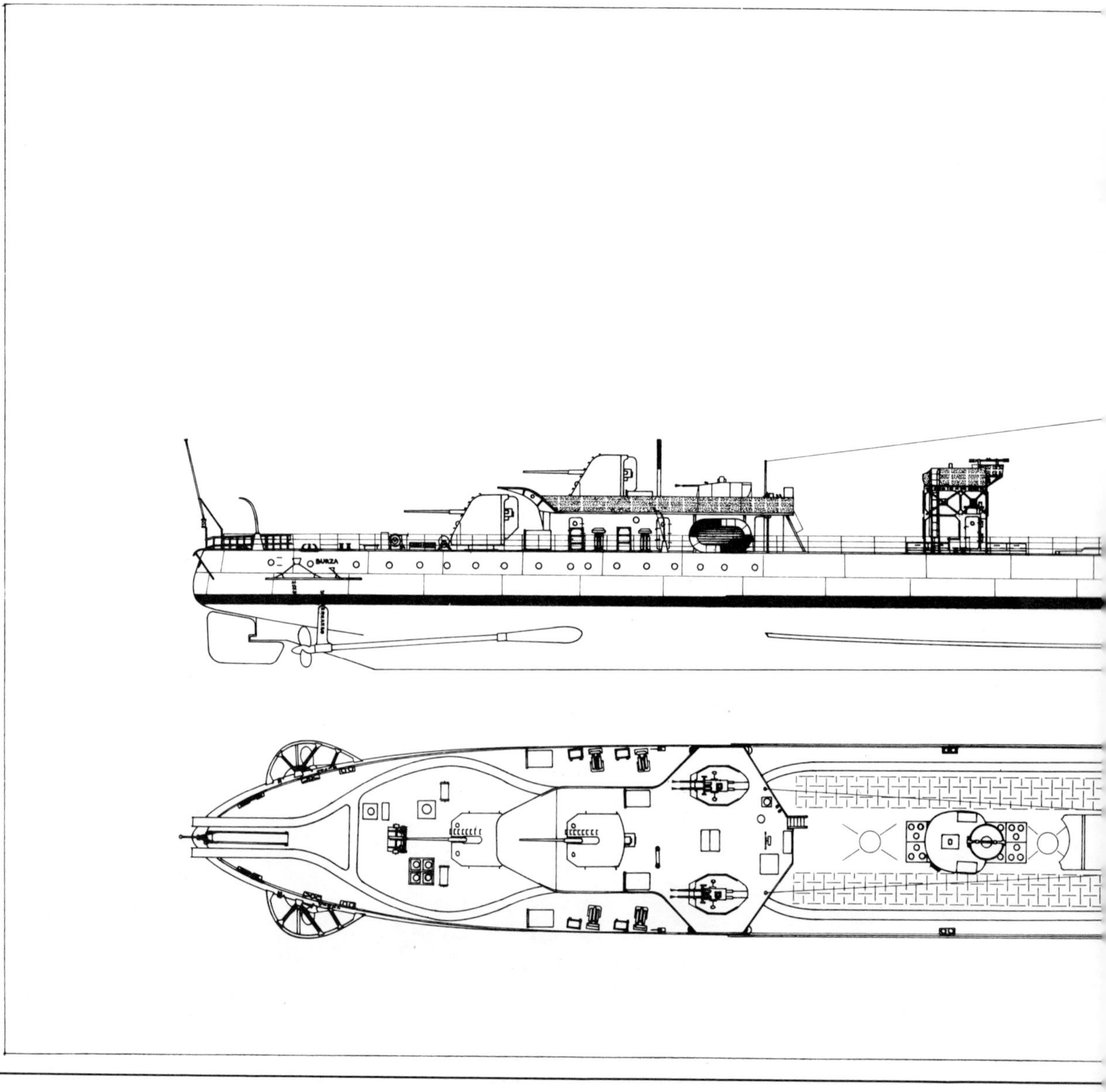

begun to take an interest in seapower. During 18–23 May 1937 *Burza* visited British waters and took part in the Coronation Review at Spithead.

By the middle of 1939 both ships had undergone short refits, which had little effect on their silhouettes. The most significant alteration was the removal of the main mast in 1935 and its replacement by an additional rangefinder. However, a significant number of minor changes had affected their stability and in April 1939 it was forbidden to add further to their weights. It was planned to modernise both ships during the early 1940s, in order to: a) bring them in line with the standardisation of the fleets ordnance; b) augment their AA defence; and c) give them a similar silhouette to the recently constructed super destroyers *Blyskawica* and *Grom.*

On 30 August 1939 the order 'Execute Pekin' (see *Warship* 4) separated this long lived couple for ever. *Burza* sailed with *Grom* amd *Blyskawica* to Britain, while *Wicher,* the best gunnery ship in the Fleet, remained to guard the Polish coast. On 1 September, accompanied by the minelayer *Gryf* and some minesweepers, *Wicher* successfully fought off an intensive attack by 32 Stukas in the first air-naval engagement of the Second World War. *Wicher* was sunk two days later in Hela harbour by three bombs from the Stukas of 4/Trägergruppe 186 (a carrier

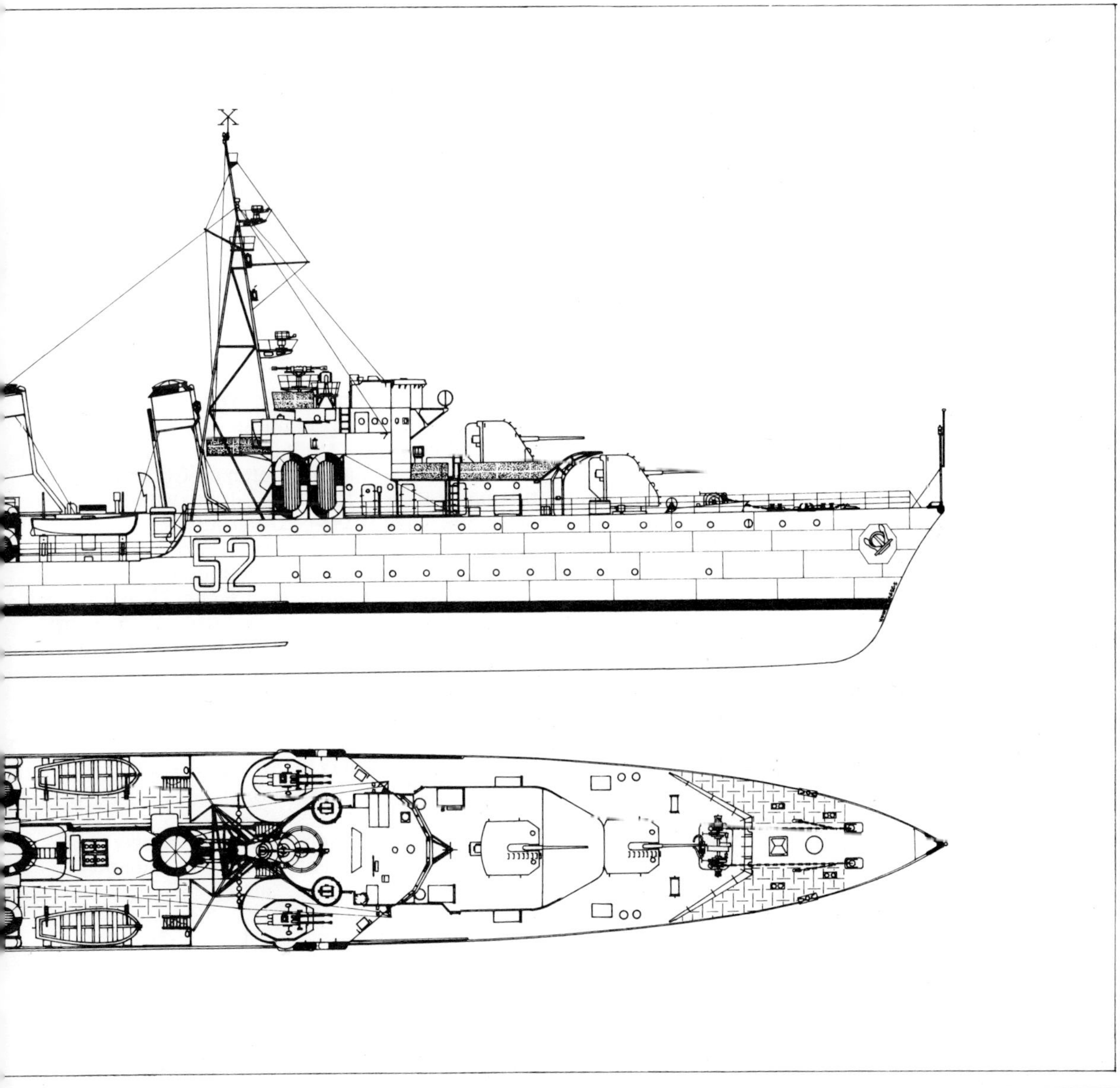

TABLE 1: COMPARISON BETWEEN THE FRENCH AND BRITISH DESTROYER DESIGNS AVAILABLE IN 1926

Name of prototype	Simoun	Amazon
Yard	Penhoët	Thornycorft
Prototype completed in	1926	1926
Standard displacement	1319 tons	1352 tons
Main armament	4–130mm	4–4.7in (120mm)
AA armament	1–75mm	2–2pdr (40mm)
MGs	4	5
Torpedo tubes	6–550mm	6–21in (533 mm)
Horsepower	31,000	39,500
Speed (trials)	33kts	37kts
Delivery time	34 months	21 months

1 The wreck of *Wicher,* photographed after the Second World War

Museum of Gdynia

2 The *Burza* as a destroyer escort in 1942. Note the reversed western-approached camouflage scheme

IWM

3 The *Burza* at sea towards the end of her war career wearing her standard western-approaches camouflage

IWM

2

group – originally ear-marked for the uncompleted aircraft carrier *Graf Zeppelin).* She was later raised by the Germans, who intended to repair her but, after the fall of France in 1940, they had more attractive opportunities to enlarge the Kriegsmarine and the wreck of *Wicher* (renamed *Seerose*) was finally towed off and sunk in the shallows of the Gulf of Gdansk. The wreck was destroyed in the 1950s, during exercises, when used as a target by Polish Naval aircraft.

WAR SERVICE OF BURZA

After her arrival to Britain, *Burza* was taken in hand for alterations to adapt her stability for service in the Atlantic and at the end of 1939 her AA defence was augmented by the fitting of two quadruple 0.5in machine gun mountings on platforms, one on each side of the third funnel. After taking part in the Norwegian Campaign and receiving bomb damage off Calais in May 1940, she was fitted with a 3in AA gun in place of the after bank of torpedo tubes, her mine rails were removed, the depth charge chutes were moved to the weather deck and she was fitted with Asdic. After this refit, *Burza* joined the Western Approaches command and took part in the attempted salvage of the *Empress of Britain* in November 1940.

Her French pattern armament and fittings gave much trouble when replacements were needed and it was proposed by the Admiralty to employ *Burza* in the planned raid on St Nazaire. Although a British destroyer was promised as a replacement, the Kier Mar Woj rejected these plans because *Burza* was undoubtedly Polish property. So instead she spent the first half of 1942 in Portsmouth Dockyard being extensively modified for escort duties.

As an escort destroyer she served under Lieutenant Commander Franciszek Pitulko with B6, B3 and EG8 escort groups respectively. After the HX217 convoy battle, *Burza* was honoured by Commander Ralph Heathcote, RN, with the famous signal, 'You are making escort history'. This and other convoy actions earned her the football nickname 'left winger' but

TABLE 2: DELIVERY TIMES OF 1926 POLISH ORDERS IN FRENCH YARDS

Name	Yard	Delivery time (months)		Delay (months)
		planned	actual	
Destroyers				
Wicher	CNF	29	49	20
Burza	CNF	42	74	32
Submarines				
Rys	Loire	29	56	27
Zbik	CNF	31	63	32
Wilk	Normand	33	71	38

her greatest achievements were her participation in the sinking of *U 606* and the rescuing of the damaged USS *Campbell* during the ON166 convoy battle. This action was even reported in occupied Poland, being widely published by the Resistance press.

In the middle of 1944 *Burza* was in need of extensive repairs but the victorious allies were not in need of such a non-standard vessel and *Burza* was decommissioned. For a while she was employed as a training ship at West Hartlepool and then became a submarine depot ship. During this time her forward bank of torpedo tubes was removed.

During her war career *Burza* covered 98,750 miles and took part in 45 patrols, 14 Atlantic and 40 smaller convoys. She was officially credited with sinking 1½ submarines (one presumably shared) and damaging 2 others.

POST WAR

In the spring of 1946 *Burza* was towed to Harwich and on 28 September was formally handed over to the Royal Navy. For five years she rusted in Harwich before being retroceded to the new Polish Government and in the middle of July 1951 she was towed to Gdynia, completing a 'round trip' which had started with Operation 'Pekin' in 1939.

The Polish Naval authority found *Burza* worth repairing and it was proposed to reconstruct her as an AS or AA frigate. The latter was approved and, after a 1952–55 modernisation and complete overhaul, she appeared with an enlarged bridge, the two after funnels trunked into a single uptake, additional controls amidships and a new armament.

During 8–11 September 1955 *Burza,* together with *Blyskawica,* visited Portsmouth. Both ships were there inspected by Admiral Sir George Creassy, who had commanded them during the war. After a short period of active service *Burza,* at the age of 31, was deleted from the active list on 24 February 1960 and began a new career as a museum ship in Gdynia.

TABLE 3: CHARACTERISTICS OF WICHER AND BURZA (AS COMPLETED)

Displacement 1320 tonnes (moulded), 1400 tons (standard), 1540 tonnes (normal), 1920 tonnes (full load)
Length 100.90m, 331.3ft (pp), 107.20m, 351ft (oa)
Beam 10.10m, 33.1ft (moulded)
Depth 6.10m, 20ft (moulded)
Draught (mean) 2.95m, 9.7ft (moulded), 3.30m, 10.8ft (normal load)
Draught (max) 3.60m, 11.8ft bow, 4.00m, 13.1ft stern (full load)
Immersion 7.74 tonnes per cm (moulded)
Machinery Parsons 2-shaft geared steam turbines in one engine room, 35,000 shp at 360 rpm = 33kts (33.8kts on trials). Three 3-drum boilers (one in the forward and two in the after boiler rooms), heating surface 930sq m, working pressure 18kg/sq cm, oil consumption 1.5 tonnes per hour at full power
Endurance 300 miles at 15kts (145rpm)
Oil Fuel Stowage 300 tonnes (max)
Armament 4 Schneider-Creusot 130mm/40cal, Model 1919; 2 Vickers 40mm/40cal AA, Mk 28; 4 (2x2) Hotchkiss 13.2mm AA MGs, Mk 30; 6 (2x3) 550mm torpedo tubes with adaptors for 21in and 450mm torpedoes; 2 DC chutes for 20 DCs; Mine rails fitted for 60 mines
Crew 155 officers and men

TABLE 4: ARMAMENT OF BURZA AFTER 1942 MODERNISATION

Guns 2-130mm, 1–3in AA, 1 quadruple pompom, 4–20mm Oerlikons
Torpedo Tubes 3–550mm (1x3) with adaptors for 21in torpedoes
ASW Hedgehog, 4 DCTs, 2 DC chutes, 110 DCs
Radars and DF Equipment Type 271, Type 292, MF/DF. A new pattern MF/DF, at the front of the bridge, and an HF/DF, on a pole mast, were fitted in 1943.
Crew 172

3

TABLE 5: ARMAMENT OF BURZA AFTER 1952–55 MODERNISATION

Guns 4–100mm AA, 8 (later reduced to 6) –37mm AA (4x2, later 2x2+2x1)
ASW 4 DCTs, 1 double DC chute, 36DCs
Mines Mine rails fitted for at least 100 mines
Radars and DF Equipment Type 292 (later replaced by Russian 'Cross Head' plus two others of Russian pattern), MF/DF
Crew 145

The *Burza* in 1955 after modernisation as an AA frigate
USN

TABLE 6: BUILDING DATES

Name	Bow Letter	Laid Down	Launched	Commissioned
Burza	B	1.11.1927	16.4.1929	18.8.1932
Wicher	W	19. 2.1929	8.7.1928	8.7.1930

Note: Turbines for both ships and the boilers of *Wicher* were subcontracted to Ateliers et Chantiers de la Loire, St Nazaire. The boilers of *Burza* were constructed by the Builder. Both ships were armed by the Arsenal in Cherbourg.

During 15 seasons in this role she was visited by over 3,700,000 people but due to the serious deterioration of her hull plating it was decided, in 1975, to replace her with *Blyskawica*. *Burza* was towed to Swinoujscie (near Szczecin) and employed there for fire fighting exercises by Polish and, occasionally, Russian seamen. This was her last service and in 1977 she was handed over to the shipbreakers.

WHO WAS RESPONSIBLE FOR THE FAULTS IN THE TEGETTHOFF CLASS?

by Ing René Greger

Reading the additional information about the faults of the *Tegetthoff* class in *Warship* 9 one may think that the designer was fully responsible for them. I would like to draw attention not to each of the faults mentioned, as some of them are questionable, but to the factors which influence the quality of the designer's work, besides those mentioned previously in the editorial of *Warship* 9.

Recently I discovered in the Military Archives in Prague parts of a bequest given to that Institution by Czech General (Ret) Vitezslav Siegfried Popper, about 50 years ago. It is little known that the designer of the first Austro-Hungarian dreadnoughts was a Czech; Popper was born in 1848 in Prague, studied there and was employed there after finishing his education at the technical school in Karslruhe. After retirement, in

René Greger is a well-known Czech naval historian who has carried out a substantial amount of original research on Eastern European navies. He has access to most of the relevant national archives and is a frequent contributor to naval journals such as *Marine-Rundschau.*

General Vitezslav Siegfried Popper in 1907
Author's collection

The *Svent Istvan* off Pola in 1917
Author's collection

The Pitzinger design of 1911 for 'Viribus Unitis Verstarkt'
Original Drawing by M Polak and R Greger

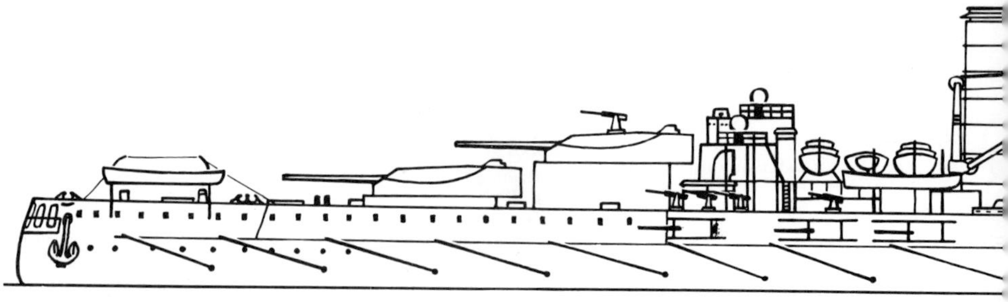

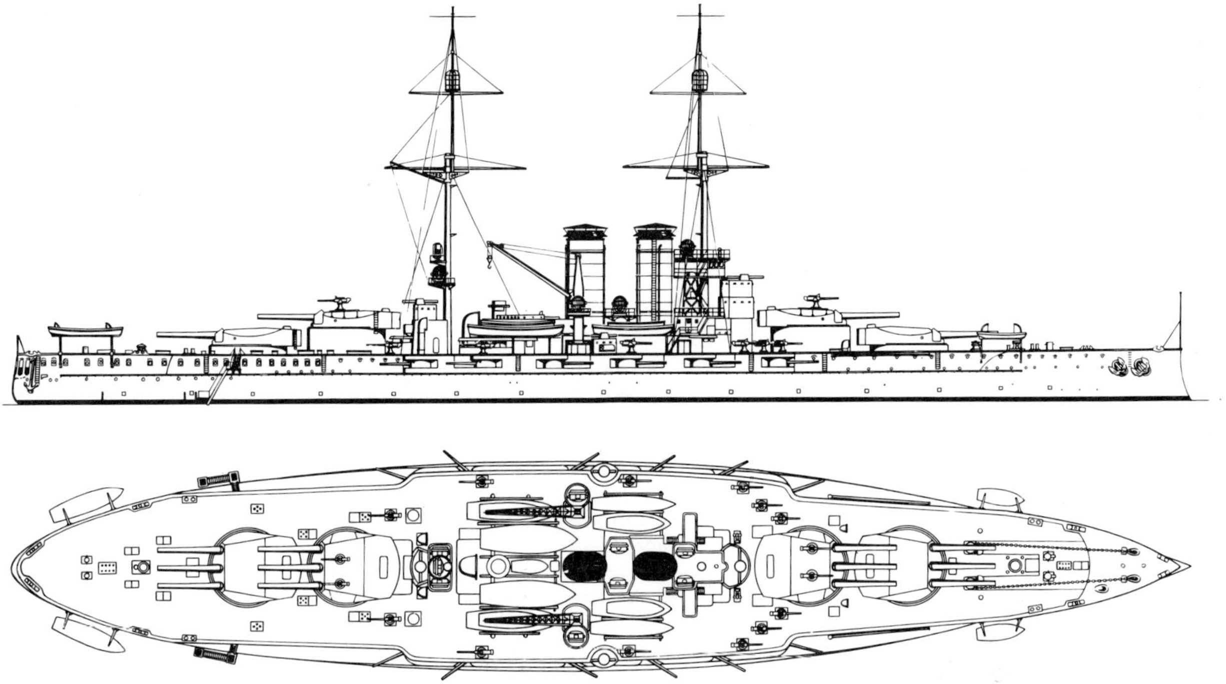

Viribus Unitis, 1918
Original drawing by M Polak and R Greger

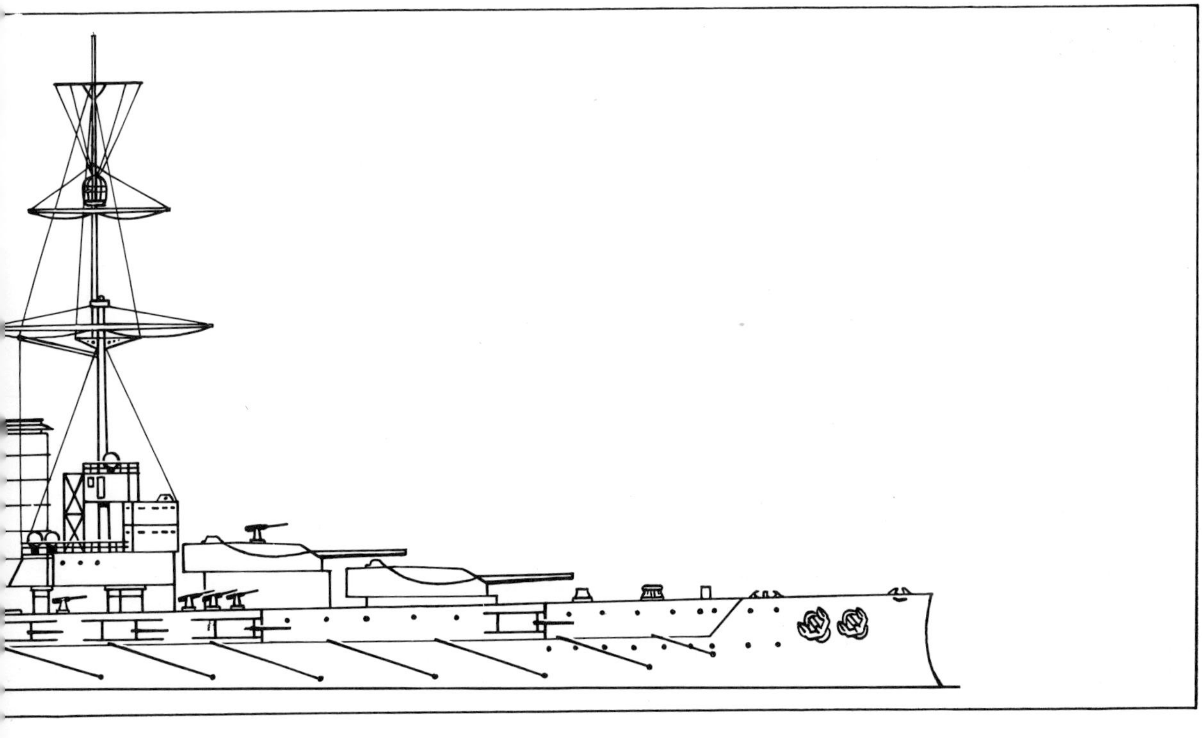

1907, he lived both in Trieste and Prague but definitely stayed in the later town after the collapse of the Habsburg Empire. Being a good Czech he asked for a position in the new Czech army and received the rank of General of Weapons Development, which was given to no other Czech officer. General Popper lost his life in an accident in the centre of Prague on 19 July 1933.

Among the documents found in Popper's bequest are several concerning the design of the first Austrian dreadnoughts. Furthermore, there are many papers relating his personal views together with details of calculations but, unfortunately, no diary and the matter cannot be reconstructed fully. It is widely known that financial problems hampered many Austrian shipbuilding projects but in this case the most important factor influencing the qualitative characteristics of the first Austrian dreadnoughts were the limits given to the designer by the Navy. In a memorandum, prepared by the Naval authority in 1906, on the necessity of strengthening the fleet, the protection of the coast line was stated as the primary mission of the Austro-Hungarian Navy. Four years later the Marine-Kommandant, Admiral Montecuccoli, asked for a special appropriation for the construction of four battleships, three cruisers, etc. He informed the Delegations (member bodies of both the Austrian and Hungarian Parliaments) that the latest naval developments dictated the need to build a new, modern type of battleship but that the strategic tasks resulting from the geography of the Adriatic did not require battleships of greater displacement than 20,000 tons!

Nobody can therefore be surprised at the discovery in Popper's documents of a note, made by him early in 1908, that the Navy was considering the construction of 18,500 ton dreadnoughts on the lines of an enlarged *Radetzky*. He called this type the 'Neues Projekt' and made a simple comparison between the *Radetzky* and the 18,500 ton ship. The derived dimensions (149.05 x 56.6 x 8.74m) he considered unacceptable and, after many further calculations, he came to the dimensions 153 x 27 x 8.12m, to meet a request for a ship carrying eight 305mm guns, but this result was stricken by him with the remark 'Zu schwasch!' (too vague). His preliminary calculations end with a simple deck plan representing a battleship carrying 5 twin main armament turrets, and 14 120mm guns as secondary armament. A table, in which this design is compared with other dreadnoughts building at that time, gives the additional characteristics of 20,500 tons displacement and dimensions of 155 x 26.5m.

Triple main armament turrets do not appear in Popper's private studies of 1908 or in official documents of that time. In official documents referring to Res Nr 2112/1.08 (ie, the proposed dreadnoughts) the techincal data and drawings of the new Skoda 305mm/50cal guns in twin turrets can be found. All of them are dated either October or December 1908 and therefore the information given by many Austrian authors that 305mm guns in triple turrets were just under development by Skoda is incorrect. It is also surprising to find that the Kuk Kriegsmarine wanted, in October 1908, 19cm guns in twin turrets as secondary armament, besides 10cm anti-torpedo-boat guns. A ship armed with these weapons would have had a normal displacement of 19,954 tons.

Popper designed a battleship to the 20,000 ton limit but this was a considerably better armed unit than the Navy originally wanted for use in the Adriatic. There is another paper in the bequest documenting the fact that the new Chief Designer of the Kriegsmarine, Ing Pitzinger, used Popper's conception in his proposal, submitted on 10 August 1911, for a battleship he called *Viribus Unitis Verstärkt*. Pitzinger was not limited as to displacement, having to respect only the shipbuilding limitations. His design represents practically a lengthened *Viribus* with the same armament but stronger hull, resulting in a much greater displacement. The protection was the same for the belt amidships, turrets, barbettes and secondary battery but the belt armour forward was removed and that aft reduced. Improvements in the underwater protection, mentioned by Pitzinger, must be considered as problematic as he chose the same system of 'Minepanzer' (mine armour) used originally by Popper. The only change was the replacement of 50(25+25)mm thick, 'special steel', longitudinal bulkheads by 40(20+20)mm bulkheads made from vanadium steel. As the bulkhead was situated in the same place, the replacement of one steel by another added nothing. Popper remarked on this document, that in 1911 no major sea power would project such a weak unit armed iwth 305mm guns and was later proved correct. In conclusion it can be said that the design submitted by Popper in 1909 was very progressive for the existing Naval conditions in the Mediterranean.

Editorial

In the world of mechanical invention there is very little that was not thought of by Victorian or earlier engineers. In many cases however, the technology of the time was insufficiently developed to put these ideas into practice and consequently the most practical had to await the modern invention of suitable processes and materials before seeing the light of day.

A case in point is the hydrofoil, discussed in this issue by David Brown. The idea of using foils and speed to lift a vessel clear of the water goes back to the 1860s but it was not until 1905 that an Italian inventor, Enrico Forlanini, actually built and tested such a craft. Proof of the theory does not necessarily mean success however, and there were still many practical problems to overcome before modern technology produced the operational hydrofoil vessels in service today.

Other types of high-speed craft have similar early histories. The hydroplane, designed to skim across the surface by using a stepped hull, was suggested to the Admiralty by a Reverend C M Ramus in 1872 and a patent for such a vessel was taken out by Thornycorft in 1877. To work however the hydroplane needed a light but powerful means of propulsion to provide speed without weighing the boat down so much that it could not lift, which was impractical with the steam engines then in existence. It was not therefore, until the introduction of reliable petrol, and diesel, engines that the development of the hydroplane became a practical proposition.

The hovercraft has even older origins having been outlined in theory as early as 1716 by a Swede, Emanuel Swedenborg. It was however, over 200 years before this idea was given a practical test, by the Finnish engineer T Kaario in 1935. The major problem once again was a suitable high power light-weight engine and it was not until the 1950s that hovercraft development began in earnest.

On a wider level the development of mechanical engineering in this century has for the most part been a continuation and refinement of the work of the Victorian engineer. In ship construction and machinery design it would not be difficult to find a multitude of similarities between modern ships and the vessels built in the last century. Hull design and construction may have changed substantially but not fundamentally. The use of welding is only a substitute for rivetting, not a change in principles; the use of high quality steel reflects only the development of material processes originated by the Victorians, and so on. In fact, to somewhat over simplify, it could be said that, so far as warships are concerned, mechanical engineering in this century has largely been concentrated on weight reduction, mainly in order to increase the load carrying capacity of ships and the power to weight ratio of their machinery.

The same arguments, but to a lesser extent, can be applied to electrical engineering for which the Victorians provided much of the ground work although considerably more of this branch of practical science was developed in the present century. A real modern revolution can however, be seen in electronic engineering, which in naval terms, began to have a serious effect with the development of radar in the 1940s. The introduction of electronic warfare has produced fundamental changes in warship design in which space, rather than weight, becomes the prime factor and in which the outcome of a naval engagement might well be decided by the efficiency of a ship's sensors rather than the power of her armament.

John Roberts

GERMAN HUMAN TORPEDOES

A 'Marder' type human torpedo in the courtyard of the Danish Army Museum, Copenhagen

Author's Collection

AND MIDGET SUBMARINES

by Erwin F Sieche

Since operational submarines first came into existance the possibilities of midget submarines have been seen as very attractive, particularly by small navies and young naval officers. Despite their advantages, however, midget submarines were always regarded as 'toys' by most naval authorities and their development in certain navies did not begin until the late 1930s.

It is interesting to learn that one of the first serious proposals for German midget submarines came from an industrialist – Dr Heinrich Draeger, the owner of the famous Draegerwerke at Lübeck – whose firm mainly produced fire-extinguishers, CO^2 exchangers and other submarine and fighter equipment. On 1 October 1941 he submitted to the German Naval High Command (OKM) a paper entitled 'Das Kleinst U Boot' ('The midget submarine') containing proposals for mass production and some sketch designs for various boat types from 7 to 120 tonnes.

At this time, however, German submarines were trying to control the high seas by oceanic warfare and Dreager's proposals were rejected in January 1942. In his paper he predicted later procedur to speed up construction times as developed for the Type XXI and Type XXIII U-boats. Draeger dic

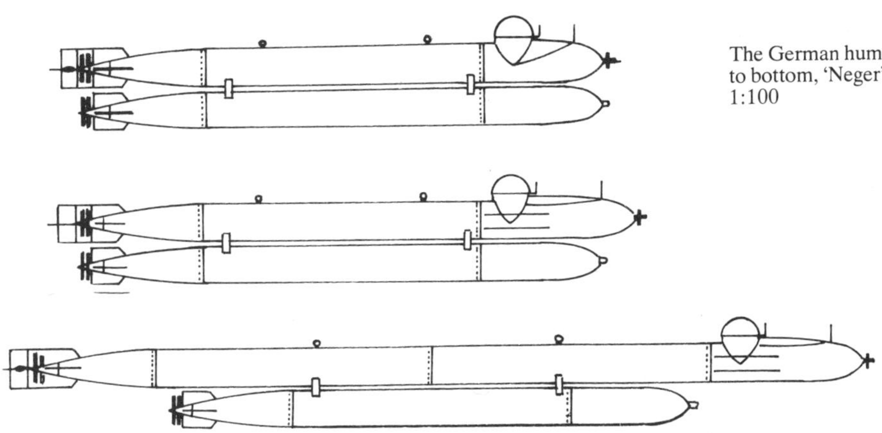

The German human-torpedo types, from top to bottom, 'Neger', 'Marder' and 'Hai'. scale 1:100

not know that Germany was already running two midget submarine projects. One was the 80 tonne Walter-propelled *V80,* an experimental vessel built around a revolutionary new air-independent turbine. (The development of the Walter turbine is, however, a different story, which will be related in a later article). The second was the so called 'Englemann-Schnellboot', a curious half-submersible speedboat of 265 tonnes, designed by Dr Engelmann, a hybrid which subsequently turned out to be a failure.

As the Allied invasion of Europe (Operation 'Overlord') became obviously inevitable, the German authorities discussed thoroughly all possible defence measures and proposed the following as basic requirements:
(a) every active defence must concentrate on the line of supply to the enemy beachhead;
(b) primary targets were therefore enemy supply ships, of every size, anchoring offshore or underway, so mobility and the highest possible explosive power would be primary requirements;
(c) owing to enemy air-superiority this must be provided by either fast surface attacking units (E-boats and their midget brothers, the 'Sprenglinsen') or midget submarines,

As the tactical, geographical and general situation in and/or around this beachhead was unforeseeable, the latter submarines must be: (a) available in great numbers; (b) easily transportable (railroad and/or truck); (c) of simple construction and easy maintenance; (d) easily constructed by unskilled workers and unskilled factories to avoid interference with the existing top-priority programmes (rocket weapons, new submarines, new jet fighters, radar development, Atlantic wall construction, V-weapons and launching sites, etc); (e) it must be possible to launch these vessels at any beach in great numbers without complex technical aid or harbour facilities.

A 'Molch' type midget submarine in the courtyard of the Danish Army Museum, Copenhagen

Author's Collection

The outcome of all these demands was inevitably a midget-submarine, or, much more primitively, a human torpedo.

To speed up the feed-back between designers and front-line

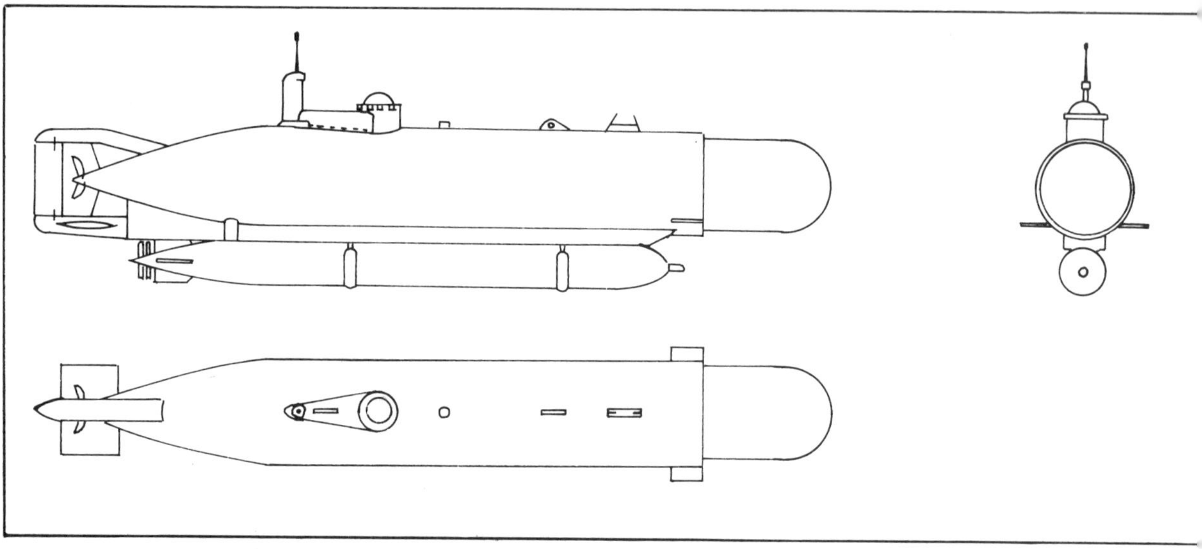

'Seehund' midget submarines being assembled in the 'Konrad' concrete shelter at Germaniaweft, Kiel on 18 May 1945. The photograph was taken by a British examination group.

Bildarchiv d Ost Nationalbibliothek

A midget submarine of the 'Biber' type

IWM, courtesy A Preston

officers, to train the future crews and to develop the attack and beaching techniques, a special operations staff was organised in the Germany Navy Group Commando North during 1943/44. From this staff later emerged the 'Small Fighting Units' ('Kleinkampfverbaende' or 'K-Verbaende') under the command of Vice-Admiral Hellmuth Heye, the later 'Ombudsman' (Wehrbeauftragter) of the Federal German Army (Bundeswehr) in the years 1961–64.

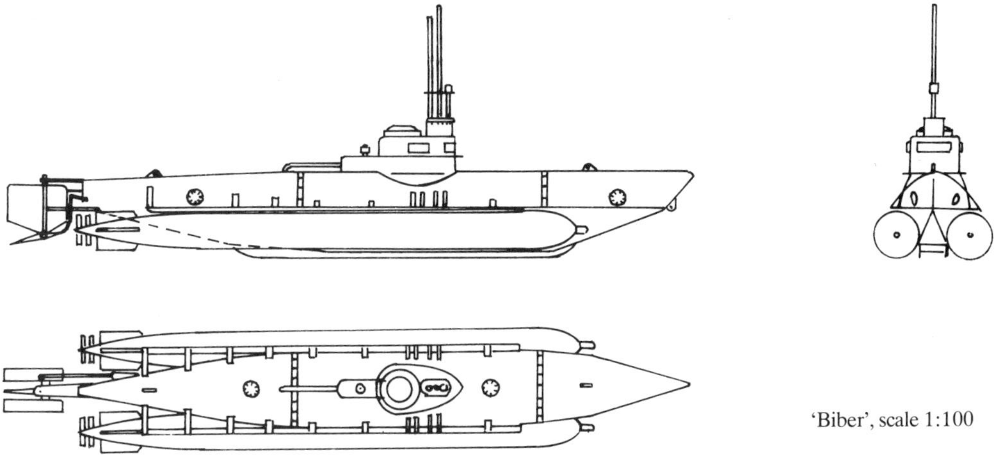

'Biber', scale 1:100

The 'Small Fighting Units' urgently required that weapons and vehicles be ready as quickly as possible so one line of development employed existing components and eventually resulted in the production of the German human torpedoes. The development of real midget submarines was set up simultaneously.

THE FIRST GENERATION: HUMAN TORPEDOES

Initially, it must be pointed out that the German human torpedoes were not suicide weapons: the operator was intended to be able to return ashore after his mission, but these men were volunteers of all ranks and services not skilled sailors and their vehicles were so crude and their navigation aids so primitive that every third mission was fatal. This was not primarily due to enemy interception, but mainly to exhaustion, stress, CO_2-poisoning, bad weather or just plain going astray. It is not astonishing, therefore, that thr real successes of these weapons were few and the percentage of losses relatively high. Their greatest success was the torpedoing of the Polish cruiser *Dragon* on 8 July 1944 by midshipman Karl Heinz Potthast (a real sailor!). The cruiser was damaged so badly that she had to be beached off Arromanches where she served as a break-water for the Mulberry harbour.

THE NEGER

This double torpedo was the brainchild of Marine-Baurat (naval engineer) Richard Mohr of the torpedo testing ground Kiel/Eckernfoerde. As 'Mohr' is a German literary expression for 'negro', the weapon was called 'Neger'. It consisted of two standard German G7e electric torpedoes. The carrying torpedo had its warhead removed, to provide space for the driver's cockpit, together with half the batteries to give sufficient buoyancy to carry the second torpedo which gave the 'Neger' its means of attack.

This weapon was designed to operate on the surface and could not dive. The driver could see out through a plexiglass cupola and lined-up his target over primitive open-sights. He was also equipped with a Draeger-type breathing mask attached to an aircraft CO_2 exchanger.

The first 'Neger' operation was ordered too hurriedly to bring a first success in time for Hitler's birthday; and at a time when tests on the equipment were incomplete. On 20 April 1944 30 'Negers' were launched at Torre Vaianica, 25km north of the Allied beachhead at Anzio but, due to a lack of special slip trailers, only 17 of them reached deep water, the others had to be blown-up on the beach on the following day. There was also no air-reconnaissance of the landing areas so the drivers found no targets and a further 3 units were lost during the mission. The Allies found one of these lost vessels drifting on the following day, the poisoned driver still in his cockpit – and the surprise was over.

On the night of 5–6 July 1944, 26 'Negers' began the second sortie, from Villers-sur-mer in Seine Bay. They sank the minesweepers *Cato, Magic* and *Pylades* together with some small transports; 9 units were lost. The third sortie took place on the night of 7–8 July, using 20 'Negers' one of which torpedoed the cruiser *Dragon,* as mentioned previously, and 16 units were lost.

Two further sorties, at the end of July and on 15–16 August, provided some small successes but the only vessel claimed as sunk was the British destroyer *Isis* which was torpedoed on 20 July 1944.

THE MARDER

This 'Pine-marten' was a slightly improved 'Neger' with diving cells, enabling the vessel to dive to 120ft. On the night of 2–3 August 1944, 58 'Marder' made their first sortie from Houlgate and the escort-destroyer *Quorn* was sunk.

There was also a plan to send 'Marders' to their targets pick-a-back on Type VIIC submarines and, for example, *U997* was fitted with supports for 4 'Marders' on the after deck. Tests in Norway were carried out during

January – April 1944 but the final order, for an attack on Murmansk, was never given.

THE HAI

This 'Shark' was an improved 'king-size Marder'. The carrying torpedo had two battery sections instead of one, which increase speed and action radius but drastically decreased seagoing and manoeuvring capabilities. It is assumed that only one prototype was ever produced.

THE SECOND GENERATION: MIDGET SUBMARINES

It is significant that Heye's drive, to give his 'K-Verbaende' the necessary weapons, resulted in most of the developments being started simultaneously. However, it naturally required more time to construct a real midget submarine than a human torpedo, although these too were built mainly of readily available components.

THE BIBER

The 'Beaver' was the brainchild of a young naval officer, Korvettenkapitän (Commander) Hans Bartels. It was based on a captured British midget of the 'Welman' type and constructed by the Flenderweke at Lübeck with notable speed: the first briefing conference took place on 4 February 1944, the first prototype (nick-named 'Adam') was ready on 15 March and, after trials, it was accepted by the Kriegsmarine on 29 March.

This one-man boat was the first German midget-submarine and had both electric propulsion and a 32hp, 6cyl 2.5 litre Opel-Blitz truck engine. It was, of course, a great risk to use a petrol engine but there was no serviceable diesel available. This primitive construction was not a real submarine, as it could not be trimmed to a given depth and could only escape from the surface by sinking to the seabed (maximum 20–30m, 100ft). The drivers head was inside a small tower, provided with sighting slots. The boats carried two torpedoes and a total of 324 of the craft were built.

The lack of submarine abilities in the 'Biber' lead to the proposed improved types 'Biber II', with two-man crew, and 'Biber II' with closed circuit propulsion (See later) but neither project was realized.

The first 'Biber' unit was K-Flottille 261 and its first sortie was made in the night of 29–30 August 1944 from Fécamp. Desipite a heavy sea all boats returned but had to be blown-up next day when the Wehrmacht evacuated Féchamp.

A second 'Biber' Flotilla operated from December 1944 until January 1945 against British shipping off the Schelde Estuary. Due to very difficult navigation in the area and personnel already under great stress, successes were few and the losses very high at 70 per cent. 'Biber' 90, which is now exhibited at the Imperial War Museum in London, was lost during one of these sorties. The midget was found on 29 December 1944 drifting 30 miles north of the North Foreland, the poisoned driver still inside. It was towed to Dover, where it sank accidentally in the harbour entrance. It was later raised and beached, subsequently being overhauled at Chatham Dockyard prior to undergoing extensive trials by the Royal Navy. From February the 'Bibers' in this area were replaced by the 'Seehunde' type but they carried out some operations on Dutch rivers. For these operations the boats were fitted with a buoyancy device, camouflaged as drift-wood, to allow snorkeling.

One of the most spectacular operations of the German midget submarines would have been the attack on the Soviet battleship *Archangelsk* (ex-British *Royal Sovereign)* anchored in the Kola Inlet off Murmansk. The submarines *U295, U716* and *U739* left Harstadt in Norway on 5 January 1945, each carrying two 'Bibers' on deck supports. The mission failed because the 'Bibers' became inoperative due to leaking petrol pipes caused by constant vibrations.

It was also planned that 'Bibers' should find the allied 'Pluto' pipeline, drill a hole into it and inject a corroding substance which would destory the engines of any vehicles using the 'contaminated' petrol. Another plan was to drop a 'Biber' into the Suez Canal, from a Blohm & Voss BV 222 floatplane, where it would attempt to block the Channel by torpedoing a ship.

THE TYPE XXVIII, 'HECHT'

The 'pike' was a development of the submarine construction bureau of the German Navy High Comman and therefore had an official type designation.

The first specification for this type was worked-out under the impression created by the British X-craft attack on the German battleship *Tirpitz* and what was asked for was a mine-carrying two-man midget. It took a long time to develop the requrested mini-gyro-compass and to adapt existing charges for use as limpet mines, but when this was finished, the requirements had changed and a complete redesign was ordered. Instead of the limpet mine, it was desired that they carry a removable bow housing three commando raiders with a rubber boat or alternatively a simple bow containing additional batteries, the two types being interchangeable.

Although it was clear to almost everybody involved that such a fundamental redesign would lead to an unsatifactory vehicle the Krupp-Germaniawerft at Kiel was ordered to build three prototypes and a production series of 50 boats under the designation Type XXVIIA. These 53 boats never became operational but proved of value as training vessels, for which purpose they had additional batteries fitted in the bow. This may be the reason why most of them survived; one o the Type XXVIIs is displayed in the Imperial War Museum.

THE MOLCH

The 'Salamander' was a one man vessel carrying two standard torpedoes. It was propelled solely by a standard torpedo motor, the driver sitting under a plexiglass cupola. The first prototype was ready for testing at Eckernförde on 12 June 1944 and subsequently a

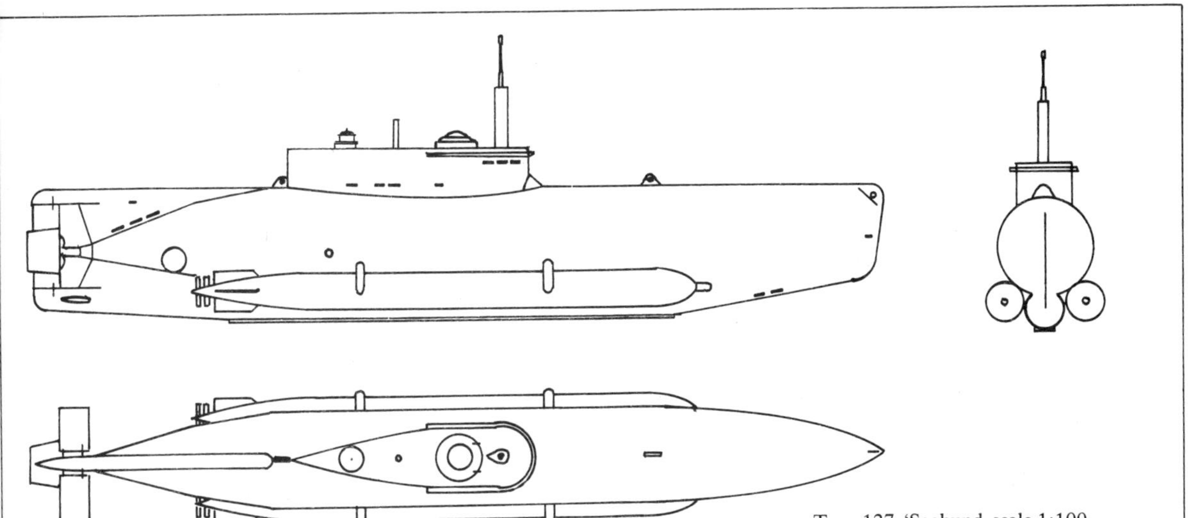

Type 127, 'Seehund, scale 1:100

total of 363 'Molche' were built.

The first operational 'Molch' unit was K-Flottille 411, which was transferred to Northern Italy in the autumn of 1944, but due to lack of experience and personnel difficulties, they achieved no success. The second unit was K-Flottille 412, which operated in December 1944 against targets off the Dutch coast. During a 12 boat sortie on 21–22 February 1945 8 vessels were lost and again on 12–13 march 1945 a 16 boat sortie suffered 14 losses. The third unit, K-Flottille 413, was also intended to operate in The Netherlands, but saw no action.

THE TYPE XXVII B5 (also Type 127) 'SEEHUND'

All the vessels previously described did not reach the minimum requirements as they were only an emergency measure to meet the neėd to work up 'something' as quickly as possible for operational service. It was soon realised, however that it was useless to develop such vessels and/or weapons by utilising available components. To meet the required abilities of a real midget submarine it was necessary to have a scaled-down version of the machinery and instrumentation of a standard submarine but with the parts carefully designed and developed specifically for the purpose.

The 'Seal' was the brainchild of two constructors, Marine-Baurdirektor (director of naval construction) Fischer and Marine-Baurat Grim. By June 1944 the pre-project Types XXVIIB1–5 had been designed. They differed in some construction details which, in themselves, were very influential on the development of such small craft.

The final version, Type XXVII B5 was a two-man midget submarine carrying two external torpedoes. The single hull boat had a surface populsion unit consisting

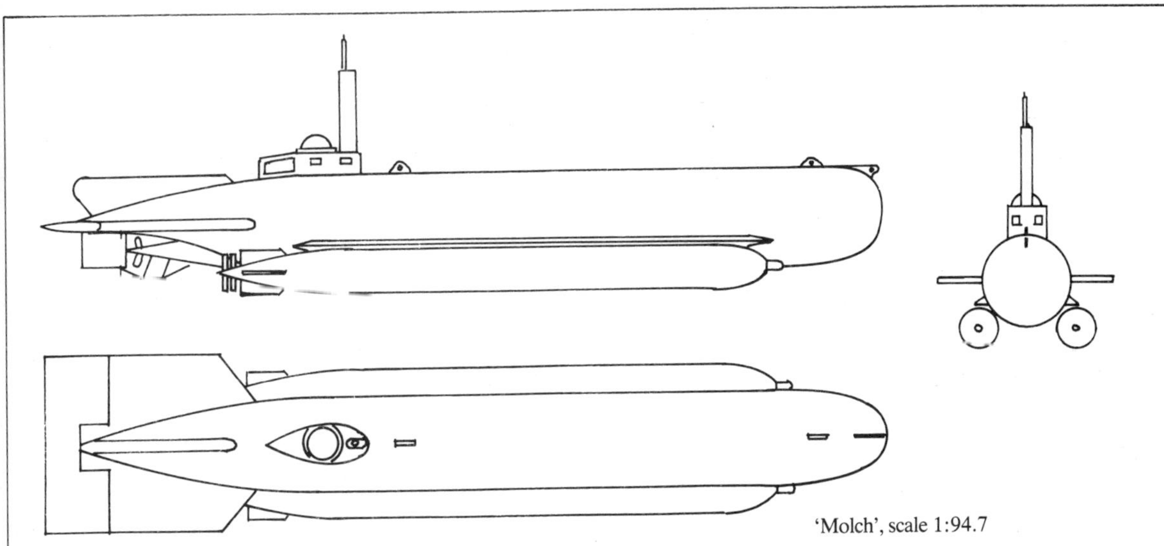

'Molch', scale 1:94.7

of one 60hp, 6cyl Büssing truck diesel and an underwater propulsion unit consisting of one 25hp electric motor. The boat was designed for a diving depth of 30m (100ft), but in action depths of up to 70m (213ft) were achieved without problems.

For operational service the 'Seehund' showed some substantial advantages: (a) The surface silhouette was so small that it was scarcely detectable by optical or electronic instruments; (b) due to its small moment of inertia is was less susceptible to depth-charges because in most cases it would simply be displayed sideways by the shock waves; (c) good seagoing capabilities up to sea state 4; (d) extremely short diving time: 2-3 seconds! The 'Seehunds' disadvantages were: (a) Bad manoeuvering abilities of the early boats, (this was later improved by surrounding the propeller with a ring (a so called Kort-jet); (b) too slow submerged to match moving targets; (c) limited action radius (this was not caused by limited fuel stowage, but by the physical stress and exhaustion of the crew who had absolutely no comfort on board – there was no possibility of sleep or use of a toilet. For this latter reason all midget submariners had to have a special diet. (d) small observation radius caused by sea-level look-out position.

As the 'Seehund' was the best of the German midget submarines it was the only one chosen for further high priority mass-production. Up to the end of 1944, 169 units were delivered and, by the end of the war, 285 boats had been completed while another 93 were found in various stages of construction in bomb-proof shelters.

The first 'Seehund' operation against the Thames Estuary in January 1945 ended in disaster; 1 losses out of the 18 boats involved The remaining crews quickly gain in combat experience, however, a the second sortie, on 17 January 1945 had no losses.

The 'Seehunds' operated relatively successfully from February to May 1945, sinking 19 freighters, totalling 18, 451 tons GRT, and damaging 3 others, totalling 18,384 tons GRT. Only ten per cent of the boats were lost. On the other hand, it is estimated that the Allies required 500 vessel and 1500 aircraft to control the North Sea and spot the 3 to 5 'Seehunds' operating there daily.

In April 1945 two special 'Seehund' missions were carried ou to supply the isolated German bas at Dunkirk with food. Instead of torpedoes the boats carried food-containers (nicknamed 'butter-torpedoes'); when returnin

A prefabricated midship section of a 'seehund' midget

IWM, courtesy A Preston

British officers examining a two-man midge of the 'Seehund' type inside the 'Konrad' concrete shelter at Kiel on 15 May 1945

Bildarchiv d Ost Nationalbibliothek

these containers were filled with the mail of the Dunkirk defenders.

THE THIRD GENERATION

When the European war ended in May 1945 a wide variety of new midget-submarines with fantastic abilities was in the experimental stage or on the drawing boards. From today's point of view some of them sound as though they were an invention for a James Bond film; rocket-propelled high speed submarines, tracked amphibious submarines with flame throwers, surface skimmers and so on. Unfortunately, most of the construction files and test prototypes were destroyed when Allied troops approached so the details of these vessels can only be reconstructed from the limited information that survives.

THE SEETEUFEL

Of all new projects the 'Sea Devil' (also code-named 'Elephant' and/or 'Projekt Loedige') was the furthest developed. The great difficulties of launching even such a small weapon as the 'Neger' resulted in Dipl Ing Alois Loedige, the leader of the development division o the torpedo testing ground at Kiel/Eckernförde, suggesting an idea for an amphibious two-man midget-sbmarine.

The first prototype was ready for testing in July 1944. As there were no diesel engines available, an 80hp petrol truck engine had to be installed. It served for both surface propulsion (8kts) and land propulsion (10km/h). The 'Seeteufel' was to be armed either with two torpedoes or mines and either machineguns or flame-throwers.

The trials, at Eckernförde Bay, showed the vessel to have remarkably good manoeuvering and handling abilities but that it was also seriously underpowered. It was therefore decided that the production series should have a 250hp diesel engine and a 100hp electric motor. It was also found that the tracks were too small.

After finalising the tests, a series of standard 'Seeteufel' were ordered at the Carl Borgward car factory at Bremen, but production never began. The only existent experimental vehicle was brought to the training camp 'Blaukoppel', near Lübeck, and blown-up at the end of the war.

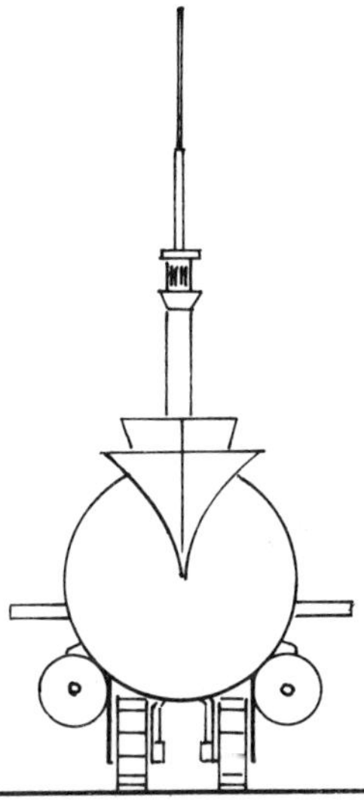

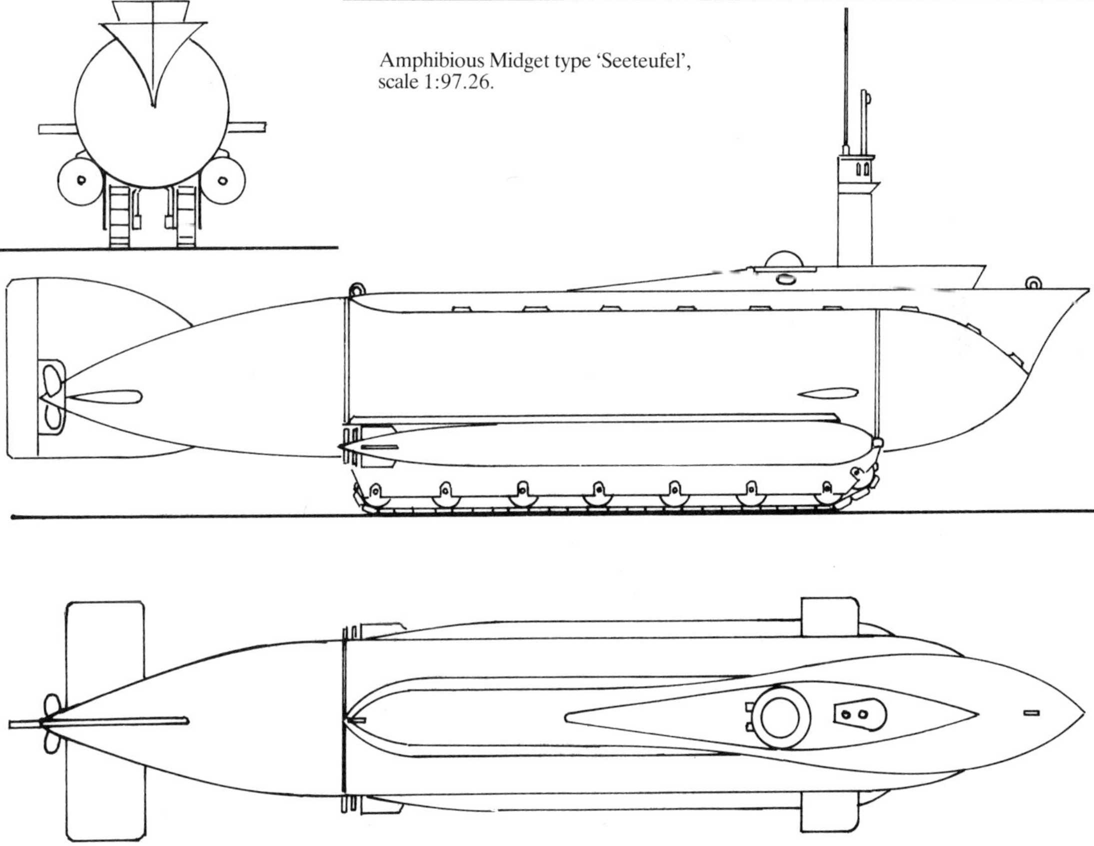

Amphibious Midget type 'Seeteufel', scale 1:97.26.

THE KREISLAUF-SEEHUND
In May 1944 Marine-Oberbaurat Kurzak proposed an air-independent diesel using normal fuel and utilising its own exhaust gases regenerated with liquid oxygen. In June 1944 tests were ordered to check the feasibility and reliability of this type of propulsion for midget-submarines. The tests being successful, Kurzak proposed the building of some experimental vessels using components of the 'Seehund'. The resultant, enlarged, vessel was called the 'close-circuit Seal' and experimental boats were ordered from F Schichau at Elbing and the Germaniawerft at Kiel but only the hulls of *U5188 – U5190* were begun, all at Kiel. Modified Büssing diesels were run on a test-bench from September 1944 to February 1945 but these were never installed in the hulls.

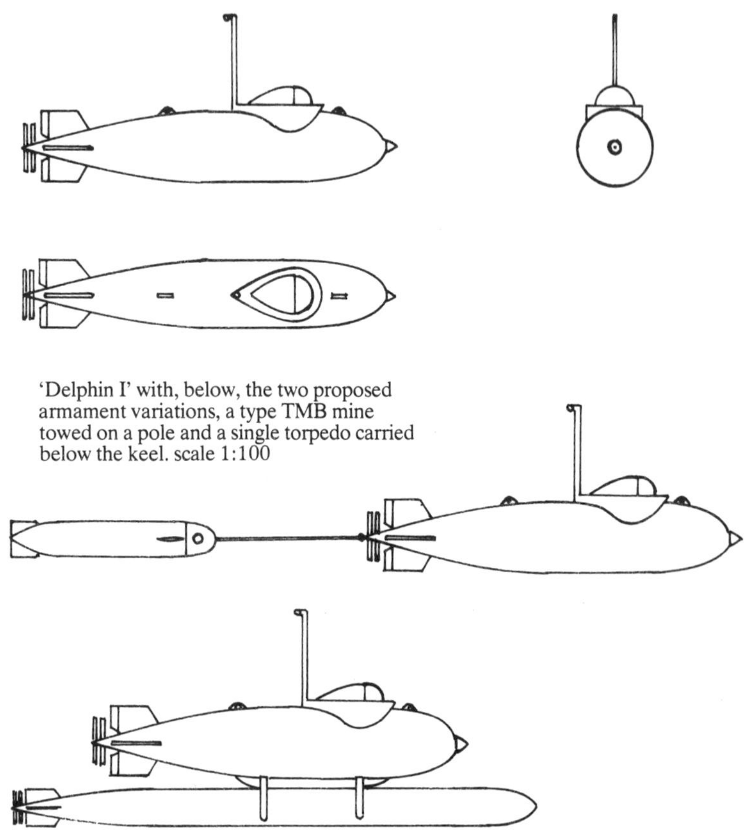

'Delphin I' with, below, the two proposed armament variations, a type TMB mine towed on a pole and a single torpedo carried below the keel. scale 1:100

THE BIBER II AND THE BIBER III
The disadvantages of the 'Biber' – the dangerous petrol engine and the physical stress of the single driver – led to the development of a two-man vessel, the 'Biber II', by the Flenderwerke at Lübeck.

When it was found that midget submarines could be propelled more efficiently by closed-circuit engines, the 'Biber II' was rejected in favour of a new version propelled by this type of machinery. In November 1944 a mock-up hull of this type, the 'Biber III,' was tested but the whole project was ordered to be abandoned because a similar vessel, the 'Kreislauf-Seehund', was in the design and development stage at the same time. Nevertheless, the command of the K-Verbaende insisted that development of the 'Biber III' continue and, in January a test-engine was running. Due to

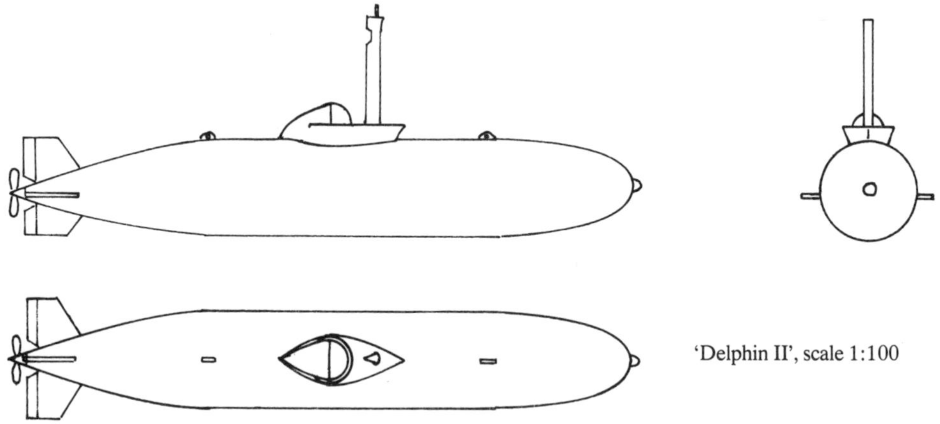

'Delphin II', scale 1:100

the military situation it was not possible to install this engine in the first 'Biber III' experimental hull. After the end of the war, all existing parts of the unassembled 'Biber III' and its machinery were collected by the British, at camp Kronshagen, and examined carefully before being scrapped.

THE DELPHIN

Operational experience showed that a 3–6kt vessel was too slow to match modern fast ships or even tide currents in the Channel. So, in the spring of 1944 Dr Ing Haug designed a streamlined, tear-drop shaped, vessel. Later model drag–tests showed that this hull had extremely good hydrodynamic coefficients and a Berlin based motor-coach bodywork manufacturer was, therefore, ordered to build 3 experimental 'dolphins'. The 'Delphin' was intended for propulsion by an 80hp, 2,5 litre Opel-Kapitän engine modified for air-independent close-circuit combustion. As this engine was, of course, not ready in time, first trials were made with a standard electric torpedo engine of only 24hp but nevertheless a submerged speed of 17kts was achieved!

Unfortunately, this prototype was damaged beyond repair during a collision on 19 January 1945 and the tests had to be abandoned. The two other vessels under construction at Berlin were evacuated on 14–15 April 1945 and blown-up at Pönitz near the Trave estuary when British troops approached.

The question of the armament of this 'tin-dolphin' was tricky, as every

A preserved 'Seehund' outside the Naval Memorial Museum, Washington DC

J Batchelor, courtesy A Preston

GERMAN MIDGET SUBMARINES

Type	Units delivered	Displacement (cu mts)	Displacement (cu ft)	Length (metres)	Beam (metres)	Crew	Horsepower surface (submerged)	Propulsion surface (submerged)	Speed (kts) surface (submerged)	Endurance (nm/kts)
Human Torpedoes										
Neger	*c*200	2.7	95.3	7.6	0.5	1	12	1 Eto	4*	48/4
Marder	*c*300	3	105.9	8.3	0.5	1	12	1 Eto	4*	48/4
Hai	1	3.5	123.6	11.0	0.5	1	12	1 Eto	4*	–
Midget Submarines										
Biber	324	6.3	222.5	9.0	1.6	1	32 (13)	1 Opel-Blitz GE (1 Eto)	6.5 (5.3)	130/6 (8.6/5)
Hecht	53	11.8	416.7	10.4	1.7	1	12	1 Eto	5.7 (6.0)	(38/4)***
Type XXVII										
Molch	390	11	388.5	10.8	1.8	1	13	1 Eto	4.3 (5.0)	50/4 (50/5)
Seehund (Type 127)	285	14.9	526.2	11.9	1.7	2	60 (25)	1 Büssing DE (1 AEG E1)	7.7 (6.0)	300 (500)/7 (63/3)
Projects										
Projected midget/submarines										
Seetaeufel	0	20	706.3	13.5	2.8	2	80 (30)	1 GE (1 EL)	10 (8)	300/10 (80/8)
Kreislauf-Seehund (Type 227)	0	17	600.3	13.6	1.7	2	100 (80) (26)	1 Daimler Benz DE (1 AEG E1)	8.0 (10.3)	340/8 (71/10, 17/4)
Biber II	0	–	–	–	–	2	(13)	1 GE (1 Eto)	–	–
Biber III	0	12	423.8	11.8	2.5	2	60 (40)	1 Daimler Benz DE	8.0 (5)**	1100/8 (100/5)
Delphin I	0	2.8	98.9	5.5	1.0	1	32	1 Opel GE	10 (17)	– (300/10)
Delphin II	0	7.5	264.9	8.7	1.3	1	100 (80)	1 Daimler Benz DE	– (18)	– (400/?)
Schwertwal I	0	14.7	519.1	11.3	2.4	2	800	(modified Walter)	– (30)	– (500/10)
Schwertwal II	0	–	–	–	–	2	–	–	–	–
Manta	0	–	–	–	–	–	–	–	–	–

KEY Eto = electric torpedo engine, GE = petrol engine, DE = diesel engine, El = electric motor, * = 6kts without torpedo, ** = 7.7 kts without torpedo, *** = 69/4 with additional batteries.

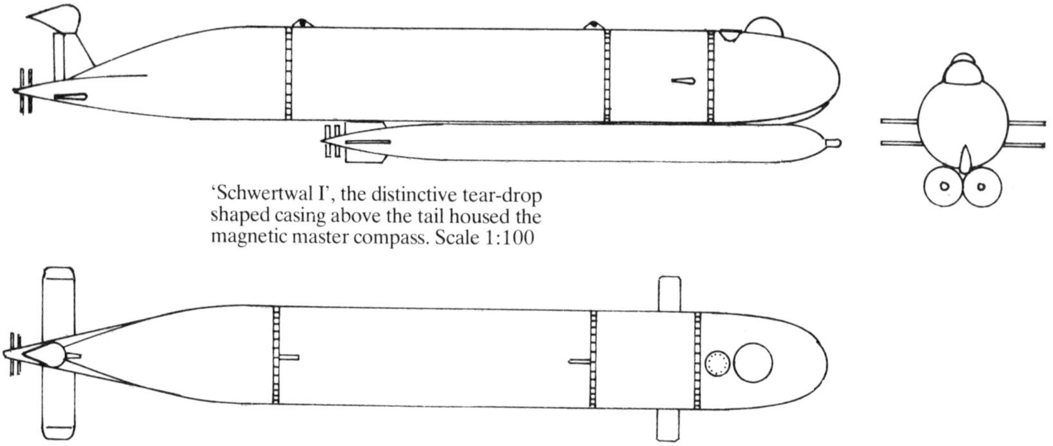

'Schwertwal I', the distinctive tear-drop shaped casing above the tail housed the magnetic master compass. Scale 1:100

weapon clamped outboard of the hull resulted in the drastic reduction of underwater speed to 4–5kts. It was suggested that it might be employed as a semi-suicide weapon containing a 500kg (1102.3lbs) internal warhead but it was quickly realised that it would be impossible to ensure the driver's escape while running at high speed underwater. Better possibilities were seen in towing a torpedo on a pole but the end of the war brought a stop to all thoughts of what to do with this promising vessel.

THE SCHWERTWAL

The 'Grampus' was a project for a high-speed midget submarine, which could operate underwater like a fighter aircraft. It was therefore designated 'submarine-fighter' part of whose function was to act as a submarine-killer. This project was still under consideration by the command of the K-Verbaende in June 1944 despite the official cancellation of the development of a Walter-turbine for midget-submarines. To achieve the high submerged speed of 30kts a sophisticated streamlined hull was developed with the use of a wind-tunnel. An 800hp Walter turbine was planned to propel this two-man vessel which was to be armed with two torpedoes.

The prototype 'Schwertwal I' was an experimental vehicle which, at the end of the war, was scuttled in Lake Plön (part of Walter's plant proving ground). In July 1945 the vessel was detected by the British, salvaged and later scrapped at Kiel. An improved vessel, the 'Schwertwal II', based on the prototype tests, was on the drawing board at the end of the war.

The 'Schwertwal' would have carried an interesting variety of armaments, including – (a) rocket-propelled anti-submarine torpedoes codenamed 'K-Butt' (standing for closed-circuit butt); (b) floating mines for defence against hunting submarine-chasers; (c) rocket tubes and flamethrowers for surface defence; (d) various types of mine.

THE MANTA:

The design of the 'sting-ray' was an example of how frustrated designers develop fantastic ideas after years of suppression and denial by officials

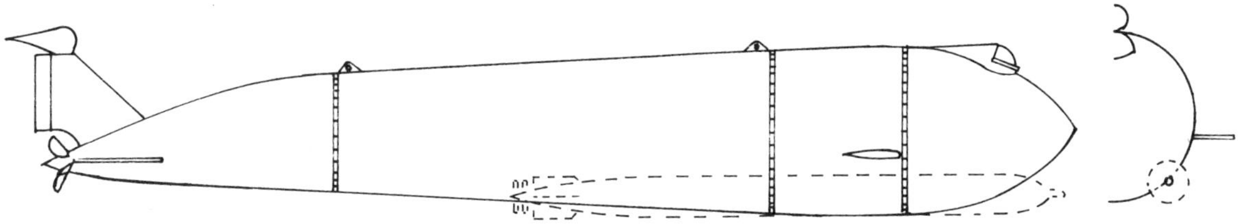

The '*Schwertnal II*' design study also carried the external master compass casing. Scale 1:100.

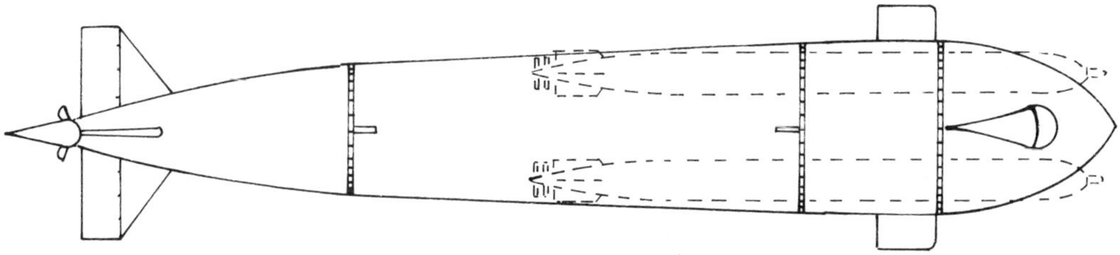

The 'Seehund' in the grounds of the Naval Memorial Museum, Washington DC

J Batchelor, courtesy A Preston

without imagination. During the breakdown of the Third Reich the technocratic control decreased and one of the most fantastic vessels ever designed was born in the minds of certain designers who had 'dropped out' of the true political/military reality – still believing in the 'Endsieg' although Allied troops were already fighting in Germany.

The 'Manta' was a curious mixture of trimaran and aircraft. Both outer hulls consisting of 'Schwertwal II' units with the centre hull housing the two-man crew and a diesel engine. These three cylindrical hulls were linked by wing-like structures, the vessel also having a pair of aircraft wheels to enable it to roll over a beach and into the water.

The 'Manta' was intended to operate in a variety of modes, namely – (a) gliding on the waves at a speed of 50kts; (b) Diesel-propelled snorkeling; (c) underwater chasing at 30kts with Walter-propulsion; (d) underwater creeping at 8kts with an electric motor. As far as known this fantastic device was never fully designed, or tested as a model.

We would ask our readers, especially those in France, The Netherlands, Belgium, Germany, Denmark and Norway to send us pictures of German small fighting units as we know that a number of them are exhibited in these countries. Please describe the exact location, as some of them are displayed at little-known museum and even at petrol stations.

Marder type human torpedo

Drüppel

THE MAGNIFICENT 7

THE ROYAL SOVEREIGN CLASS OF 1889 PART 1

by R A Burt

The British naval manoeuvres of 1888 revealed a number of defects in the design of the fleet vessels and in particular emphasized a major problem with the battleships. The low freeboard turret ships and the barbette ships of the 'Admiral' class suffered badly in a seaway and in some cases rolled and pitched so violently that they would have been unable to fight their guns efficiently, if at all. In the turret vessels the problem was worsened by the fact that the main armament was positioned so close to the waterline that it was not unusual for the muzzles of the forward guns to be dipping into the foam shipped over the forecastle.

As a result of a detailed study of these problems the Board of Admiralty, largely influenced by the Director of Naval Construction (DNC), Sir William White, decided that a new design of battleship should be adopted, combining the best features of earlier vessels with the high freeboard necessary for good performance in a heavy swell. Other new ideas were also introduced and the resulting design, which became the *Royal Sovereign* class, was the first to be fitted with the 6in quick firing (QF) gun, nickel-chrome steel armour (for the upper belt) and casemate protection to the secondary armament. The

The turret ship *Hood* viewed from ahead showing the layout of the flying bridge with the conning tower beneath it. Note the QF guns in the tops and the crest on the stem head.

MoD

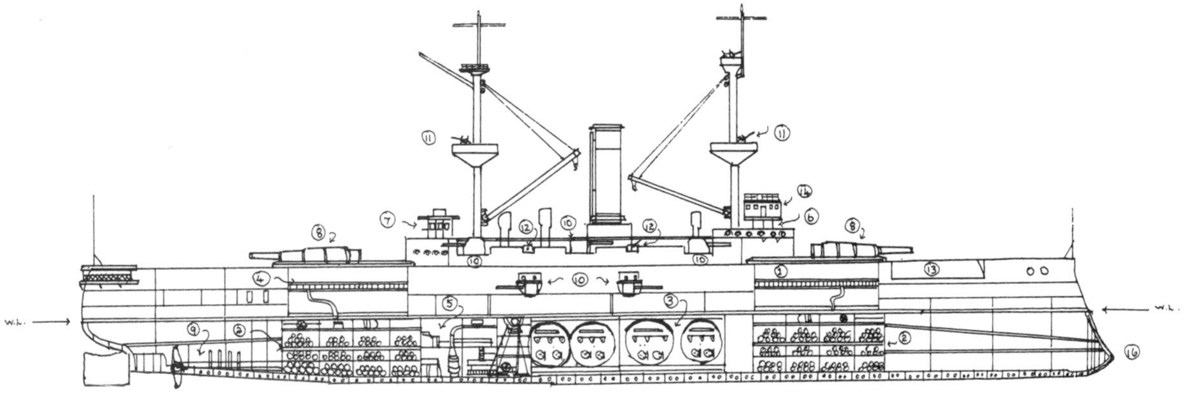

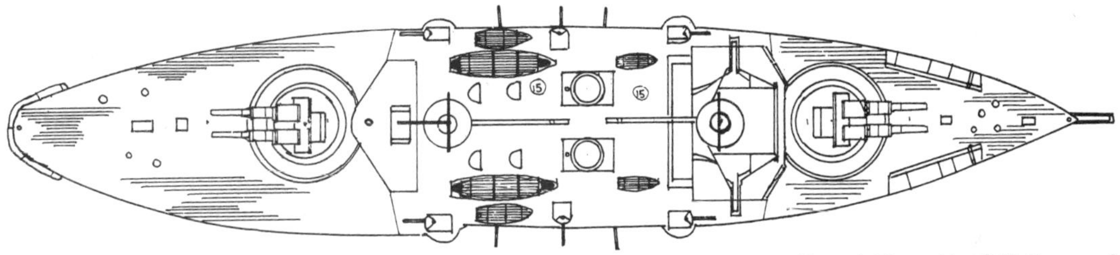

Key – 1. Turntable of 13.5in guns; 2. Shell rooms and magazines; 3. Boiler rooms; 4. Turntable mechanism; 5. Engine rooms; 6. Conning tower; 7. After conning tower; 8. 13.5in guns; 9. Shafts and steering gear; 10. 6in secondary armament; 11. 3pdr guns; 12. 6pdr guns; 13. anchor beds; 14. Bridge; 15. boat stowage; 16. ram bow.

DNC also wanted to provide them with a sloping armoured deck but the Board would not approve this feature and it was not adopted until a few years later (in the 2nd class battleship *Renown* of 1892).

As completed the *Royal Sovereigns* represented a great advance on earlier ships and initiated the gradual phasing out of the low freeboard type. Although one of the class, *Hood,* was designed and completed as a low freeboard turret ship at the insistence of the First Naval Lord, Admiral Hood, she soon proved inferior to her high freeboard half sisters. They were fine ships, strong and well built with excellent accommodation and a simple but effective layout. They could probably have taken-on any of their contemporaries with an almost certain chance of winning and for a few years enjoyed an unchallenged reputation, but developments in gun

The two after 13.5in guns of *Royal Sovereign.* Note that these weapons are entirely above the level of the barbette where they were not only vulnerable to damage themselves but could act as shell bursters, sending splinters down inside the barbette. Limited splinter protection was however provided by covering the top of the barbette with thin metal plates. A sighting hood, in this cover, is visible to the right of the nearest gun and behind that, standing on the cover, a machine-gun land-carriage. The chequered plates, on the deck in the foreground, are blast plates intended to protect the wood deck-planking when the guns were fired

IWM

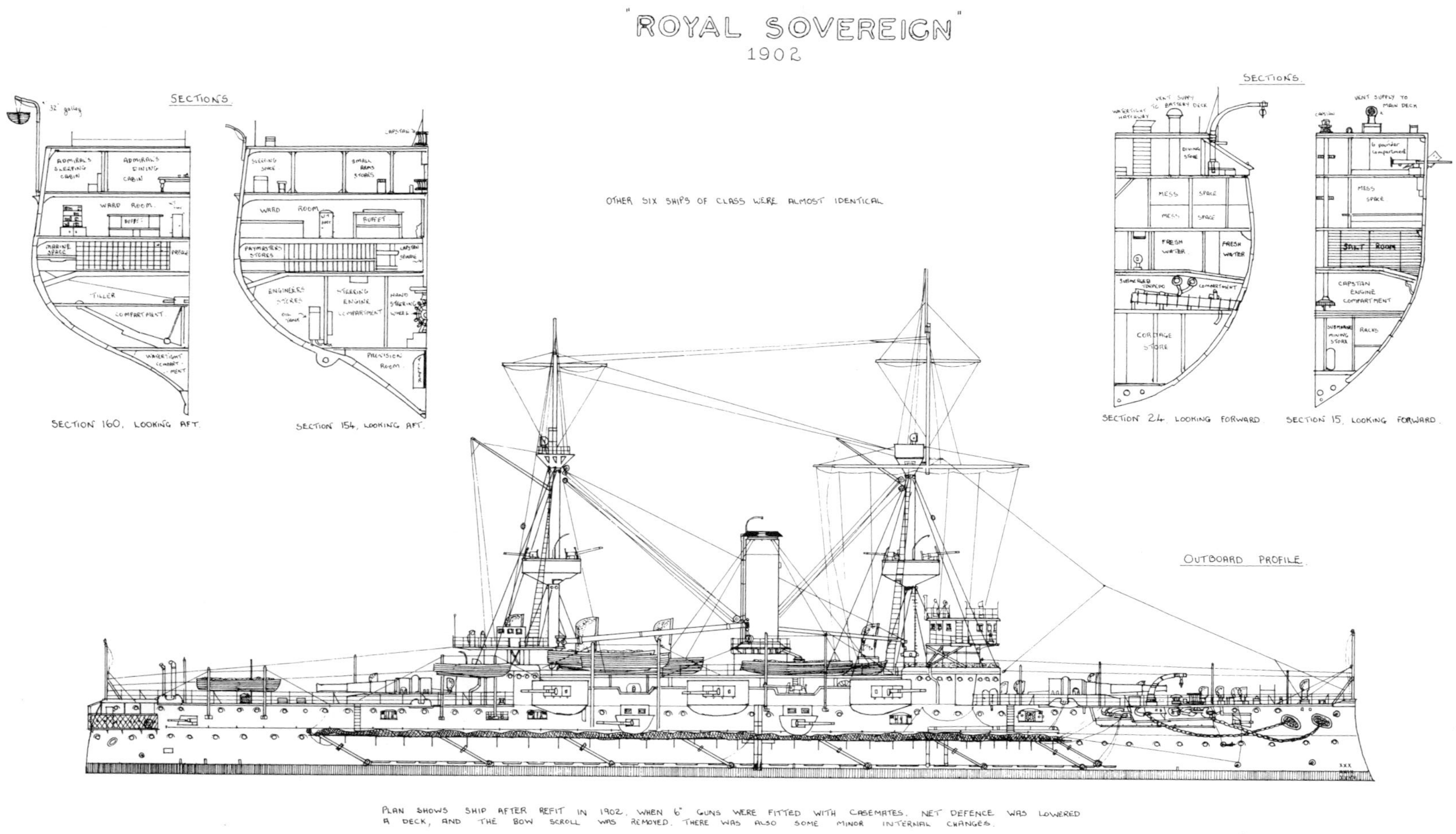

All drawings by the author

and armour manufacture were advancing very rapidly and they were quite obsolete within a decade of their completion. The general features of their design were copied by all the major naval powers and surely no finer tribute than this could be given to William White's revolutionary ships which had greatly advanced the evolution of the battleship and placed Britain well ahead of her rivals – a lead she was to enjoy for many years to come. The *Royal Sovereigns* never had the opportunity to prove themselves in action, all except one, the *Revenge,* being scrapped prior to the outbreak of the First World War but had they been engaged there can be little doubt that they would have given a good account of themselves.

DESIGN

On 17 August 1888, a meeting of the Board of Admiralty took place at Devonport dockyard, during the Boards annual inspection of the Yard, to decide on the specification and dimensions of the battleships to be ordered under the estimates for

The *Royal Sovereign* in the Solent in about 1898 or 1899, compare this with the view of *Hood* taken from the same angle

Beken of Cowes

The *Hood* some time during 1894-95. Note her low freeboard and turret mounted guns – the only obvious external differences between her and her half sisters of the *Royal Sovereign* class

IWM

"Internal armour comparison"

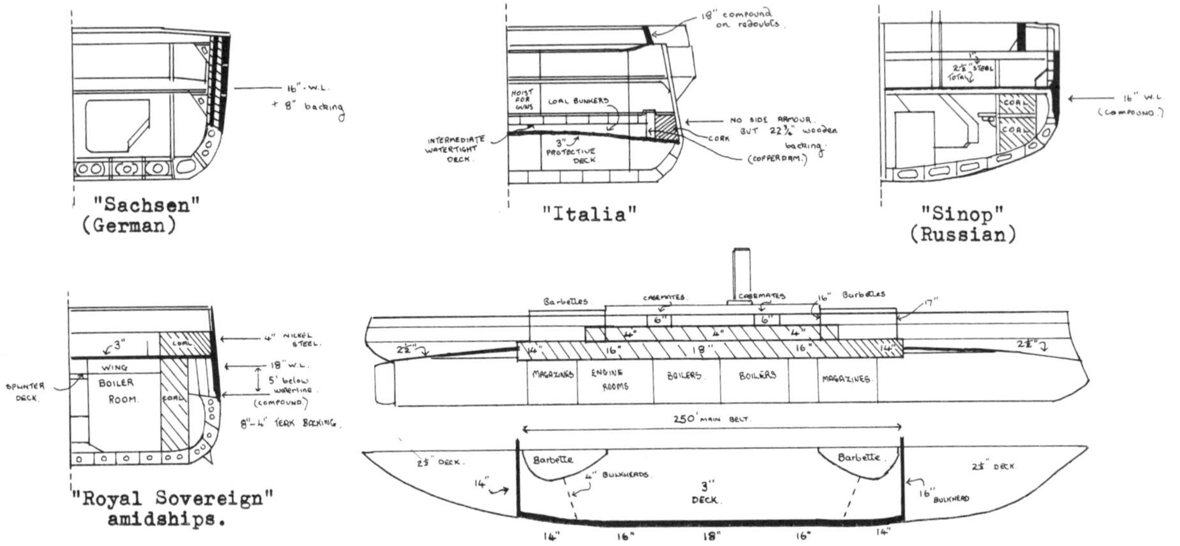

Layout of the "Royal Sovereign"

1889. The members of the Board had papers and plans detailing various battleship types and in the course of discussions reviewed the merits of the various features of these vessels. The main topics were: a) whether the new ships should have a high or low freeboard (the former only possible with a barbette mounted main armament); b) the number, distribution and mounting arrangements of the main armament; c) the disposition of the secondary armament; d) the arrangements and thickness of the armour protection; and e) the main machinery. After a lengthy debate the following points were agreed by all present:

1. There should be four heavy guns in two widely separated protected stations, each pair of guns having arcs of fire of at least 260° and both being capable of firing on either side of the ship. After considerable discussion it was agreed that these weapons should be of 13.5in calibre, which would bring them into line with the newest battleships then completing (*Nile* and *Trafalgar*), and that they should be mounted fore and aft.

2. The secondary armament was to consist of ten 6in guns, the main part of which should be placed in a long broadside central battery between the main gun positions with light protection where possible. The remaining secondary guns should be placed one deck higher to reduce the mutual blast interference between one gun and another and limit the danger of several guns being put out of action by a single hit. This latter resulted from consideration of recent developments in high explosives and mediuum calibre QF guns.

3. The main armour belt was to be of the same ratio to length as in the *Trafalgar* class. The main belt was to be at least 18in thick with a 5in (later reduced to 4in) upper belt and a 3in protective deck.

There was also discussion on the citadel (a complete armoured box above the main belt) and redoubt (heavy armoured protection restricted to bases of main armament mountings above main belt) systems of armouring, figures of 14in and 12in thickness being given for citadel armour. No decision was made as to the type of ship to be adopted but it was made clear that whatever was chosen would be subject to the same cost limitation, although ample funds were allowed for the final design. From these requirements the DNC prepared several outline designs for both turret and barbette ships which were the subject of considerable discussion and some controversy. The DNC forcefully expounded on the advantages of the barbette ship, with redoubt protection and, particularly, high freeboard, while the First Naval Lord, Admiral Hood, preferred the more extensive protection available in the turret type which, due to considerations of top weight, also required low freeboard. To resolve this difference of opinion the Board decided to seek the opinions of officers with recent sea-going experience and to this end they invited Admiral Sir W Dowell KCB, Admiral Sir R Vesey Hamilton KCB, Vice Admiral Sir F Richards KCB, Vice Admiral J K E Baird and Captain Lord Walter Kerr RN, ADC, to a meeting in the First Lord's Room at the Admiralty which took place on 16 November 1888 (Admiral Sir George Tryon KCB was also invited but was unable to attend). Also present were the DNC, the Director of Naval Ordnance (Captain John Fisher CB – later Admiral Lord Fisher) and the members of the Board – The Right Hon Lord George Hamilton MP, Admiral Sir Arthur Hood KCB, Vice Admiral Sir A H Hoskins KCB, Rear Admiral J O Hopkins (Controller of the Navy), Rear Admiral C F Hotham CB and A B Forward MP Parliamentary and Financial Secretary).

The officers were asked to comment on various aspects of the outline designs prepared by the DNC, the particulars of the more important of which are given in Table 1. After lengthy debate all present agreed that the speed was adequate within the ships limit of size and that the proposed distribution of armament was satisfactory. They considered the redoubt system of protection better than the citadel system but would have liked the upper belt to be thicker if possible and requested that protection be given to the secondary armament. In the case of freeboard the DNC's views were confirmed and all agreed that for a sea-going battleship capable of maintaining high speed and fighting power in a seaway the high freeboard and the high placed guns of a barbette ship were preferable to the low freeboard and low guns of a turret vessel. Although finding almost everybody disagreeing with his point of view Admiral Hood recommended that at least one ship of the programme should be a turret vessel and the DNC was therefore asked to work out in detail both the barbette design (6) and the turret design (3) which ultimately became the *Royal Sovereign* and *Hood* respectively.

There was some justification for Admiral Hood's stand as the open topped barbette mounting left the guns unprotected, and the mechanism and crew beneath them vulnerable to shells bursting against the guns and light weapons fired from the tops of enemy vessels. There was also more unprotected structure in the barbette design but the greater efficiency of the type as seagoing vessels greatly outweighed these disadvantages. If proof were needed, it was provided when the *Hood* completed for she soon proved inferior to the other ships of the class and she was consequently the last low freeboard battleship built for the RN. The major disadvantages of the protective

TABLE 1: OUTLINE DESIGNS

	Trafalgar (as designed)	Citadel turret ship	Redoubt turret ship	Redoubt turret ship	Barbette ship Admiral type (as completed)	Barbette ship	Barbette ship
Length (ft)	345	380	380	350	330	380	350
Beam (ft)	73	75	75	70	68½	75	70
Draught (mean) (ft)	27½	27½	27½	27	27-10	27½	27½
Displacement (tons)	12000	14000	14000	12000	10550	14000	12000
Freeboard (ft)	11¼	11¼	11¼	11¼	10¼	18	18
Height of heavy guns above water (ft)	15	17	17	17	20	23	23
Horse power (forced draught)	12000	13000	13000	12000	11000	13000	12000
Speed (forced draught) (kts)	16½	17	17	17	16¾	17	17
Speed (natural draught) (kts)	15	15	15	15	15	15	15
Coal (tons)	900	900	900	900	900	900	900
Armament – main	4 67 ton BL	4 67 ton BL	4 67 ton BL	4 45 ton BL	4 67 ton BL	4 67 ton BL	4 45 ton BL
– Secondary	8 5in BL	10 6in QF	10 6in QF	10 6in QF	6 6in BL	10 6in QF	10 6in QF
	8 6 pdr	10 6pdr	10 6pdr	10 6pdr	12 6pdr	16 6pdr	16 6pdr
	10 3pdr	8 3pdr	8 3pdr	8 3pdr	7 3pdr	8 3pdr	8 3pdr
Length of belt (ft)	230	250	250	230	150	250	230
Armour – belt (ins)	20 to 14	18 to 14	18 to 14	14	18	18 to 14	14
– Turrets (ins)	18	15	18	13	–	–	–
Armour – citadel, redoubts, or barbettes (ins)	18	14	17	13	14	17	13
side between bases of barbettes or turrets (ins)	16	12	5	4	Unprotected	5	4
Armour tube, or base of barbettes (ins)	–	–	–	–	12	17	13
– Conning tower (ins)	14	14	14	14	12	14	14
– secondary armament	unprotected	protected	protected	protected	unprotected	protected	protected

TABLE 2: PARTICULARS OF ROYAL SOVEREIGN CLASS

Name	Builder	Machinery	Laid down	Launched	Completed	Cost (£)
Royal Sovereign	Portsmouth Dky	Humphreys & Tennant	30. 9.1889	26. 2.1891	5.1892	912,986
Revenge	Palmers	Palmers	12. 2.1891	3.11.1892	22. 3.1894	954,825
Resolution	Palmers	Palmers	14. 6.1890	28. 5.1892	5.12.1893	953,817
Repulse	Pembroke Dky	Humphreys & Tennant	1. 1.1890	27. 2.1892	25. 4.1894	915.302
Ramillies	Thompson (Clydebank)	Thompson	11. 8.1890	1. 3.1892	17.10.1893	980,895
Empress of India	Pembroke Dky	Humphreys & Tennant	9. 7.1899	7. 5.1891	11. 9.1893	912,162
Royal Oak	Lairds (Birkenhead)	Laird	29. 5.1890	5.11.1892	12. 6.1894	977,996

Displacement: 14,150 tons (legend), 14,050 *(Empress of India)* to 14,420 *(Ramillies)* tons (actual), 15,220 tons (full load) (see also Table 4) *Ramillies* – 10,464 tons light at 21ft 8in draught, 13,290 tons at 25ft 11in draught, 14,620 tons at 28ft draught, 15,220 tons deep at 28ft 11in draught

Sinkage: 52.5tons/in at deep displacement, 51.9 tons/in at light displacement

Length: 380ft (pp), 410ft (oa)

Beam: 75ft

Draught: 27ft 6in (mean), 29-30ft (max) (see also Table 4)

Armament: four 13.5in, 67 ton BL; ten 6in QF; sixteen 6pdr QF; twelve 3pdr QF; seven 18in torpedo tubes (two broadside submerged; four broadside and one stern above water)

Shell stowage: as approved 30 June 1891 – 80rpg 13.5in (20 AP, 12 Palliser, 39 common, 10 Shrapnel); 150rpg 6in (10 AP, 38 Palliser, 128 common, 24 shrapnel) as completed – 80rpg 13.5in, 200rpg 6in, 500rpg 6pdr, 5400rpg machine guns.

Armour: main belt 18–16in and 14in (250ft long x 8½ft deep) with 4in to 8in teak backing; upper belt 4in (150ft long x 6½ft deep); total height of side armour 15ft (5½ft below and 9½ft above load water line); bulkheads 16in (forward) 14in (aft), 3in at ends of upper belt; barbettes 17in, 16in and 11in sides, 1in roof and floor; casemates 6in face and sides (roof and floor unprotected); conning tower 14in – 12in and 8 in; after conning tower 3in; decks – middle 3in (over length of main belt), lower 2½in (outside citadel)

Machinery: Two sets 3 cyl, vertical, triple expansion engines; 9000 ihp = 15.5kts (natural draught), 11000ihp = 16.5 kts (forced draught); 8 single-ended cylindrical boilers, working pressure 155lbs/sq in; coal stowage 900 tons (legend), 1100 tons (normal) 1490 tons (maximum); radius of action 2780 miles at 14kts, 3086 miles at 13 kts, 4720 miles at 10 kts, engine room complement 138

Boats: Approved stowage as designed (1888-89) – one 40ft steam pinnace, one 53ft steam vedette boat, one 56ft wooden torpedo boat, one 42ft launch, one 36ft pinnace, one 30ft cutter, one 30ft gig, one 27ft whale gig, one 25ft gig, two 23ft gigs, one 26ft jolly boat, one 16ft skif dingy, one 14ft punt; as fitted in *Empress of India* (1893) – two 56ft steam pinnace, one 40ft steam pinnace, one 42ft launch, one 36ft pinnace, two 30ft cutters, one 26ft cutter, one 24ft whaler, one 13½ft balsa raft, one 32ft gig, one 28ft gig, one 24ft gig, one 16ft dingy

Anchors: Two 118cwt Inglefield (bower) one 117cwt Inglefield (sheet), two 55cwt Martins, two 43cwt Martins

Masts: Foremast – housing from heel to deck 15ft x 37in dia, deck to head 70ft 6in x 37in dia, overall height of mast 85ft 6in; fore topmast – 48ft x 12in to 6in dia; lower yard – 53ft x 10in dia, yardarms 3ft 2in x 5in (dia at ends); signal gaff – 25ft 4½in x 6½ dia (heel), 4½in dia (head); upper yard – 24ft 6in x 5½in dia, yardarms 1ft x 3in dia. (main mast dimensions identical to foremast)

Complement: 670-700; *Empress of India* as a flagship (1903) 692, *Resolution* 1903) 672, *Revenge* as private ship (1904) 673, *Revenge* as gunnery training ship (1906) 466.

arrangements of the *Royal Sovereign* class were in any case largely solved in later ships, mainly due to the adoption of hard faced steel armour, which allowed a greater area to be protected without increase in weight, and the addition of armoured hoods (later known as turrets) to the tops of barbettes.

CONSTRUCTION

The design work being complete, seven battleships of the *Royal Sovereign* class were ordered under the Naval Defence Act of 1889, three of the ships going to Royal Dockyards and the remainder to private contractors. They represented a great success for William White, who had clearly seen the faults of existing battleships and how to solve them, and pushed through his ideas against serious opposition. The *Royal Sovereign*s broke a long lasting size limitation on British battleships and were the largest such vessels so far built for the Royal Navy. They were also the largest group of battleships built to a single design since the introduction of the ironclad and heralded a decade of uniformity for the British fleet in which homogeneous battle squadrons once again became possible.

While under construction some modifications were made to the design and incorporated in the ships. These included increasing the height of the funnels, which added 13 tons, increasing the fighting tops adding 4 tons, and increasing the thickness of the boiler tubes, adding 20 tons. The complement was also enlarged adding a further 10 tons,

"ARMOURED BELT LAYOUT"

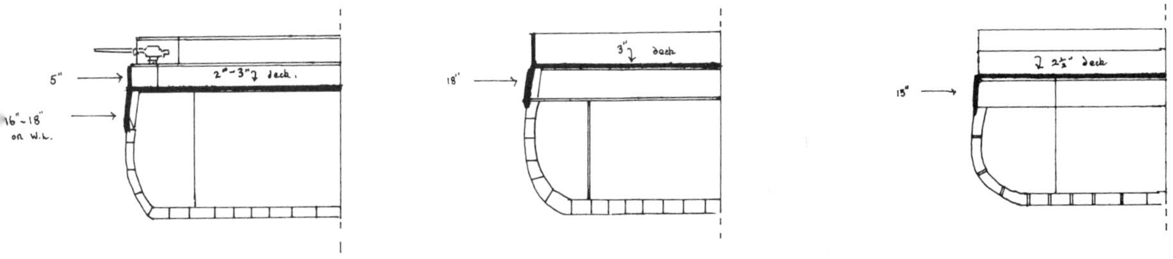

"TRI-SVITITELIA" "INDIANA" "WORTH"

ıree more 3pdr QF guns were
.tted and the carpenter's stores and
ıe number of boats increased.
'hese, with various other minor
dditions, amounted to a total
alculated weight of 137.5 tons
vhich was taken from the Board
ıargin.
The ships completed during
892-94 and soon proved highly
uccessful and, at the time, were
›robably unequalled in all-round
ighting efficiency. They were also
›f impressive and elegant
ppearance and could easily be
dentified by their twin,
.thwartships, funnels, high
reeboard and open topped
›arbettes. The *Royal Sovereigns*
ame as a welcome change and all
oncerned with their design and

The *Ramillies* fitting-out at the Thompson Yard on the Clyde in the summer of 1893

Scottish Records Office

construction were more than pleased with the result of their labours. No better description can be given to these ships than that provided by Dr Oscar Parkes *(British Battleships)*, 'White . . . provided the Navy with the finest group of fighting ships afloat, which sat in the water with majesty and distinction. For the first time since the *Devastation* set a new standard for unsightliness, a British battleship

A close up of *Ramilies* during fitting-out

Scottish Records Office

TABLE 3: WEIGHTS

Legend Weights (tons)

Hull	5,075	35.7%
Vertical armour & backing	3,460	32.2%
Horizontal armour & backing	1,100	32.2%
Main armament	910	10 %
Secondary armament	500	10 %
Main machinery, engineers stores, and auxiliary machinery	1,115	7.9%
Warrant officers stores and boats	100	4 %
Crew	110	4 %
Water and provisions	125	4 %
Masts, rig and net defence	110	4 %
Anchors and cables, etc.	125	4 %
Coal	900	6.4%
Board margin	520	3.7%
Displacement at load draught	14,150	

Armour weight breakdown	**As designed (tons)**	**As completed (tons)**
Main belt and bulkheads	1,350	1,310
Upper belt and battery sides	450	–
Upper belt	–	225
Casemates	–	150
Barbettes	1,385	1,345
Conning tower	85	80
Wood backing to sides and bulkheads	80	75
Wood backing on barbettes	50	40
Deck plating and shell proof gratings	1,010	1,060
Total	4,410	4,285
Hull	4,890	5,320
Total hull and armour	9,300	9,605

Note: The total weight of hull and armour given in the legend estimates was 9640 tons which included 250 tons for the main armament turntable structure which is not included in the above armour weights.

TABLE 5: ESTIMATED COSTS

	£
Hull, propelling machinery, fittings and equipment	263,000 (dockyard labour)
	445,000 (dockyard material)
	125,000 (contractors work)
Stores for first fitting	12,000
Total	£845,000
Gun mountings and torpedo gear	60,000
fittings and equipment	5,000
Armour: main 18in armour Belt (dockyard est)	121,800
barbettes (official est)	130,000
conning tower (official est)	4,800
casemate fronts (official est)	4,900
lighter armour (dockyard est)	12,100
Total	£273,600*

*£275,000 was allowed for this item in the final estimate, less 5% contract discount.

Auxiliary machinery (prices accepted from Humphrey & Tennant, 13 November 1889):

Steering engines	825
Air compressing machinery	3,000
Loading machinery	2,800
Boat hoisting engines	781
Total	£7,406/ship

TABLE 4: STABILITY PARTICULARS

Royal Sovereign (based on inclining experiment of 16 April 1892)
Legent displacement: 14,060 tons at 27ft 6in draught
Load displacement: with 1100 tons coal on board, boilers full to working height and feed tanks full, GM = 3.6ft, angle of maximum stability, 27°, range 63°
Deep displacement: with 1450 tons coal on board, GM = 4ft (almost)
Light displacement: at 25ft 5in draught, GM = 3.55ft

Royal Sovereign (based on inclining experiment of 21 November 1903)
Load displacement: with 200 tons coal in upper bunkers and 700 tons in lower, draught 27ft 7in, GM = 3.4ft, angle of maximum stability: 36°, range 62°
Deep displacement: with 1490 tons coal onboard, draught 28ft 11in, GM = 3.7ft, angle of maximum stability, 35°, range 64°
Light displacement: draught 25ft 10in, GM = 3.2ft

Ramillies (based on inclining experiment of May 1902)
Load displacement: with 200 tons coal in upper bunkers and 700 tons in lower, draught 28ft 2in, GM = 3.3ft
Deep displacement: with 1480 tons of coal onboard, draught 29ft 8in, GM = 3.6ft
Light displacement: draught 26ft 6in, GM = 3.2ft

Revenge (based on inclining experiment of 24 May 1902)
Load displacement: 14,635 tons with 900 tons coal on board, draught 28ft, GM = 3.35ft
Deep displacement: 15,535 tons with 1410 tons coal on board and 128 tons reserve feed water, draught 29ft 5½in, GM = 4.65ft
Light displacement: 13,560 tons, draught 26ft 4in, GM = 3.2ft

Revenge (based on inclining experiment of 14 October 1915)
Load displacement: 14,799 tons, draught 27ft 6in
Deep displacement: 16,011 tons, draught 29ft 5½in, GM = 3.31ft
Extra deep displacement: 16720 tons, draught 30ft 6in, GM = 3.19ft
Light displacement: 14036 tons, draught 25ft 10in
Note: the above test was carried out after (removable) anti-torpedo bulges had been fitted to the ships hull.

SECTIONS. AS "FITTED"

"ROYAL SOVEREIGN"

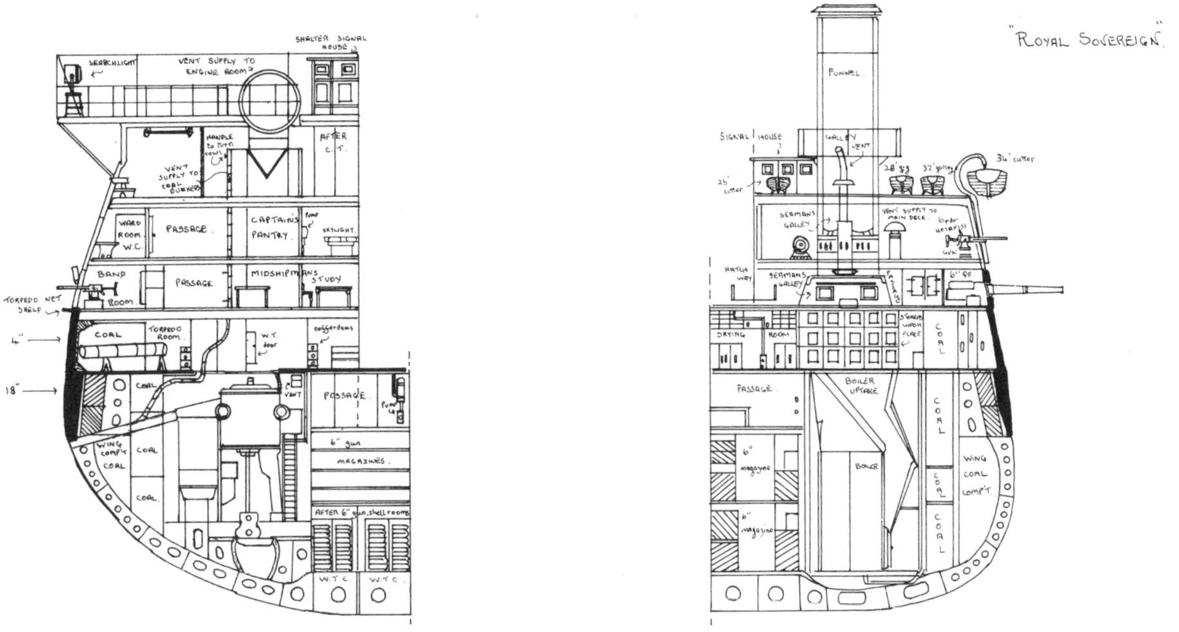

SECTION 119. LOOKING AFT. (ENGINE ROOMS.)

SECTION 77. LOOKING FORWARD. (BOILER ROOMS.)

THERE WERE SOME SMALL CHANGES THROUGHOUT THE LIVES OF THE SHIPS, BUT REMAINED MUCH THE SAME INTERNALLY, ALTHOUGH SOME OF THE TORPEDO TUBES WERE REMOVED.

A view of the *Royal Sovereigns*'s after bridge and 13.5in guns

IWM

1

2

3

4

5

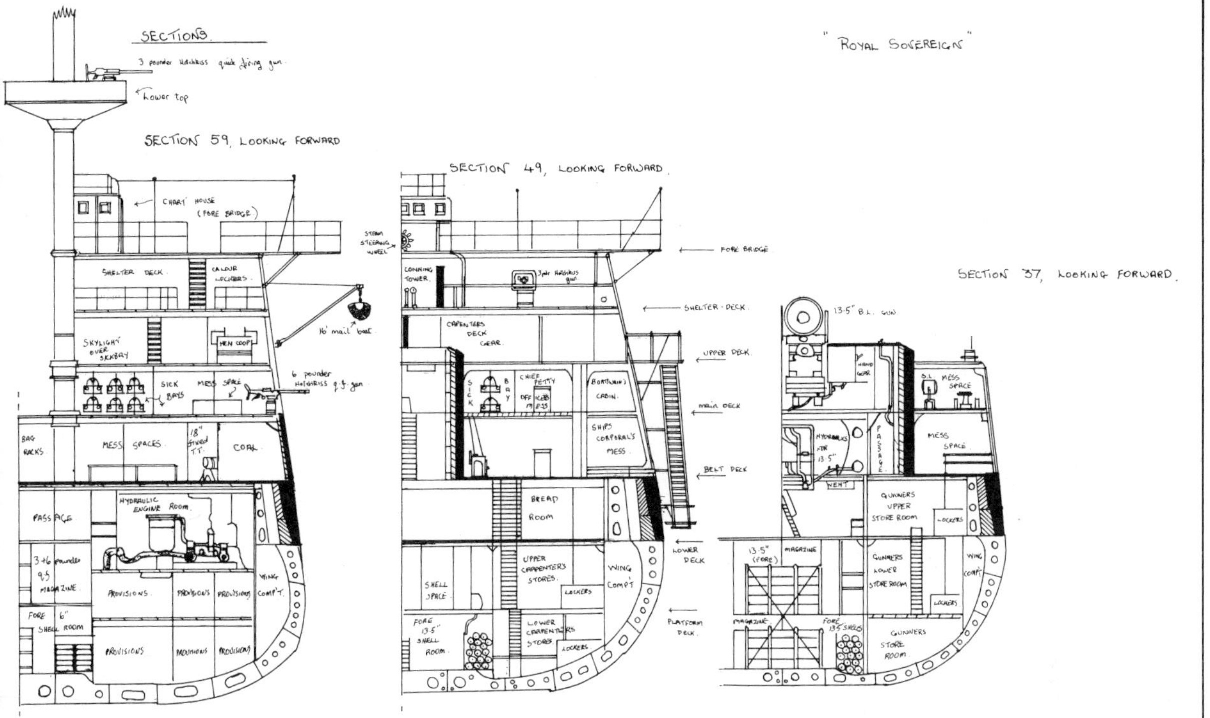

presented a proud, pleasing and symmetrical profile which was unmatched by any other warship afloat, initiating a new era of Volcanic beauty, after two decades of sullen and mishapen misfits.'

A broadside view of *Ramillies* at the time of her trials, note the absence of guns which, in practically all vessels built for the RN during the Victorian period, were fitted in one of the Royal Dockyards after the completion of builders trials.

Scottish Records Office

The *Royal Sovereign* in a heavy swell, possibly during her trials in 1892. Note that although her head is well down no green sea's have been shipped over her forecastle.

IWM

The *Ramillies* departing for her preliminary builders trials in 1893

Scottish Records Office

The *Resolution* shortly after completion, 1894. Note the 3pdr QF guns in the fighting tops and the absence of a bow scroll. Although there were several small differences between the ships of the class they were very difficult to tell apart!

IWM

The *Royal Sovereign* in about 1896

IWM

TRIALS

On completion the ships were put through a lengthy series of steam trials which were studied closely but the records of these were not, unfortunately, preserved apart from a few details. Only the *Royal Sovereign,* the first of the class to run trials, had her boilers forced to the design limit and consequently she achieved one of the highest recorded speeds. During her full power trials at natural draught, which were run in Stokes Bay, she achieved 16.43kts with 9661ihp at 99rpm (mean) during the 8 hour trial and 16.77kts (mean) during the 4 hour trial. On the full power forced draught trials in the Channel, which were run in deep water when possible, she produced mean figures of 18kts with 13,360ihp at 106.3rpm on a displacement of 14,200 tons over a 3 hour period. It was observed during this latter trial that sections of the boiler tubes were cracking and leaking and it was subsequently decided that neither *Royal Sovereign* or her sisters were to be forced beyond 11,000ihp to provide the boilers with a margin of safety against similar breakdowns.

While *Royal Sovereign* was on a voyage from Plymouth to Gilbraltar the opportunity was taken to test her machinery over an extended period. The trial was run in fine weather over a 72 hour period during which time she produced mean figures of 15kts with 8180ihp – an excellent result. When she left Plymouth her displacement was 14,650 tons at a draught of 28ft 6in and it was calculated that 484 tons of coal were consumed during the 72 hour run giving a consumption figure of 1.84lbs/hp/hour – a record low for battleships that had never before been equalled. The ship was also economic at cruising speeds using only about 1 ton of coal per hour at 7.6kts; the ihp for 10kts was 2500.

The *Royal Sovereign* class were the first British battleships to exceed 17kts and, at the time of completion were rated as the fastest such vessels in the world although some foreign vessels had paper speeds which equalled this figure. Other members of the class achieved similar results to *Royal Sovereign;* the *Empress of India* made 15.25kts (natural draught), the *Ramilies* 17.25kts with 11,571ihp, the *Repulse* 17.8kts

TABLE 6: PARTICULARS OF GUNS

13.5in MkII BL, 67 ton gun
Weight of gun (excluding breech mechanism): 67 tons 1cwt 1qtr
Overall length: 433in (36ft 1in)
Weight of breech mechanism: 1 ton 3qtr
Calibre: 13.5in
Length of bore: 30 cal (405in)
Chamber: 18in x 66.5in, 17,100cub in capacity
Rifling: 34 grooves, uniform right hand twist, 1 turn in 30 calibres
Weight of charge: 630lbs SBC (later 187.8lbs cordite)
Weight of shell: 1250lbs (burster = 84$\frac{1}{2}$ lbs)
Muzzle velocity: 2061 ft/sec
Muzzle energy: 35,230 ft tons
Rate of fire: 1 round/2 min
Penetration: 28in wrought iron at 1000 yds, 11in Krupp steel at 3000yds
Cost: £10,859 each
Maximum elevation: 13$\frac{1}{2}$°

6inch MkIV QF gun
Weight of gun: 5 tons
Overall length: 173.5in
Calibre: 6in
Length of bore: 26 cal (160in)
Diameter of chamber: 8in
Rifling: 24 grooves
Weight of charge: 42lbs primatic brown
Weight of shell: 100lbs (burster = 9$\frac{1}{4}$lbs)
Length of shell: 26.5 in
Muzzle velocity: 1960 ft/sec
Muzzle velocity: 2665 ft tons
Penetration: 12.5in wrought iron at muzzle, 8in wrought iron at 200yds.
Weight of cradle: 1 ton 17cwt
Weight of carriage: 2 ton 8cwt

and the *Royal Oak* a remarkable 18.27kts with 11,608ihp. The majority of these trials were run in the designed load condition (14,050 tons). The subsequent fitting of bilge keels produced no discernable effect on the ships' speeds.

As first completed the ships of the class were found to roll rather heavily in heavy seas and it was pointed out that this would greatly reduce their efficiency as sea-going fighting ships. While on a voyage to Gibraltar in *Resolution,* Captain W H Hall made notes on the ship's performance and at one time recorded roles of 23°. To correct this problem bilge keels 200ft long and 3ft wide were fitted in all the ships of the class and an immediate improvement resulted. The performance of *Resolution,* with bilge keels, was monitored while she was operating with the Channel squadron. Those present noted a loss in speed but this was later put down to the heavy swell then running which varied in direction from ahead to abeam. It was concluded that had the ship not been fitted with bilge keels she would have rolled quite violently but as it was she behaved very well and rolled very little. One unexpected side effect of the addition of bilge keels was the improvement of the ships' manoeuvrability. The turning circle was found to reduce from 680 to 500yds with both engines ahead and from 450 to 300yds with one ahead and one astern (see Table 7).

While on a voyage in *Ramillies* in 1895, Commander John Jellicoe (later Admiral of the Fleet, Lord –) stated that he had never seen a ship turned out so well from a builder's yard, especially a contractor. He went over the ship with a fine-tooth comb and all he found to criticise were a few minor details of the design and nothing regarding the fighting capabilities of the vessel. The only major point he raised was that the ship was very stuffy in the accommodation spaces, both for officers and men, due to a lack of good ventilation. There were plenty of cowl vents in the areas concerned but their arrangement did not

TABLE 7: TURNING TRIALS

	Empress of	**Ramillies**	**Repulse**	**Resolution**	**Revenge**	**Revenge**	**Royal Oak**	
With or without bilge keels	without	without	without	without	with	without	with	without
Date	–	22.8.93	13.2.94	2.11.93	–	15.1.94	12.3.95	28.1.94
Draught	–	27ft 1$\frac{1}{2}$in	24ft	26ft 7in	–	26ft	27ft 11$\frac{1}{2}$in	25ft $\frac{1}{2}$in
Tactical diameter ahead at at 14kts (yards): Port	655	660	670	690	–	685	525	660
: Starboard	624	675	650	670	–	670	550	690
Tactical diameter with one engine ahead and one astern (yds): port	–	410	465	485	290*	455	330	415
: starboard	–	430	425	415	270*	435	350	440
Tactical diameter on half power at 8kts, ahead both engines (yds) : port	619	655	685	680	456*	665	510	645
: starboard	609	665	640	660	497*	655	530	650

*at 10kts

TABLE 8: COMPARISON WITH FOREIGN SHIPS

Ship	*Sachsen* **(Germany)**	*Italia* **(Italy)**	*Sinope* **(Russia)**	*Magenga* **(France)**
Launched	1877	1880	1887	1890
Displacement (tons)	7400	13900	10800	10580
Length (ft)	298	400	300	323
Beam (ft)	60	72.75	69	65.6
Draught (ft)	19.33	31.25	26	27.25
HP	5600	18000	13000	7000
Speed (kts)	14	17–18	16–17	16
Coal (tons)	700	1650	886	600
Belt thickness	10in	0	16in	18in
Armament	6–10.2in	4–17in 8–6in	6–12in 7–6in	4–13.4in 17–5.5in

ensure that air in sufficient quantities reached all compartments.

Vice Admiral Walter Kerr, aboard *Empress of India* in 1895, said the ship was 'A1' in heavy weather, having been caught out in a real gale one night she had rode it out magnificently. He added that he had never before found it possible to praise a ship in this manner.

COMPARISON WITH FOREIGN SHIPS

When the specification of the *Royal Sovereigns* was under consideration the details of existing foreign vessels as well as those of British ships were studied and taken into account. The particulars of four of these vessels, all barbette ships, are given in Table 8, and an examination of these ships reveals that while superiorities can be found in some respects none possess the all round capabilities of the *Royal Sovereigns*. Account must however be taken of the fact that these foreign vessels were contemporaries of the British 'Admiral' class and the *Royal Sovereign*, being later, was able to take advantage of newer technology and in all but one case *(Italia)* larger size.

All four vessels took advantage to varying degree, of the high freeboard possible with barbette mounted guns. The best in this respect is *Italia* with the axis of her main guns 30ft above the water line, compared with 23ft in *Royal Sovereign,* and a freeboard of 25ft compared with 18ft. The *Sinope* and *Magenta* had comparable freeboards but the German *Sachsen* had a low freeboard amidships with a high forecastle and poop joined by a longitudinal superstructure incorporating the barbettes. This latter vessel was however a coast defence ship and not an ocean-going vessel.

The best protected of the four ships was *Sinope* with a 16in belt and 2½in compound (1in steel on 1½in wrought iron) deck which was quite adequate. The vulnerable structure was also less than in the other vessels and she had 2in thick hoods over her barbette guns which, although only providing protection from splinters and light weapons, was an improvement on the flat splinter plates laid across the top of the *Royal Sovereigns*' barbettes. The *Sachsen* had a well protected citadel amidships but, apart from a protective deck, no armour outside this and *Magenta* a narrow 18in waterline belt with no protection above it apart from that of the barbettes. The *Italia,* with no side armour, was the least well protected, having only a protective deck and a shallow elongated barbette amidships. At the time this ship was designed, in the late 1870s, her system of protection had substantial advantages and she caused something of a stir but subsequent developments in ordnance, particularly the introduction of the QF gun, rendered such vessels, with large unprotected areas of structure, too vulnerable for serious consideration.

In the case of armament the apparent advantage held by *Italia,* with her large calibre guns, was largely negated by their very slow rate of fire – one round every 4 to 5 minutes. The *Sachsen* had comparatively light weapons while *Sinope* had guns with limited performance owing to the development of faults in the barrels. The *Magenta* is the only member of the group with an armament equal to that of *Royal Sovereign* but she could only bring three to bear on each broadside as she had four single barbettes in a diamond arrangement. All four designs could theoretically bring more guns to bear on fore and aft bearings (forward only in case of *Sachsen*) but, with th exception of *Magenta* broadside fire was equal to that in *Royal Sovereigns*.

Only *Italia,* which for her time was a very fast battleship, could approach the *Royal Sovereign* in speed, the remainder having the moderate speeds common to the majority of battleships built during the 1880s.

Comparisons with the contemporaries of *Royal Sovereign* reveals that the original design and the discussion thereon, of which the British made no secret, did not result in an immediate following abroad. Table 9 gives the particulars of four foreign battleships all of which were laid down in 1891.

The German *Worth* was well thought of at the time of her completion, having six guns in her main armament and a good all-round balance of features within a fairly restrictive displacement. She was low freeboard barbette ship but had a high forecastle and the advantage of armoured hoods over the guns. She did however lack the speed and seakeeping qualities necessary to match the British ships, being little more than a large coast defence vessel.

TABLE 9: FOREIGN CONTEMPORARIES

Ship	*Indiana* **(USA)**	*Tri Svititelia* **(Russia)**	*Worth* **(Germany)**	*Jauréguiberry* **(France)**
Completed	November 1895	1895	October 1893	February 1897
Displacement (tons)	10288	12540	10060	11824
Length (ft)	348	357.5	340.9	364
Beam (ft)	69.25	72.25	64.75	72.8
Draught (ft)	24	27	25.33	27.75
Complement	473	582	591	597
Armament	4–13in	4–12in	6–11in	2–12in
	8–8in	14–6in	8–4in	2–10.8in
				8–5.5in
HP	9000	10600	10000	15800
Speed (kts)	15	18	17	17.5
Armour (in): belt	18	16	15	17.75
decks	2.75	3	2.5	2.75
turrets	15	16	11	14.5
CT	10	12	11.75	9
Radius of action (miles/kts)	3000/10	3980/10	4500/10	3920/10

The USS *Indiana* was a direct reply to the *Royal Sovereign*s and had similar protection and main armament but again the size was limited, which is reflected in the very low speed. The Americans also retained a turret mounted main armament and the low freeboard that went with it makeing her a poor seaboat. In addition, she did not handle well but did have the theoretical advantage of a heavier secondary battery. Again, this ship was little more than a large coast defence vessel and was no real match for the *Royal Sovereign* class.

The Russian *Tri Svititelia* was much closer to the *Royal Sovereign* in armament and protection but followed more closely the design of the British *Trafalgar* class having a low freeboard and turret mounted guns. Like most such vessels she was well protected with a 250ft, 7ft 10in deep, belt of 18in to 16in thickness covering her vitals but, like *Trafalgar,* and the later *Hood,* would have experienced difficulty i maintaining speed and fighting her guns in a seaway.

The French ship, *Jauréguiberry,* was a good seaboat which combine a turret mounted armament with a high freeboard. This latter was achieved by adopting a steep tumblehome and no protection to

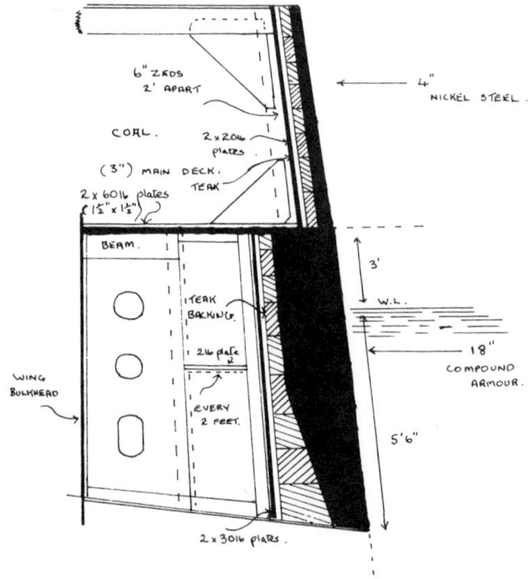

Section of side protection system in *Royal Sovereign* showing teak backing and frame supports behind the armour belt. Some of the plate sizes are given in lbs/sq ft – plates of 40lbs/sq ft are 1in thick, other weights being in proportion

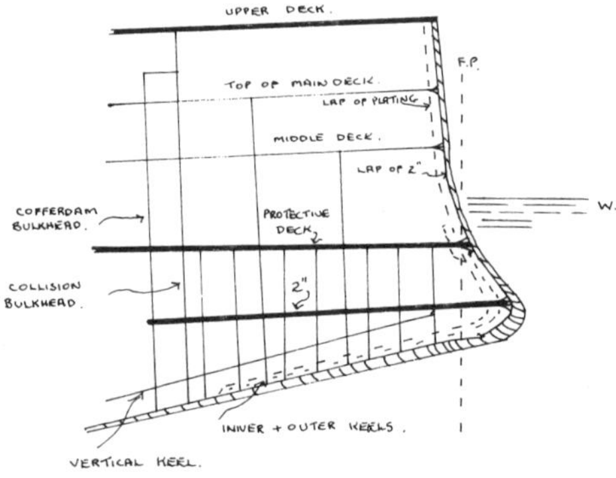

Profile of bow section of *Royal Sovereign* showing internal construction and ram stem

Viewed from the port quarter, the *Royal Oak* in 1896 *MoD*

The *Ramillies* in 1894 *NMM*

the hull above the water line belt. This arrangement produced a vessel somewhat vulnerable to superstructure damage and with a limited reserve of stability and while the paper qualities of the British and French ships could be debated at length it seems unlikely that she would have survived a straight fight with her British counterparts. It is also worth noting that *Jauréguiberry* was very late in completing by which time she was bordering on obsolescence.

The *Revenge* at Portsmouth in the summer of 1894

MoD

The *Repulse* as completed, 1894

NMM

HISTORIC HYDROFOILS OF THE ROYAL NAVY

by D K Brown

The hydrofoil MTB *101* on preliminary trials in 1939

MoD

The success of the coastal motor boats (CMB–stepped hull planing boats) during the First World War led to some continuing interest in fast torpedo-boats after peace returned. A few CMBs were retained in service for trials and considerable attention was paid to Dr Graham Bell's successful test of a ladder-foil craft, driven by aircraft engines and propellers, on Lake Bras d'Or in Canada during 1919.

The Admiralty Experiment Works (AEW), Haslar, initiated a prolonged correspondence with Bell and, at the same time, began a series of model tests on foil sections and configurations. In June 1920 an Admiralty party, consisting of Mr Childs (Constructor) and Commanders Dobson and Mann, visited Bell at Baddeck and spent 3 months studying the performance of his craft. In 1921 Bell offered the Admiralty a 67ft hydrofoil with a guaranteed speed of 40kts but money was short and the proposal was turned down.

EARLY EXPERIMENTS

In 1924 a high speed, towed, target was built based on knowledge of the Bell craft and on the AEW tests. Instead of four ladders in a diamond arrangement the target omitted the bow ladder, leaving two forward and one aft. When completed, it was towed behind a destroyer off Portland at about 30kts and proved to be a complete failure due to instability. In a seaway the hydrofoil would alternately leap out of the water and then crash back and dive deep with a period of about four seconds. The target weighed about 5 tons and was towed behind a long, heavy, wire with a breaking load of 40 tons. The unsteady loading soon snapped the cable and, as might be expected, the development was abandoned.

During the years of financial crisis and disarmament there was little interest in hydrofoils, either in the Royal Navy or elsewhere and the next step did not come until 1936 when Commander Hampden, working with J Samuel White of Cowes, produced an 18ft 'runabout' hydrofoil. It was a dinghy, fitted with an improvised step to which three sets of ladder foils had been added. One set was fitted each side, about one third of the length back from the bow, and the other hung

Not an RN vessel but nevertheless an interesting example of the early experiments with hydrofoils. This is the *Crocco and Riccaldone,* an experimental hydrofoil built for the Italian Army by Cantieri Baglietto in 1906

Courtesy Cantieri Baglietto, Varazze

TABLE 1

Craft	Length (ft)	Displacement (tons)	Speed (kts)	bhp
Hydrofoil Dinghy	18	1.34	33	130
Fast Dinghy	16	1.12	24	50
Fast motor-boat	24½	1.78	24	100

from the transom. There were 5 foils in each ladder supported on a single vertical strut at mid span and arranged with a 3in gap between the foils. The chord was a uniform 4in, while span and angle of attack varied from 24in and 8° at the top to 18in and 1° at the bottom.

This 18ft craft was tried off Cowes in May 1936. Driven by Cdr Hampden it reached about 33kts with 130bhp at a displacement of 1.34 tons. The boat ran with the hull a clear 6in out of the water. No formal seakeeping trials were run but White said that behaviour was 'satisfactory' in roughish water. A comparison was made with contemporary hard chine motor boats as shown in Table 1.

MTB101

Samuel White then put forward a number of proposals, based on this craft, which led to the design for a 67ft motor torpedo-boat, driven by three Isotta-Franschini engines of 3000bhp total. Speeds of up to 60kts were expected. MTB *101*, as the craft was to be known, was ordered from White under the 1936 Estimates at an estimated cost of £34,000. She was of stepped-hull layout and constructed from an aluminium alloy. Ladder foils were arranged as in the 18ft boat but

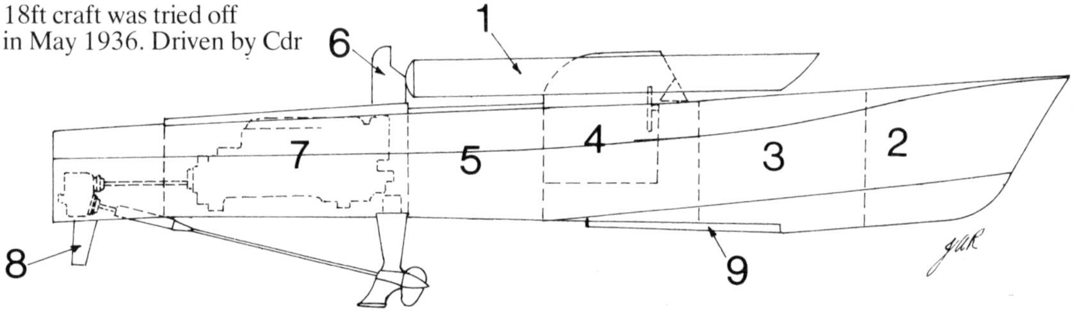

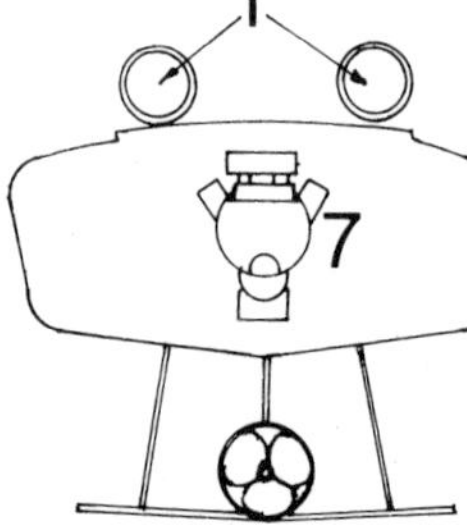

GENERAL ARRANGEMENT OF MTB 109
(with section of foils looking aft)
1. 18in torpedo tubes
2. Fore peak
3. Forward compartment
4. Cockpit
5. Fuel space
6. Air intake
7. Engine room
8. Twin rudders
9. Step raked back to centre line.

The US Navy hydrofoil *Pegasus* constructed by Boeing during 1973-77

additional vertical struts were incorporated to take the heavier load.

Preliminary trials were very unsuccessful. In an attempt to obtain higher speeds new propellers were fitted, as were new foils with a sharper leading edge, and further trials were run in May 1940 with the results shown in Table 2.

TABLE 2

Displacement (tons)	**Speed (kts)**	**Rise on foils (ins)**
22.8	41.35	16
28	34.23	7

It was found that severe cavitation arose both on the backs

HMS *Speedy* under construction at Boeing's Renton works early in 1979

Boeing Marine Systems, courtesy 'Defence'

The *Speedy* ready for launching on 9 July 1979, in Renton, Washington

Boeing Marine Systems, courtesy 'Defence'

of the foils and on the struts as speed increased and that, once this condition had developed, any further increase in power or r.p.m. had no effect on rise or speed. With this incurable problem, MTB *101* was found unsatisfactory and she was not accepted into service. Some records state that she was lost due to hydrofoil failure but it is most likely that this is a misinterpretation of the above facts.

The next development came from the well known Dumbarton firm of Denny who, in 1939, had proposed a semi-hydrofoil consisting of a stepped hull with a single, fixed, fully submerged foil at the after end. Very high speeds were anticipated with the boat riding on the step and the after end lifted right out of the water by the foil.

Model tests in Denny's ship tank (the oldest surving tank in the world) were promising and a 20ft prototype was built by McGruer. The engine was right aft, driving forward to a 'V' drive box in the bow. The single foil was immediately abaft the propeller with a deep rudder behind it. On trials in the Gare Loch, at a displacement of 19 cwts (with 2 men), the boat got up well and reached 23.7kts. The stability was less satisfactory since the boat was inclined to roll from side to side at speed. This was thought to be due to excessive dihedral angle on the foil (15°).

MTB 109

On the whole, this small craft was a success and Sir Maurice Denny proposed a high speed motor torpedo-boat, comparable to Thorneycroft's stepped hull boat, *105*, with two 18in torpedo tubes and a single Rolls Royce 'Merlin' engine. Denny expected that the speed would be 55–60kts at full power and full load in calm water. The Admiralty approved the proposal and the craft was completed and tried, as MTB *109*, in September 1944. The results were 32.6kts at 2000 rpm (max continuous rating) and 45.9kts at 2500 rpm (max continuous rating). On firing trials at Portland it was found that torpedoes could only be launched safely at speeds of 25kts or less. Stability at high speed was unsatisfactory and it was necessary to reduce speed before turning. Thi lack of lateral stability was considered to be so serious that the boat was not accepted. The end of the war in Europe made further development of this type of craft unnecessary and no more hydrofoil were built for the RN for many years.

Some testing work did, however, continue. At the end of the war the Admiralty acquired several ex-German craft: a 27ft 6in runabout and the 46ft VS7, both designed by Professor Tietjens, and the VS6, 51ft 6in in length, designed by Sachsenberg. The damaged remains of VS10 were also available for study. Neither VS6 or 7 were successful having too little power for effective foil borne operations.

MODERN DEVELOPMENTS

In 1952 the Royal Canadian Navy ordered a hydrofoil from Saunders Roe, Anglesey. The British Admiralty contributed funs for tank testing and the research committee was chaired by the Superintendent AEW. After various studies, this craft was completed in 1957 as the *Bras d'Or*.

Since then Admiralty officers have had the opportunity to visit and in many cases carry out instrumented trials on the new, larger, *Bras d'Or,* USS *Tucumcari* and several other hydrofoils.

Early in 1977, the Boeing Jetfoil *Flying Princess* became available for trials, just at the time when serious consideration was being given to such craft for offshore patrol. Another happy chance allowed trials on *Condor V,* an RHS 160 commercial craft, built by Rodrigues of Messina to a surface piercing design. Discussions were also held with Grummans on the feasibility of adapting their *Flagstaff II* design and with the US Coast Guard on their experience with hydrofoils in fisheries' law enforcement. As a result, an order was placed with Boeing for a Jetfoil which is about to enter service as HMS *Speedy*.

The *Speedy* nearing completion

Boeing Marine Systems, courtesy 'Defence'

The *Sverige* (Sweden) and her two sisters, *Drottning Victoria* and *Gustav V* were the last and largest of a series of armoured coast defence ships built for the Swedish Navy. The first vessel of this type was the monitor *John Ericsson* launched in 1864 and during the following fifty years a further 15 armoured vessels, intended solely to defend Swedish territorial waters, were added to the fleet. Each succeeding group was larger than the first and the comparatively small monitors built initially were superceeded by vessels which could well be described as miniature battleships.

The *Sverige* was both heavily protected and heavily armed for her size, although speed was limited to about that of existing battleships. She was originally ordered in 1911 and then cancelled, following a change of Government, but, by asking the Swedish people for voluntary contributions to built the ship, sufficient funds were raised for both *Sverige* and the preliminary costs of her sisters. She was eventually laid down by Gotaverken in Gothenberg in 1912, launched in 1915 and completed in 1917. The *Drottning Victoria* and *Gustav V* were built to a slightly larger design incorporating 6 boilers and a speed of 23 knots.

The *Sverige* was refitted with a heavier foremast in 1926 and, during 1931-33, her forward funnel was remodelled to rake aft (to keep funnel gases clear of the bridge), her torpedo tubes and 75mm guns were removed and four 75mm AA added. She was modified again during 1938-40 when the two amidships 15cm turrets were removed and four 40mm and four 25mm AA guns added A further two 40mm and three 20mm AA were added before the ship was removed from the effective list in 1953. She was eventually scrapped in 1957.

The drawing shows *Sverige* as originally built and is based on official drawings.

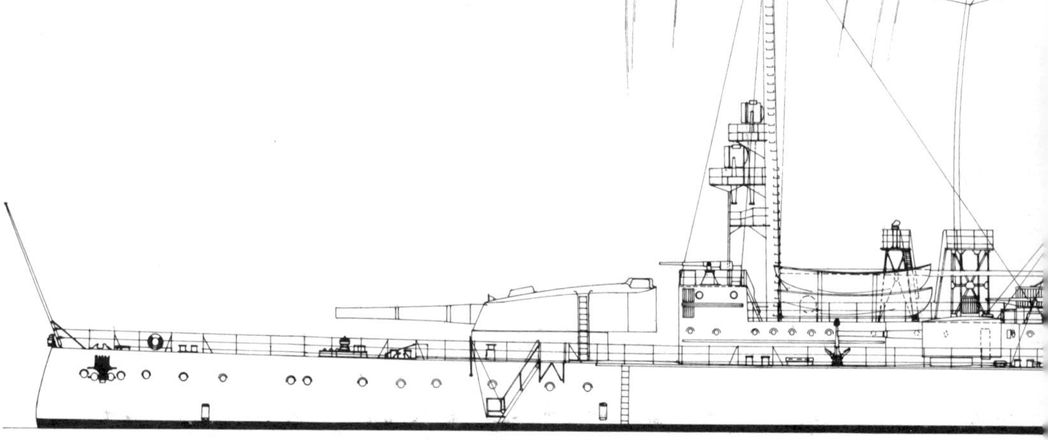

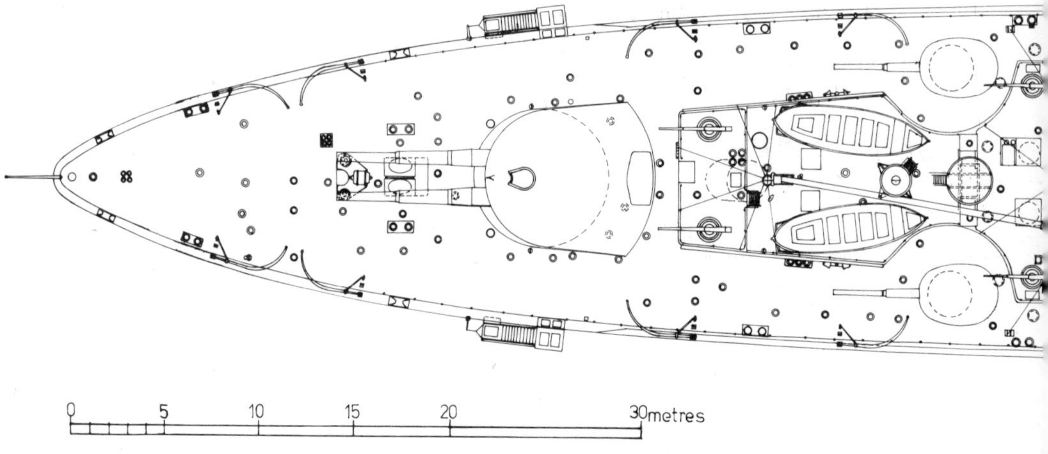

SVERIGE Drawn by John Roberts

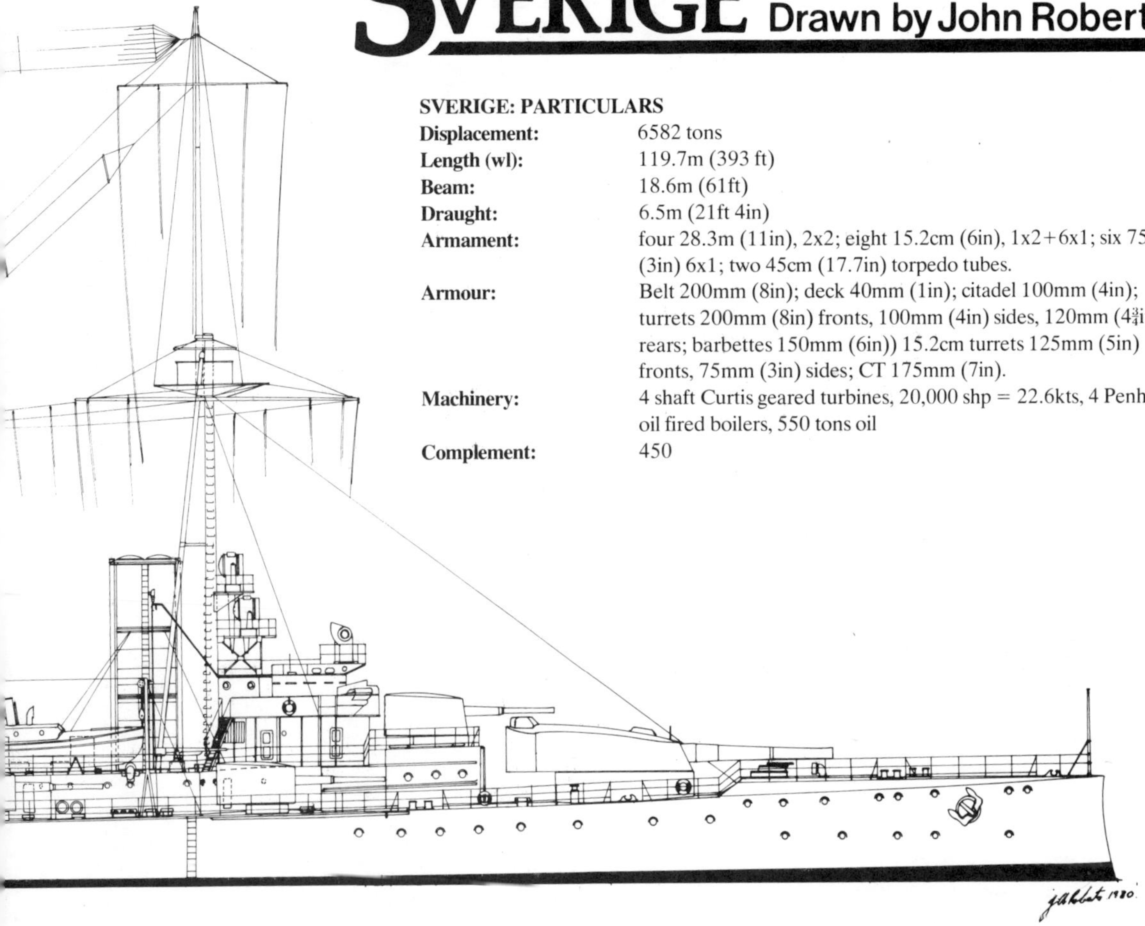

SVERIGE: PARTICULARS

Displacement:	6582 tons
Length (wl):	119.7m (393 ft)
Beam:	18.6m (61ft)
Draught:	6.5m (21ft 4in)
Armament:	four 28.3m (11in), 2x2; eight 15.2cm (6in), 1x2+6x1; six 75mm (3in) 6x1; two 45cm (17.7in) torpedo tubes.
Armour:	Belt 200mm (8in); deck 40mm (1in); citadel 100mm (4in); turrets 200mm (8in) fronts, 100mm (4in) sides, 120mm (4¾in) rears; barbettes 150mm (6in)) 15.2cm turrets 125mm (5in) fronts, 75mm (3in) sides; CT 175mm (7in).
Machinery:	4 shaft Curtis geared turbines, 20,000 shp = 22.6kts, 4 Penhoet oil fired boilers, 550 tons oil
Complement:	450

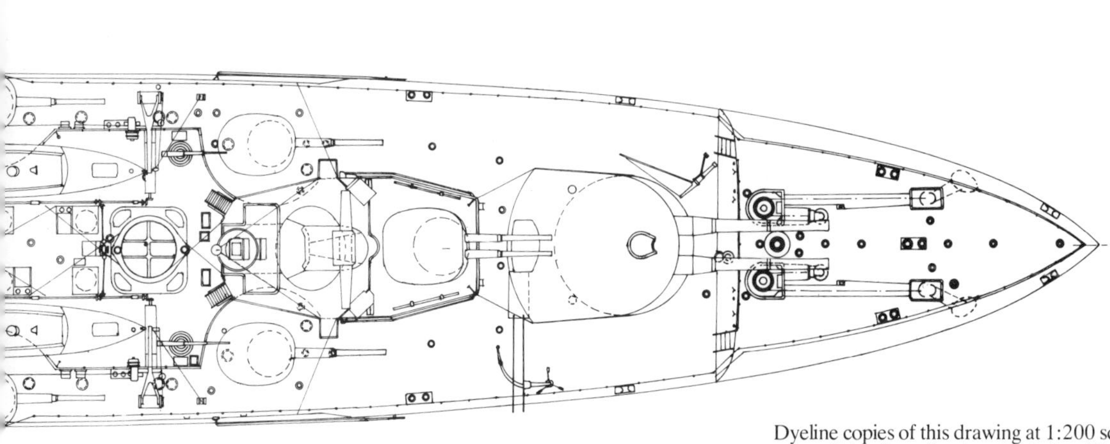

Dyeline copies of this drawing at 1:200 scale are available from Sambrook Marine (Plans), 84-86 Broad Street, Teddington, Middlesex, TW11 8QY. Price £1.75 plus 30p post and packing.

GERMAN DESTROYER MINELAYING OPERATIONS OFF THE ENGLISH COAST (1940–41)

by Pierre Hervieux

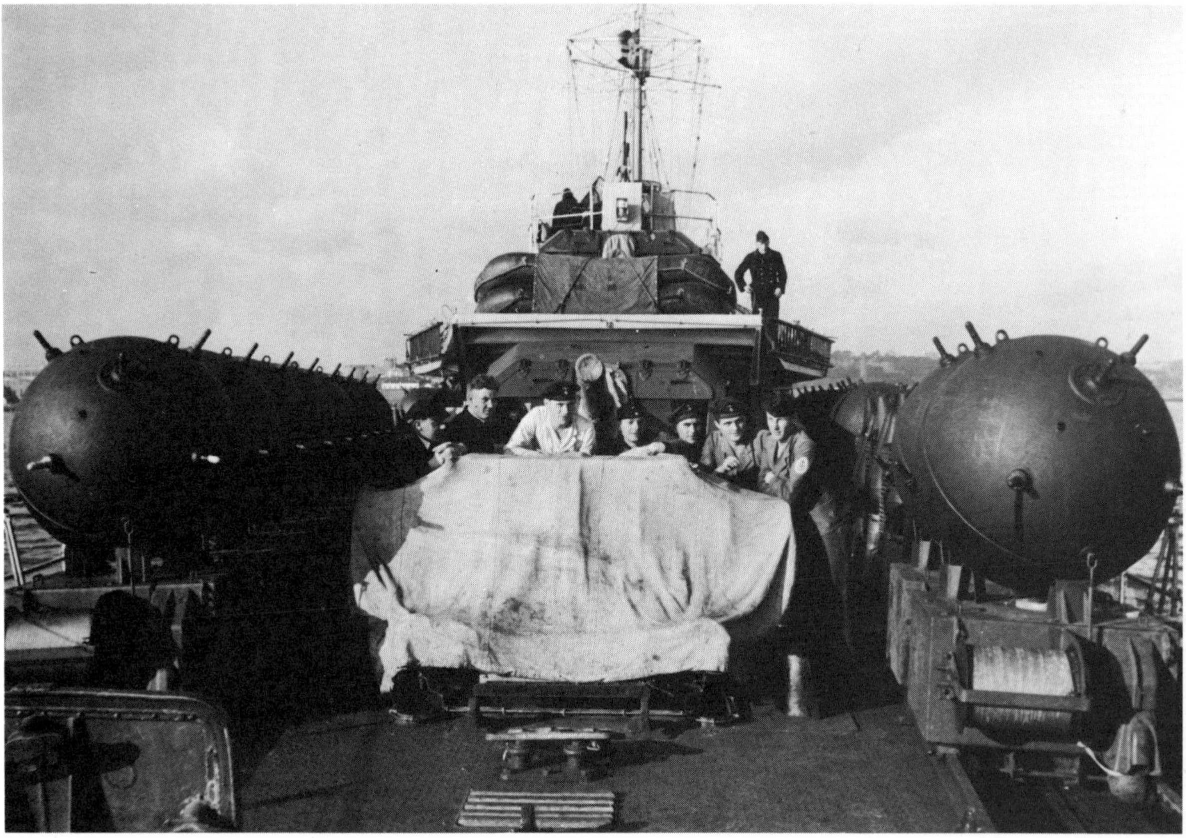

On board a German *Maass* class destroyer at Brest in the autumn of 1940

Between October 1939 and February 1940, German destroyers under the command of Commodore Paul Friedrich Bonte made eleven minelaying sorties close to the English coast without being discovered by the British. They laid 1800 mines – half moored contact mines and half magnetic ground mines – on the trade routes between the Thames Estuary and the Tyne. During these operations, which always took place on very dark nights with little or no moon, they suffered no losses despite having to pass through narrow gaps in their own minefields. The success of these operations depended to a large extent on the accuracy of the intelligence provided by the Naval Staff regarding British East coast minelaying activities.

A small quantity of mines was also dropped in this area by German seaplanes and it is of value to consider the relations between the German naval and air forces to clarify the results of the situation as, in questions of mine warfare, the Luftwaffe and Navy did not see eye-to-eye. The Luftwaffe regarded this area as its own responsibility and wished to delay the start of a minelaying offensive until sufficient stock had been accumulated for mass laying, which would not be until the spring of 1940 when 5000 aerial mines would be available

A *Leberecht Maass* class destroyer under-way in Brest harbour – autumn 1940

ECP Armées

compared with only 120 ready for use in November 1939. The Naval Staff, however, was unwilling to wait so long and minelaying from U-boats commenced on the outbreak of war and from destroyers a month later. In November the Luftwaffe followed the Navy's lead in a rather dilatory way by dropping a mere 68 mines and by mid December the waters of the seaplane bases had frozen over so the minelaying aircraft could no longer take-off.

It seemed certain that some considerable time would elapse before the British developed a counter to the magnetic mine but, fortunately for the British, two such weapons fell on mudflats in the Thames Estuary during the first air mining operations and mine warfare experts from Portsmouth were able to recover them at low water. A study of these magnetic mines enabled the British to devise suitable counter measures and quickly produced not only an effective magnetic sweep but two methods of reducing the magnetic field of a ship.

Nevertheless, the German mining offensive of October 1939-February 1940, resulted in the loss of 67 merchant vessels representing 238,467 tons, plus 3 destroyers and 9 naval auxiliaries. Most of these ships, if not all, must have fallen victim to those mines laid by the destroyers because there were 1800 of them compared with the Luftwaffe's total of 68 (ships sunk by U-boat mines are not included in the above total). This success was never repeated and it must be said that even greater results could have been achieved if the Luftwaffe's idea of suddenly laying very large quantities of mines in the relatively narrow channels, to which British coastal shipping was confined, had been adopted. However, it is questionable, in view of the inadequate navigational training of the Luftwaffe air crews, whether aircraft could have laid these mines with the accuracy required. In contrast, the mines laid by the destroyers and U-boats were usually positioned exactly where required. After March 1940, the nights became too short for German surface vessels to approach the English coast unobserved but the fact that the British were forced to allocate large numbers of small craft to minesweeping, which could

Viewed from the port side amidships the after section of a *Leberecht Maass* class destroyer showing the after bank of quadruple torpedo tubes, No 4 5in gun and, in the foreground, a few mines stowed on the port rails

ECP Armées

otherwise have been employed on anti-submarine work, meant that the effect of these operations lasted throughout the war and assisted to some extent the U-boat campaign. In fact it is debateable whether a postponement of the minelaying operations, as suggested by the Luftwaffe, would have produced any better long term results.

The Admiralty was surprised by the high losses on the East coast and assumed that U-boats and aircraft were responsible, the possibility of surface ships being involved was not even considered. Of the 22 destroyers in the Kriegsmarine suitable for such operations no less than 17 took part. These ships included: *Friedrich Eckoldt* (Captain Schemmel – 5 operations), *Hermann Künne* (Captain Kothe – 4 operations), *Wilhelm Heidkamp* (Captain Erdmenger – 4 operations), *Richard Beitzen* (Captain Von Dvidson – 3 operations), *Bruno Heinemann* (Captain Beger and Captain Langheld – 3 operations), *Karl Galster* (Captain Bechtolshein – 3 operations) and *Erich Steinbrinck* (Captain Johannesson – 3 operations). A list of the minelaying operations is given below.

Mines board a *Leberecht Maass* class destroyer in Brest harbour – autumn 1940. This view, from the port side, was taken from just abaft the second funnel

ECP Armées

MINELAYING OPERATIONS

1. On 17/18 October 1939 in the first of these operations the *Wilhelm Heidkamp, Hermann Künne, Friedrich Eckoldt, Diether Von Roeder, Karl Galster* and *Hans Lündermann* under the command of Rear Admiral Lütjens laid a minefield off the Humber estuary which sank 7 ships totalling 25,825 tons.
2. Mines laid off the Thames Estuary, by *Karl Galster, Wilhelm Heidkamp, Hermann Künne* and *Hans Lüdemann* under the command of Captain Bonte, sank the destroyer *Blanche* and 13 ships totalling 48,728 tons. The force was met on its return by the light cruisers *Königsberg* and *Nürnberg* and four *Möwe* class torpedo boats.
3. On 17/18 November 1939 the *Hermann Künne, Bernd Von Arnim* and *Wilhelm Heidkamp,* under the command of Frigatten Kapitan Hartmann laid a field in the central area of the Thames Estuary which sank the destroyer *Gipsy,* one trawler and 7 ships totalling 27,565 tons. The force was met near Terschelling Bank by the light cruisers *Leipzig* and *Nürnberg* and three *Möwe* class torpedo boats.
4. On 18/19 November 1939 the *Erich Steinbrinck* (flag Fregattenkapitan Bey), *Friedrich Eckoldt* and *Hans Lody* laid a field off the Humber estuary which sank 7 ships totalling 38,710 tons (this included the Polish passenger ship *Pilsudski*). On return the destroyers were met by the *Leipzig* and four vessels of the *Möwe* class.
5. On 6/7 December 1939 the *Erich Giese* and *Hans Lody* laid mines off Cromer which later sank 2 ships of 5286 tons and damaged one other. After laying these mines, lookouts on the German destroyers sighted two ships at about 8000 metres (8750 yards). It was 02.54 (German time) and shortly afterwards the two vessels were identified as British J class destroyers. At 03.15, with the range down to 5300 metres (5800 yards), the *Erich Giese* launched 4 torpedoes at the second ship while *Hans Lody* similarly launched torpedoes at the first. One torpedo, from *Erich Giese,* found its target –

-IMS *Jersey* – which was badly lamaged and was later taken in ow. The second British ship lropped depth charges, obviously hinking that her companion had een torpedoed by a U-boat and ubsequent Admiralty nnouncements confirmed this ssumption. The Germans, seizing he opportunity to keep their lestroyer operations secret, also confirmed the U-boat attack and, ollowing direct orders from the Ministry of Propaganda, it became ront page news in the German ress!
5. On 12/13 December 1939, *Hermann Künne, Friedrich Ihn, Erich Steinbrinck, Richard Beitzen* and *Bruno Heinemann,* under the command of Commodore Bonte, aid mines in the area off Newcastle which later sank eleven ships otalling 18,979 tons. On return hey were met by the cruisers *Nürnberg, Leipzig* and *Köln* but, on he same day, the first two were orpedoed and damaged by the British submarine *Salmon* and on he following day (14 December) he escort vessel *F9* was sunk by the ubmarine *Ursula.*
7. On 6/7 January 1940, *Friedrich Eckoldt, Erich Steinbrinck* and *Friedrich Ihn* under the command of FregattenKapitan Berger, laid mines in the Thames Estuary which ater sank the destroyer leader *Grenville* and 6 ships totalling 21,617 tons and damaged one other.
8. On 10/11 January 1940, *Wilhelm Heidkamp, Karl Galster, Anton Schmitt, Friedrich Eckoldt, Richard Beitzen* and *Friedrich Ihn* under the command of Commodore Bonte laid mines off the coast near Newcastle which claimed only one trawler of 251 tons.
9. Also on 10/11 January 1940 the *Bruno Heinemann, Wolfgang Zender* and *Erich Koellner,* under the command of Fregattenkapitan Bey, laid mines off Cromer which sank 3 ships totalling 11,155 tons.
10. On 9/10 February 1940 the *Friedrich Eckoldt, Richard Beitzen* and *Max Schultz*, under the command of Fregattenkapitan Berger, laid mines in the Shipwash area which later sank 6 ships,

VICTIMS OF GERMAN MINES

Identified Warships and Merchant Ships sunk by Mines laid by German Destroyers off the East Coast during 1939-40 (unless otherwise stated ships are British).

Date	Name	Type	Tonnage	Position
22.10.39	Whitemantle	Cargo ship	1692	5 to 6 miles east of Witherinsea Light
29.10.39	Varangmalm (Norwegian)	Cargo ship	3618	off the Humber.
3.11.39	Realf (Norwegian)	Tanker	8083	53°48N/0°360
4.11.39	Sig (Norwegian)	Cargo ship	1342	Off the Humber
13.11.39	HMS Blanche	Destroyer	1360	Thames Estuary
13.11.39	Ponzano	Cargo ship	1346	51°29N/01°25E
13.11.39	Matra	Cargo ship	8003	1 mile east of Tongue LV
15.11.39	Woodtown	Cargo ship	794	¾ mile NE of NE Spit Buoy, near Margate.
18.11.39	Blackhill	Cargo ship	2492	145° 7½ cables from Longsand Head Buoy.
19.11.39	Torchbearer	Cargo ship	1267	025° 2 miles from Shipwash LV
20.11.39	HMS Mastiff	Armed trawler	520	Thames Estuary
21.11.39	HMS Gipsy	Destroyer	1350	Off Harwich.
21.11.39	Geraldus	Cargo ship	2495	3 miles WNW of Sunk LV
22.11.39	HMS Aragonite	Armed trawler	315	Off Deal
22.11.39	Lowland	Cargo ship	974	2 miles ENE of NE Gunfleet Buoy
23.11.39	Hookwood	Cargo ship	1537	3½ miles ENE of Tongue LV
24.11.39	Mangalore	Cargo ship	8886	288° 1½ miles from Spurn LH
26.11.39	Pilsudski (Polish)	Passenger	14294	Off the Humber estuary
28.11.39	Rubislaw	Cargo ship	1041	1-1½ miles ENE of Tongue LV
30.11.39	Sheafcrest	Cargo ship	2730	51°32/01°26E
1.12.39	Dalryan	Cargo ship	4558	2 to 2½ miles SW of Tongue LV
2.12.39	Sancalisto	Tanker	8010	2½ miles NNE of Tongue LV
8.12.39	Merel	Cargo ship	1088	Nr Gull LV, off Ramsgate
8.12.39	Corea	Cargo ship	751	2½ miles NE ½N of Cromer Light
10.12.39	HMS Bay of Hope	Armed drifter	98	Thames Estuary
12.12.39	King Egbert	Cargo ship	4535	4 miles SW of Haisboro LV
13.12.39	RFA W Hallett	Armed tanker	202	Off the Tyne
14.12.39	RFA J Ludford	Armed tanker	506	Off the Tyne
14.12.39	Inverlane	Tanker	9141	55°05N/01°07W
15.12.39	Ragni (Norwegian)	Cargo ship	1264	Off the English east coast
15.12.39	H C Flood (Norwegian)	Cargo ship	1907	near Tynemouth
15.12.39	Strindheim (Norwegian)	Cargo ship	321	near Tynemouth
16.12.39	RFA Sedgery	Armed tanker	520	Off the Tyne
16.12.39	Amble	Cargo ship	1162	54°55N/01°03W
20.12.39	RFA Napia	Tug	155	off Ramsgate
31.12.39	Box Hill	Cargo ship	5677	53°32N/00°24E
6. 1.40	ETA	Fishing vessel	81	6 miles N ½W of Outer Gabbard LV
7. 1.40	Cedrington Court	Cargo ship	5160	2 miles NE of North Goodwin LV
7. 1.40	Towneley	Cargo ship	2888	1 mile ENE of NE Spit Buoy, near Margate.
9. 1.40	Dunbar Castle	Cargo ship	10002	51°23N/01°34E

11. 1.40	Lucida	Fishing vessel	251	55°00N/00°53W
12. 1.40	Granta	Cargo ship	2719	53°13N/01°21E
15. 1.40	Newhaven	Fishing vessel	162	18 miles SSE of Lowestoft (est)
19. 1.40	HMS Grenville	Destroyer	1485	Thames Estuary
2. 2.40	Portelet	Cargo ship	1064	3 to 4 miles SW by W of New Smiths Knoll
9. 2.40	Agnes Ellen	Cargo ship	293	On voyage from Holyhead to Workington
13. 2.40	British Triumph	Tanker	8501	53°06N/01°25E
17. 2.40	Baron Ailsa	Cargo ship	3656	53°17N/01°12E
23. 2.40	RFA Benvolio	Armed tanker	352	Off the Humber
24. 2.40	Clan Morrison	Cargo ship	5936	53°07N/01°22E
24. 2.40	Jevington Court	Cargo ship	4544	53°08N/01°22E
27. 2.40	Ben Attow	Fishing vessel	156	7 miles E $\frac{1}{2}$S of May Island
2. 3.40	Albano	Cargo ship	1176	$128\frac{1}{2}$° 7.6 miles from Coquet Light
3. 3.40	Cato	Cargo ship	710	51°24N/03°33W
8. 3.40	Counsellor	Cargo ship	5068	280° 6 miles from Bar LV Mersey
9. 3.40	Chevychase	Cargo ship	2719	53°18N/01°13E
11. 3.40	Halifax	Fishing vessel	165	3 miles SE of Aldeburgh LV
12. 3.40	Gardenia	Cargo ship	3745	53°04N/01°33E
15. 3.40	HMS Peridot	Armed trawler	550	off Dover
28. 3.40	Burgos (Norwegian)	Cargo ship	3219	near Cromer Knoll LV

LV = Light Vessel, LH = Light House

totalling 28,496 tons, and damage one other.

11. Also on 9/10 February 1940 th *Bruno Heinemann, Wolfgang Zenker* and *Erich Koellner,* under the command of Fregattenkapitan Bey, laid mines in the area of Haisborough (Cromer Knoll) whic resulted in the sinking of 3 ships totalling 11,885 tons.

For comparisons with the above the first air mines were laid by German seaplanes off the east coa on 20 November 1939, 41 mines being dropped in three sorties. A further 27 mines were dropped from aircraft in the Humber and Thames Estuaries and in the Dowr on the night of 6/7 December 193

After the destroyers stopped the minelaying sorties in February further operations were undertake

At anchor in Brest harbour a view looking a from forward of the second funnel on the *Leberech Maass* class destroyer. Note that t two nearest mines have been taken off their sinkers and are stowed on the mine rails

ECP Armées

A destroyer of the *Leberecht Maass* class, loaded with mines, in Brest harbour in the autumn of 1940

ECP Armées

ɔy other vessels although not with he same intensity. The next ɔperation was carried out by the ıuxiliary minelayer *Schiff II* which, ɔn 9 March 1940, laid a mine ɔarrage east of the North Foreland. Ɔerman aircraft recommenced their ɔffensive on 17 April 1940, when hey laid air mines in the Downs ınd Edinburgh Channel. These ɔperations were followed by many nore of similar type and, later, orpedo-boats were also employed or this work.

During these early war ɔperations the German destroyers ıad operated with immunity, to the considerable cost of the British, but he wheel of fortune was soon to urn. Between the end of February ınd the first half of April 1940 no ess than 12 destroyers were lost to he Kriegsmarine. Two were lost as ı result of an error by German ıircraft which attacked their own ships and forced them into a ninefield during Operation Wikinger and a further 10 during he two Battles of Narvik. These osses were a terrific blow the German Navy which, in August 1940, after the Norwegian Campaign, had only 7 destroyers in service and 3 under refit. These latter vessels were *Richard Beitzen* (due to return to active service on 21 September 1940), *Bruno Heinemann* (due to return on 4 October 1940) and *Hermann Schoemann* (due to return on 15 October 1940).

CHANNEL OPERATIONS

The 7 active ships, plus *Richard Beitzen* fresh from the dockyard, moved to Brest, with a stop over at Cherbourg, in September. Despite their losses the spirit of the flotilla was still high and as early as the night of 28/29 September the 'officer commanding destroyers' Captain Bey, took the *Hans Lody, Karl Galster, Paul Jacobi, Erich Steinbrinck* and *Freidrich Ihn* on an offensive minelaying operation into Falmouth Bay. The operation was given distant cover by the destroyers *Friedrich Eckoldt* and *Theodor Riedel.* The mines claimed five ships totalling 2026 tons but only two of these have been identified: the cable ship *Lady of the Isles* of 1660 tons, sunk about 3 miles east of St Anthony point, near Falmouth, on 3 October 1940 and the cargo ship *Jersey Queen* of 910 tons, sunk 1½m miles from St Anthony point on 6 October 1940.

On the night of 17/18 October 1940 another sortie took place, once again under the command of Bey in the *Hans Lody* together with *Karl Galster, Friedrich Ihn* and *Erich Steinbrinck,* with the 5th torpedo boat flotilla, consisting of

Ꟃrom left to right the German destroyers *Theodor Riedel, Friedrich Ihn* and *Friedrich Eckoldt* at Cherbourg in September 1940

Ɔrüppel

six units of the *Möwe* class, in support. They steamed towards the western exit of the Bristol channel but were intercepted 30 miles southwest of the Scilly Isles by the British cruisers *Newcastle* and *Emerald* and 5 destroyers and no mines were laid. Little damage was received on either side although *Erich Steinbrinck* later claimed to have hit one of the cruisers with a torpedo.

A second operation, planned for October, in which the destroyers were to carry out a sweep from the Bay of Biscay towards Cape Finnisterre was called off due to machinery failures. Of the 8 destroyers available in the west at this time 5 went into dockyard hands for major overhaul which left only 3, *Karl Galster, Hans Lody* and *Richard Beitzen,* available for the next two sorties.

Both these operations, again under the command of Captain Bey, were raids into the waters around Plymouth in which no mines were laid. In the first, on the night of 24/25 November 1940 two ships were sunk by gunfire about 8 miles south east of Wolf Rock, the Dutch motor tanker *Apollonia* (2086 tons), proceeding in ballast from Plymouth to Avonmouth, and the Belgium motor trawler *Marguerite Simonne* (70 tons). The second raid, on the night of 28/29 November, resulted in another two ships sunk by gunfire, 8 miles off Start Point – the tug *Aid* (134 tons) and the barge *BHC No 10 (290 tons)* – and a short action with British destroyers in which HMS *Javelin* was torpedoed and damaged.

By the end of the year two more of the German flotilla were due for extensive overhaul and only *Richard Beitzen* remained fully operational. On the night of 23/24 January 1941, together with two *Möwe* class torpedo-boats, she sailed for her last offensive mining sortie in the west – operation Weber. Under the command of Korvettenkapitan Von Davidson she successfully laid her mines off the south coast of England which probably caused the sinking of two cargo ships, the *Ben Rein* (156 tons) off Falmouth on 17 February and the *Shoal Fisher* (698 tons) in position 50°10N/04°50W on 23 February.

Thus ended the flotillas operations in the Channel. Compared to the results obtained in their earlier North sea sorties the achievements were negligible, only a few small vessels being sunk. There were several reasons for this off the East coast during a 4 month period the ships had carried out eleven minelaying raids while during a similar period in the Channel they had only carried out operations, of which only two involved minelaying. Circumstance had also changed, the British were operating more minesweepers and patrols, the convoy system was becoming more effective and, of course, the German flotilla had fewer ships. German torpedo-boat continued to be active in the Channel but this is another story which I hope to cover in a future article concerning the operations o the *Möwe* class.

A *Leberecht Maass* class destroyer passes th breakwater of Brest harbout at the start of a minelaying Sortie in the autumn of 1940. (D any of our readers know of any variations in the appearance of the vessels of this class which would enable a differentiation of one ship from another? I have noticed that the vessel illustrated above has her propeller guard lower than in the ships in the other photographs and her foremost boat, and its crane, to starboard rather than port of the forward funnel – apart from this they appea almost identical – Editor)

ECP Armées

THE DEVELOPMENT OF RADAR

IN THE ROYAL NAVY (1935-45) PART 2 by Alastair Mitchell

'he bridge of the aircraft carrier *Indomitable* ı 1943, with a Seafire taking-off in the ›reground. Mounted on the front of the ridge is the aerial lantern of Type 271 with, bove it, the frame aerial of the MF/DF. On ıe bridge itself are three pom-pom directors arrying the aerials of Type 282 and the main irector carrying the aerials of Type 285. The ›remast carries a homing beacon on the :arfish and at the topmast head, not visible in ıis photograph, the aerials of Type 281B, 'hile the mainmast carried the aerial for 'ype 79B

'PL W/14/001

THE FIRST CENTIMETRIC RADARS

From the first days of radar the aim had been to achieve workable systems at high frequencies, mainly in order to overcome the problems of 'lobing' which restricted low-angle and surface range on metric sets. In their 23cm work during 1936–37 the SS (Signal School) had used a 'split-anode magnetron' oscillator to obtain the required high frequency and a Dr Woods had actually sketched an idea for a multiple-cavity magnetron but, unfortunately, it was not developed at that time and it was left to scientists at Birmingham University to solve the high-frequency problem by producing the first resonant-cavity magnetron in February 1940. It

'he first ship to be fitted with the Type 271 :entemetric surface warning radar, the :orvette *Orchis,* seen here in December 942. The radar lantern is mounted ımmediately above the radar office at the rear f the upper bridge

1oD

oscillated at 9.8 centimetres (3000MHz) and gave a power output of 1 kilowatt. From then on work went ahead rapidly and soon the TRE (Telecommunications Research Establishment) at Swanage had a set working that could detect a submarine at 7 miles. With the U-boat menace increasing, ASE (Admiralty Signal Establishment) formed a new section and, taking a replica of the TRE equipment, developed a set for fitting on all AS escort vessels. Known as Type 271 it was ready by February 1941 and fitted in the corvette *Orchis* for trials. So urgent was the need that 150 sets were ordered two weeks before the tirals even began.

Type 271 operated on 10cm and initially had a power output of 7kw peak, contained in a 1.5 microsecond pulse transmitted 500 times a second. The receiver was also novel in that it contained a crystal detector which was much more efficient than previous valve types. To counteract the high power losses in coaxial signal cable at high frequencies, certain transmitter and receiver stages were mounted directly on the back of the aerial. The aerial system itself was enclosed in a perspex dome, to

The escort sloop *Woodcock* in June 1943 carrying a standard mid-war warning outfit of Type 271Q (aerial on lattice tower, aft) and 291 (aerial at foremast head). Gunnery radar Types 285 (aerials on main HA/LA director/rangefinder) and 282 (aerials on twin Hazemeyer Bofors mountings amidships) are also carried for main and close-range armament control. The lattice tower for the Type 271Q aerial allowed the set to 'see' ahead to some extent but was superseded b the much taller lattice foremast in ships fitte later

MoD

Radar Transmitters consisted basically of 3 units:

(1) Trigger producing the basic timing and repetition rate.
(2) Modulator controlling the operation of the oscillator.
(3) Oscillator controlling the burst of radio energy which was fed to the aerial.

The main features of the radar transmitter were:–

Peak Pulse power – measured in kilowatts was the energy contained in the pulse. The average power was much less, eg if the transmitter radiated 2 microsecond pulses, 500 times a second, it was only on for 1000 microseconds or one thousandth of the total time.

Pulse length – The length of the pulse in microseconds, determined by the modulator Long pulses were used for long range to ensure good detection capabilities. Pulse length determined range discrimination.

Pulse shape – Normally the leading edge was the important feature as it determined to some extent the range accuracy. A square wave was the ideal but could only be obtaine by employing a wideband receiver, to pressure the shape, giving rise to high noise problems.

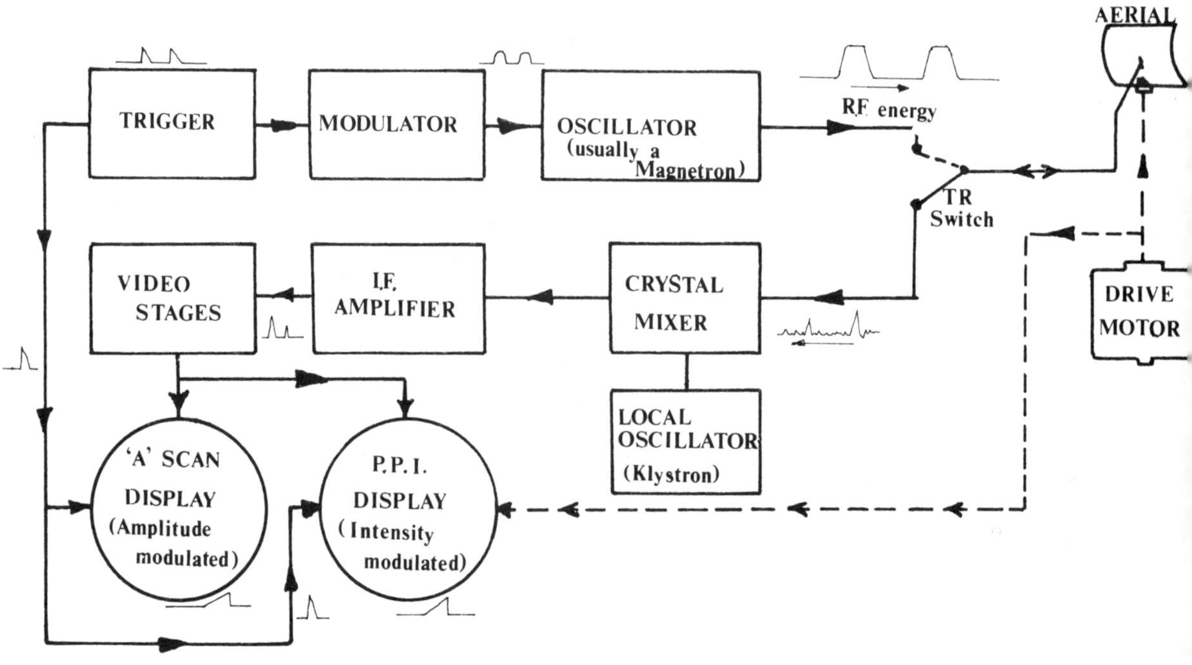

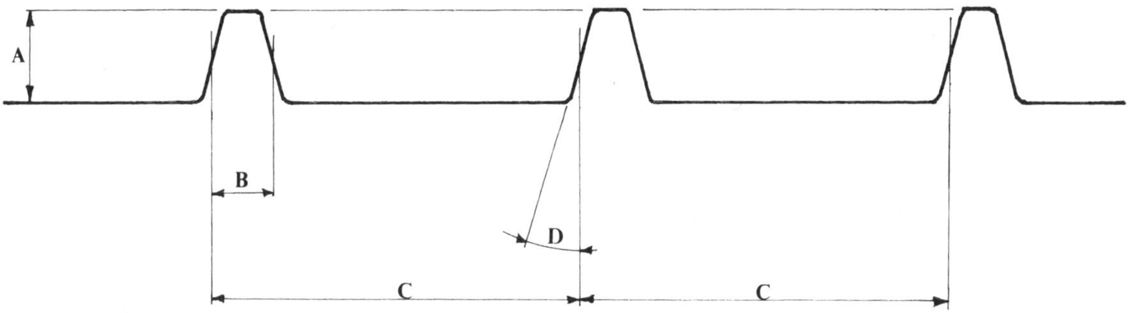

. Pulse height determined by peak power. leasured in kilowatts (kW)

. Pulse Width measured in microseconds.)etermines range resolution, ie ability to istinguish two targets on the same line of earing. A one microsecond pulse = 63.9yds. Long pulses of 3–10 microseconds ere employed for long-range sets and short ulses of 0.2 to 1 microseconds were mployed for precision sets.

. Pulse spacing, or number of pulses per econd (PRR – pulse repetition rate) varies etween 50 and 2000. Number of pulses overned by range required and width of ulse. Normally the longer the range the wer the PRR.

. The rise time or angle of the leading edge. s all the timing circuits are related to this dge for perfect accuracy it would need to be ertical. The actual angle is however overned by transmitter modulator circuits nd receiver bandwidth.

protect the transmitter from weather effects, mounted on top of a cabin housing the equipment and usually fitted at the rear of the bridge. The 'cheese' array was chosen because of its fan shaped vertical beam which allowed the radar to illuminate the surface in any sea state. The coaxial cables restricted the rotation of the cheese to plus and minus 200° and the echoes were displayed on an 'A' scan borrowed from Type 286.

The trials showed detection of a surfaced submarine at 5000yds and a periscope at 1300yds. Almost as important, other ships could be detected at 6 miles or more, simplifying convoy escort duty at night enormously.

Fitting of the new sets went ahead rapidly and by the end of 1941 over fifty ships were equipped. On large ships, with much greater stability than corvettes, a twin 4½ft diameter parabolic dish aerial was employed, still enclosed in a radome. The battleship *Prince of Wales* had this type, numbered 273, fitted in August 1941 giving her a comprehensive fit for that period which, however, did not save her four months later.

In September 1941 Birmingham University introduced the second major advance in magnetron design by 'strapping' or linking the individual cavities. This gave single-mode oscillation and resulted in much higher output powers on the desired frequency. Output peak powers of 100kW were soon attained and this led to a redesign of the Type 271 transmitter. Coaxial signal cables could not carry this level of RF (Radio Frequency) energy at 3000MHz so another innovation, the rectangular waveguide, was introduced. Measuring 3in x 1in in section the waveguide could handle powers up to 10mW (10,000kW) with a signal loss one twentieth of that experienced with cables and, literally, freed the aerial system from the transmitter. This allowed for the remote positioning of the aerial so that the office could be low

The 'Battle' class destroyer *Barfleur* was the first ship to be fitted with Type 275, the twin nacelles of which can be seen on the Mk VI director above the bridge. She is seen here in June 1949 with a standard destroyer fit of Type 293P with its cheese aerial on the foremast, and the associated Type 242 interrogator aerial behind, air warning Type 291, with aerial on the short pole mast aft, and Type 282 fitted to the twin 40mm mountings. The aerial at the fore-topmast head is for the HF/DF

Wright and Logan

down in the ship to meet considerations of top-weight and space availability while the aerial was placed high up to give the best coverage.

Both inventions led to the development of Type 271Q in 1942 which could give surface ranges of up to 46,000yds depending on the target size. In addition the larger ship's Type 273Q provided a useful detection range of 15 miles on low-flying aircraft. The Q version also introduced the continuously rotating aerial and the PPI (Plan Position Indicator) display which revolutionised radar plotting. For the first time a complete picture of the surface within a 10–20 mile radius was presented and continuously updated although, like all new developments, it highlighted the need for radars with even shorter pulses and narrower beam widths to give a high-definition picture.

The first sets entered service in 1943 and gradually supplanted the earlier Type 271–3 versions. Several destroyers were then fitted with lattice foremasts and sported the Type 272Q version with its cheese aerial enclosed in the now familiar radome at the mast head. In service it was soon found that Type 273Q outranged the Type 284 gunnery set and facilities for using it to provide gunnery ranges beyond 30,000yds were incorporated in the ship's transmitting station. On small ships without directors a ranging outfit, RTA, was fitted to provide facilities for laying the guns on to a target and an outfit, RTC, with high accuracy to 3000yds, was added for AS work.

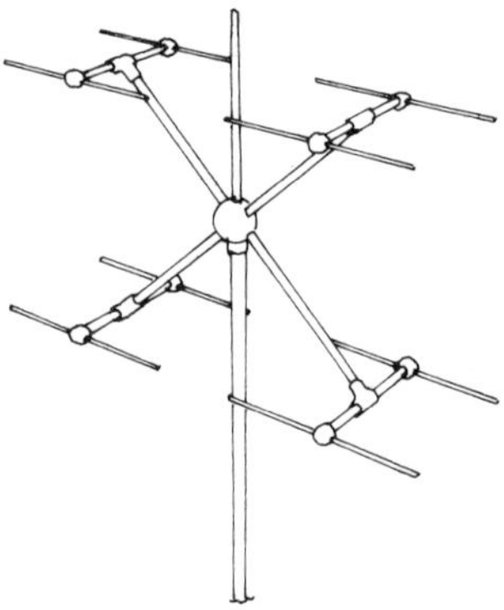

The aerial of radar Type 291

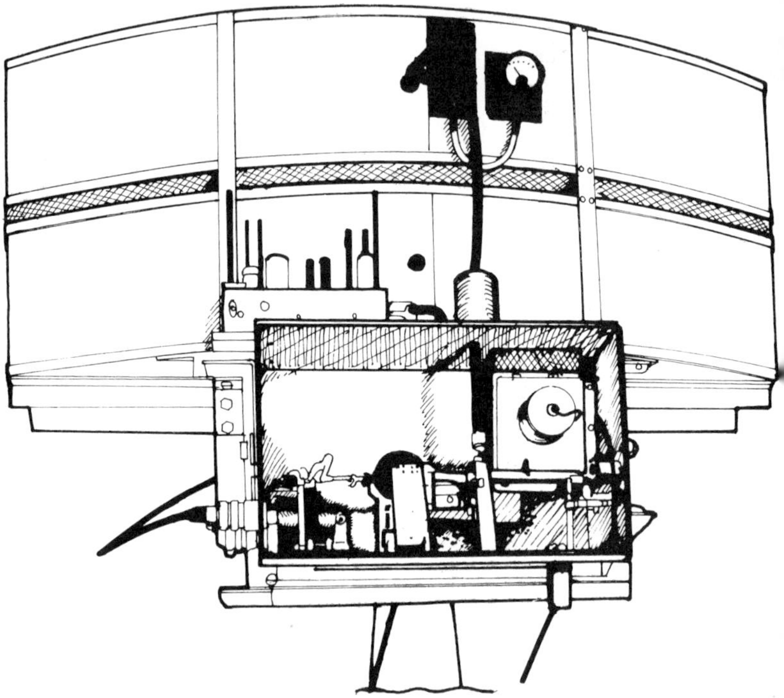

The 'double cheese' of radar Type 271, viewed from the rear, showing those parts o the transmitter/receiver fitted directly to the aerial

THE METRIC RADARS – FINAL PHASE 1942–44

In 1941 the three metric-wavelength radars in service were the Type 79/279 and its variants, Type 281 and Type 286. All suffered from various deficiencies, none more so than the 286. Although replaced in the surface role by Type 271 there still existed a requirement for air-warning on ships of escort size and for a general-purpose set on coastal craft too small to merit the centrimetric set. To meet these requirements a modified aerial-system consisting of two Yagis, one above the other, was developed for the Type 286 and a new 100kW transmitter was produced for larger ships. An aerial system of four horizontal dipoles, with reflectors similar to the Type 281, was devised only this time they were supported by an 'X' frame rather than a rectangular one and for the first time common aerial working with TR (Transmit-Receive) switching was used. A beamwidth of approximately 40° was achieved and detection ranges of up to 35 miles on aircraft were claimed. The set, numbered Type 290/291, entered service in 1942 and by 1944 was fitted in virtually every destroyer and escort, other than Lend-Lease ships, in the fleet. In its final version, the Type 291 had a peak power output of 80kW, with a 1.1 microsecond pulse transmitted 500 times a second. Th aerial system, initially hand rotated was power driven and a PPI display was provided. Its compact size made it suitable for fitting in ships down to minesweeper size and it also became a standard fit in submarines. Despite the suitability of the frequency band for long-range warning the RN abandoned all work on 200Mhz after Type 291, and never again designed a small ships air-warning radar, the small ships simply grew i size until they could accommodate the standard air-warning set!

In larger ships the introduction o Type 281 was not without problems. The intended use of Types 281 and 279 for gunnery control was not popular in the fleet as it inhibited all round search, and

The *LFD 13,* a standard LST (2) converted to an RAF fighter direction ship. Equipment fitted included the RAF GCI set, Type 8F, near the bows, which had a range of 60 miles and a height-findind capacity; a Type 11 amidships, operating on 50cm wavelength and effective against low-flying aircraft; and an assortment of VHF, R/T and DF aerials. A small navigational radar is fitted aft, with its aerial enclosed in a radome on a short lattice tower. The LFD's were employed off the Normandy beachheads in 1944 until the RAF became established ashore.
IWM

A close-up of the bridge and foremast of the carrier *Sydney* off Korea early in 1952. The rectangular waveguide feed to the Type 293 cheese aerial is clearly visible and even the flare in the guide for matching it to the aerial can be seen. Similarly the feed to the Type 277Q aerial is shown and the end-on view gives some idea of the high weight of the installation. The two Type 242, Mk III IFF, interrogator aerials are mounted on the brackets extending from the forward side of the mast with their, somewhat temperamental, drive motors visible at the base. The aerial on the bridge front is the receiver for the FM12 MF/DF.
IWM

RADAR SETS IN SERVICE 1939-46

1. Long-Range Air-Warning

Wavelength	Type Number
metric 7m	79Y, 79Z, 279, 79B
4-6m	280
3.5m	281, 281B, 281BQ
3.5m	960
1.5m	SK (US type)

2. Air-Warning

metric 1.5m	291, 291M, 291P/Q
1.5m	SA (US type)

3. Combined Surface/Air-warning/Height finder

centimetric 10cm	277, 277M/P
10cm	SM (US type)

4. Surface-Warning

metric 1.5m	286M, 286P/Q, 286U, 286W
	290
centimetric 10cm	271, 271M/P, 271Q
	272, 272M/P
	273, 273Q
	276
	293, 293M/P
	970 (RAF type)
	SL (US type)
	SG (US type)
3cm	268
	267 (submarine set 268 & 291)

5. Gunnery

decimetric 50cm	282, 282M/P
	283
	284, 284M
	285, 285M/P
centimetric 10cm	274
	275
3cm	262, 262M

6. IFF/Beacons

metric 1.5m *interrogators*	242, 242M/P
243	243
responders	251, 251M/P
	252
	253, 253M/P
beacons	254, 255, 256, 257, 258

twin open-wire feed and a powered aerial which allowed fitting of PPI displays. The short pulse for gunnery requirement was removed from the BQ type.
Note: The 281 series were superceded post war by Type 960.

Type SK (US)							
SK	1943/44	195	200	5	120 at 10,000ft (surface)	20°H 20°V	3°

Aerial: Mattress array, 18ft x 17ft reflector behind 36 dipoles (6x6). Power rotation from 0 to 5rpm. Weight 2400lbs
Displays: 5in 'A' scan and 12in master PPI with provision for 16 repeaters.
Note: SK2 and SK3 versions were not used by RN.

AIR-WARNING

Type	Date	Frequency (MHz)	Power (kW)	Pulse (micro-seconds)	Range (miles)	Beamwidth H=horizontal V=vertical	Accuracy (plus and minus)
Type 291, 291M, 291P, 291Q, 291U							
291	1942	214	80	1.1	8–9 surface; 30 on air-craft at high angles	35°	

Aerial: Type 291, 291M/P/Q – Single aerial consisting of 4 parallel dipoles (2x2) with reflectors. Manual rotation on 291, power rotation on remainder. Type 291U – 2 Yagi arrays (1 transmit/1 receive) mounted above each other on pole. Manual rotation.
Displays: 'A' scan on all types. PPI displays on final versions with power rotation.
Modifications: M, P and Q versions had powered aerials and PPI displays. U version fitted on small ships.
Note: Type 291 series remained in service until 1952 on destroyers and smaller vessels. It was replaced in coastal craft by Type 268.

Type SA (US)							
SA	1942	175–225	140	5	30 at 10,000ft, 5–8 on surface targets	30°H 50°V	5°

Aerial: Mattress array, 5ft 4in x 8ft 8in wide, 6 dipoles (2 high x 3 wide). Power rotation at 2.5 and 5 rpm. Weight 520lbs.
Displays: 5in 'A' scan, PPI fitted retroactively.
Note: Fitted in some *Captain* and *Colony* class frigates.

Notes on Terms Used in Tables
1 Dates given are those for entry into operational service.
2 Frequency is the number of cycles per second of a wave and is measured in megahertz (MHz) old term megacycles. NB 1 million cycles is 1 MHz.
3 Frequency x Wavelength = 3 x 10^8 metres/second (300,000,000m/s)
4 Power is peak power (power contained in the pulse) and is measured in kilowatts (kW)
5 Pulse length is measured in microseconds, (1 millionth of a second).
6 Range is average reliable detection distance. Actual operational ranges depended on several factors including target size, weather, tuning of the radar set and ability of the operator.
7 Beamwidth as given is the width of the beam at the half power points. (3db down on maximum).

PARTICULARS OF RN RADARS 1938-45
LONG-RANGE AIR-WARNING
Types 79Y, 79Z, 279 and 79B

Type	Date	Frequency (MHz)	Power	Pulse (micro-seconds)	Range (miles)	Beamwidth (H=horizontal, V=vertical)	Accuracy (plus and minus)
79Y	1938	43	15	8–30	50 at 10,000ft	75°	5°
79B	1942	43	70	8–30	65 at 10,000ft 90 at 20,000ft	75°	

Aerial: Separate transmit and receive aerials for Type 79Y, 79Z and 279 each consisting of 2 parallel dipoles with reflectors – manual rotation through 400°.
Display: 'A' scan on 5in tube calibrated in miles. No PPI was ever fitted.
Modifications: Type 79Z introduced in 1939 had power output increased to 70kW to increase range. Type 279 introduced in 1940 had ranging unit GL1 fitted for accurate ranging in AA gunnery control within 7 miles range. Type 79B introduced common aerial working and one aerial was removed – it was fitted in carriers as a vertical gap filler for Type 281BQ.

Types 281, 281B and 281BQ

Type	Date	Frequency (MHz)	Power	Pulse (micro-seconds)	Range (miles)	Beamwidth (H=horizontal, V=vertical)	Accuracy (plus and minus)
281	1941	90	350 or 1000	15 or 2–3	80–100 at 25,000ft	35°H	0.5 beamswitched
281BQ	1944	90	350	5	100–120 at 25,000ft	35°H	5°

Aerial: Type 281 – rectangular framework supporting 4 parallel dipoles with reflectors in 2x2 configuration. One set on foremast for transmission, one on mainmast for reception. Receiver aerial beamswitched. Manual rotation. Type 281B – single aerial, beamswitching removed. Type 281BQ – power rotation at 2 and 4 rpm. Weight 1344lbs.
Displays: Type 281 and 281B – 'A' scan with GL1 ranging unit. Type 281BQ – PPI
Modifications: Type 281B introduced common aerial working with TR switching, the beamswitching facility being removed. Type 281BQ had coaxial feed to aerial in place of

COMBINED SURFACE/AIR-WARNING/HEIGHT FINDER
Types 277, 277M, 277P, 277Q

Type	Date	Frequency (MHz)	Power (kW)	Pulse (micro-seconds)	Range (miles)	Beamwidth H=horizontal V=vertical	Accuracy (plus and minus)
277	1943	3000	400	0.7 or 0.9	15–20 surface, 30 at 15,000ft on aircraft, height finding out to 25 between 4° and 40° elevation.	4.5°H 4.5°V	

Aerial: Outfit AUK – parabolic dish array stabilised in elevation and azimuth. Power rotation continuous from 0 to 15rmp. Elevation stepped from 0 to 40°.
Displays: PPI/HPI with feeds to additional PPIs and Skiatrons. Common 277/281BQ PPIs, and bearing information to Type 293 PPI to indicate bearing of 277 aerial for height finding on 293 target.
Modifications: Type 277M – elevation range increased to 75°. Type 277P – improved receiver design which increased reliability and gave overall range increase of 15%. Type 277Q – post war development with new aerial.

AIR WARNING/FIGHTER DIRECTION
Type SM-1 (US)

Type	Date	Frequency (MHz)	Power (kW)	Pulse (micro-seconds)	Range (miles)	Beamwidth	Accuracy
SM-1	1944	2780–2830	1000	1	25–30 surface, 41–70 on aircraft at 10,000ft	2.7°V 2.7°H	500ftV at 30 miles, 0.5°H

Aerial: Fully stabilised (in three planes) 8ft diameter parabolic dish with offset, spinning, waveguide feed giving conical scan at 5° coverage. Power rotation at 2 or 6 rpm or sector scan. Elevation from plus and minus 75° to plus and minus 3°. Weight 4400lbs.
Displays: PPI with repeaters, elevation dial with altitude meters.
Note: 26 sets ordered by RN but only ship fitted appears to have been carrier *Indomitable*.

SURFACE WARNING

Types 286M, 286P, 286PU, 286W, 286PW, 286PQ

Type	Date	Frequency (MHz)	Power (kW)	Pulse (micro-seconds)	Range (miles)	Beamwidth H=horizontal V=vertical	Accuracy (plus and minus
286M (ex RAF ASV Mk2)	1940	214	10	1	7 on large ships, 1 on submarines	120° arc forward cover	

Aerial: Outfit ATQ – non-rotating fixed rectangular frame supporting forward facing Yagi array for transmission, 2 Yagi arrays angles outwards for beamswitched reception. Type 286PQ had same aerial as Type 291, with manual rotation, giving a beamwidth of 35°. Type 286PU had one forward facing Yagi at top of mast for transmission and two angled Yagis below for beamswitched reception giving a fixed forward cover beamwidth of 140°. Type 286W/PW had a frame aerial mounted on a telescopic mast between the periscopes in submarines.
Displays: All types had 'A' scan.
Modifications: Type 286M was first operational naval version, P and Q variations had improved transmitters and directional aerials. PU was developed for trawlers and PW for submarines.

Types 271, 271M, 271P, 271Q, 272, 272M, 272P, 273, 273Q

271	1941	3000	7	1.25	6 on ships, 5000 yds on submarines	5°H	
273Q	1942	000	100	2 or 0.7	20-23 surface, 12–15 on aircraft	4.5°H	

Aerial: Type 271/272 – double cheese enclosed in perspex radome, manual rotation plus and minus 200°. Type 273 – twin 4.5ft diameter parabolic reflectors excited by dipole in 273 and by waveguide nozzle in 273Q, enclosed in radome, manual rotation.
Displays: 'A' scan.
Modifications: M and P variations had a ranging unit fitted. Q versions were fitted with strapped magnatron giving higher power and longer range and were fitted with waveguide feed in place of coaxial cable to cope with extra power. In Type 271 the transmitter final stage was mounted on the back of the aerial. Type 272, a development of 271, employed a cable feed between the modulator and the magnatron oscillator on the back of the aerial which allowed the aerial to be fitted up to 40ft from the main equipment – aerial usually mounted on short lattice tower or on lattice foremast in destroyers.

Types 293, 293M, 293P

293	1944	3000	500	0.7 or 1.9	12 surface, 12 air	3.2°	

LOW ANGLE GUNNERY

Types 284, 284M and 274

Type	Date	Frequency (MHz)	Power (kW)	Pulse (micro-seconds)	Range (yards)	Beamwidth H=horizontal V=vertical	Accuracy (plus and minus)
284	1940	600	25	2	24,000 (plus and minus 120)	8°H	0.5°
284M	1942	600	150	1	30,000 (plus and minus 25)	8°H	0.085° beam-switched
274	1944	3296	400	0.5	40,000 (plus and minus 25)	2°H	3 min of arc, beam-switched

Aerials: Type 284 – 2 sets (1 transmit/1 receive) of 24 dipoles with a parabolic trough reflector on director, 21ft long x 2.5ft high. Due to director limitations sometimes reduced to 12 dipoles with an 11ft trough. Type 284M – 1 set of 24 dipoles and trough. Type 274 – 14ft aperture double cheese array mounted on director. Beamswitching of receiver incorporated.
Modifications: In Type 284M the power was increased for greater range and pulse length reduced for better range discrimination (164yds). The adoption of common aerial working, with TR switching and beamswitching, had the incidental advantage of reducing topweight. In addition a new display panel and ranging unit were introduced giving improved range accuracy while the beamswitching facility improved bearing accuracy.

HIGH ANGLE GUNNERY

Types 283, 285, 285M, 285P and 275

283	6/1941	600	–	–	–	–	–
285	1940	600	25	2	18,000 (plus and minus 150)	18°H 43°V	
285P	1942	600	150	1	18,000 (plus and minus 25)	9.5°H	15 min of arc. Beam switched
275	1944-5	3100	500	0.5	30,000 (plus and minus 25)	5°H 5°V	10 min of arc in both planes. Beamswitched

Aerials: Type 283 – twin Yagi with reflector. Type 285 – 6 Yagi arrays with 3 reflectors (3 transmit/3 receive) mounted on HA director. Type 285 M/P – as 285 but with common aerial working. Type 275 – 2 4ft diameter parabolic dishes in nacelles either side of Mk. 6 director. Separate transmit/receive dishes with receiver offset to produce conical scan when spinning.
Displays: Type 283 – ABU (Auto Barrage Unit). Type 274 – course and fine 'A' scan for range, side by side 'A' scan (with HPI for elevation search) for bearing/elevation.
Modifications: 285M/P – as for Type 284M plus fitting of ABU.

Aerial: Type 293M/P outfit AQR – cheese aerial with 8.5ft sweep diameter, waveguide feed and common TR. Power rotation at 5, 10 or 15 rpm. No stabilisation. Weight 951lbs.
Displays: PPI with sector selection of 'A' scan for target indication. Associated with Type 277 on large ships to put 277 height-finder on target.
Modifications: in M and P variations aerial AQR replaced the smaller AUR of the original 293. This gave a narrower beamwidth and greater bearing accuracy. They also had an improved receiver design giving better reliability.
Note: Type 276 is believed to be an intermediate type between 272 and 293 – it had a common TR aerial with waveguide feed. The aerial system was a small cheese.

Type 268

268	1944	9320–9475	30	0.75	10–12 on large ships, 5–6 on small ships.	4°H 17°V	

Aerial: Small cheese array, no stabilisation, power rotation at 22rpm.
Displays: PPI and 'A' scan with ranging unit RTB.
Note: Canadian design widely fitted during 1945-52. Replaced by Decca Type 974.

Types SL, SG (US)

SG	1942-43	3000	50	2	15–22 surface, 10 on submarines	5.6°H 15°V	2°
SL	1943	2945–2967	120	1.5	13–20 surface, 10 on submarines.	6°H 12°V	0.5°

Aerials: SG – cut parabola 15in high x 48in wide. Power rotation at 4, 8 or 12 rpm. Weight 340lbs. SL – truncated parabolic shield 20in high x 42in wide with wooden housing. Waveguide feed. Power rotation at 18rpm. Weight 300lbs.
Displays: SG – 5in 'A' scan and 9in (later 7in) PPI with repeaters. SL – 7in PPI with repeaters.

CLOSE-RANGE ARMAMENT CONTROL

Types 282, 282M, 262

282	1940	600	25	2	5000 (plus and minus 50)	40°H 43°V	
282M	1942	600	160	1	6000 (plus and minus 50)	29°H 43°V	1.5°H beamswitched
262	1946	10,000	30	0.5	7000 (plus and minus 25)	5°H 5°V	

Aerials: Type 282 – 2 Yagi arrays with reflectors (1 transmit/1 receive) mounted on pompom director. Type 282M – as 282 but with common aerial working. Type 262 – spinning parabolic dish with offset feed to give conical scan, mounted on 40mm STAAG mounting.
Display: Types 282 – 'A' scan with mechanical cursor.
Modifications: Type 282M as for Type 284M. Panel L22 replaced original ranging panel and L12 panel added giving improved ranging accuracy and reliability.
Notes: Type 262 gave auto follow in range and angle at 5000 yards – it searched an arc of 30° in two planes then locked in when the target was located.

PARTICULARS OF IFF MK 3 AND BEACONS
INTERROGATORS, 240 Series

Type	**Date**	**Frequency** (MHz)	**Power** (kW)	**Pulse** (microseconds)	**Beamwidth**
241	–	–	–	–	–
242	1943	179-182	1	6	360° or 60°
242M/P/Q	1943	159-189	2 or 10	4	88°
243	1943	159-189	–	–	–

Aerials: Type 242/243 – outfit ASD with Types 291M and 168, outfit ASW with Type 275. Type 242M – outfit ASS for Types 276, 277 and 293. Type 242P – outfit ANR for Type 277P. The ASS aerial consisted of four broadband vertical dipoles with power rotation.

RESPONDERS AND BEACONS, 250 series

251, 251M/P	Beacon operating on 173–179 MHz, horizontal polarisation.
252	– Employed in association with radar Types 79, 286, 290 and 291.
253MW/P	– IFF transponder operating on 157–187 MHz. All round radiation. Aerial outfit ASH. Vertical polarisation.
257	– Blind-approach landing system (BABS)

254, 255, 256, 258 – Beacons.
Note: IFF transponder replied with pre-selected coded reply to interrogations from 242, 243, etc. Beacons normally replied to any transmission on a specific frequency.

the beam-switching of the receiving aerial for Type 281 so reduced the range that on occasions enemy aircraft could overfly the bottom lobe. In addition Type 281 suffered severe loss of range if the two aerials were very different in height, as for instance in the fleet carriers.

Various steps were taken in 1942 to overcome these problems. First, beam-switching was dropped, as the gunnery requirement had diminished with the introduction of gunnery sets, and secondly TR switching was finally adopted. The principle of TR switching had been clear for some time and indeed the method had been in use in the United States since 1937 and in the UK from 1940. Unfortunately the ASE had initially attempted to use 'gas-gap' systems with little success but finally in 1942 achieved success with a diode system. Implementation began in 1943 and ships coming in for long refit had their Type 79 and 281 sets converted to single-aerial working. Single cable feed using coaxial cable and power rotation were also incorporated, the set number being amended with suffix 'B' added to indicate that the modifications had been done. In the final modification, the short pulse option on the transmitter was also removed and the sets used solely for air-warning. On aircraft carriers the opportunity was taken to fit both Types 79B and 281B as their different vertical lobe structures gave gap-free cover from about 1° elevation upwards. With battleships and cruisers the requirement for single aerials was not so urgent and quite a few carried on well beyond the war with their twin aerial systems.

As with Type 291, the RN took no steps during the war to design a replacement for Type 281 but simply confined themselves to a tidying up operation that appeared in 1946 as Type 960.

THE INTRODUCTION OF IFF

IFF (Identification Friend or Foe) was not used by the RN initially; instead the fleet relied on keeping track of all the friendly vessels in the area of operations. The RAF had found that this was not practicable with aircraft so had started development early on a method which allowed friendly aircraft to respond to radar signals. This was acceptable initially but proved inadequate as radars became widespread so a new system, involving the use of another transmitter to interrogate the aircraft, came into use. The aircraft fit, comprising a receiver and a transmitter and called the transponder, replied to the

Troops transferring from a transport to a motor launch off Normandy on 6 June 1944. The ML is one of those fitted as a navigational leader for the Normandy invasion and the lattice mast carried the aeria for a navigational radar (probably the RAF type H2S – RN number Type 971)

IWM

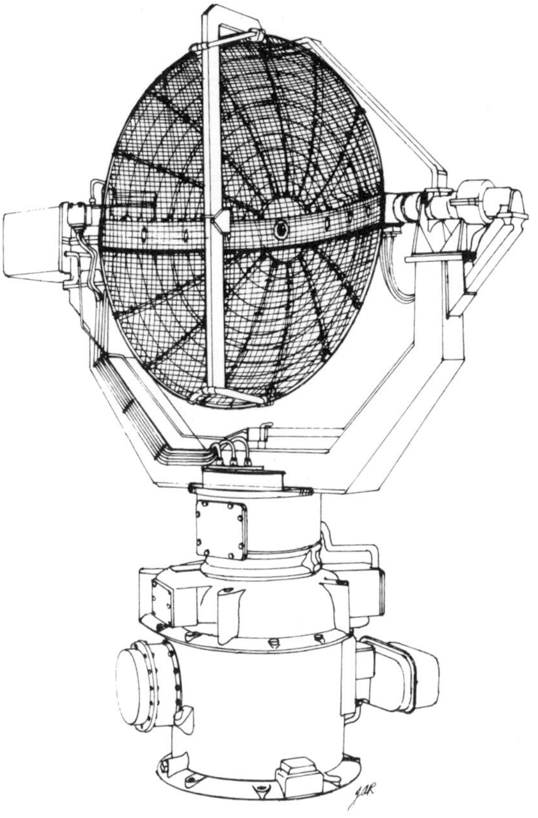

The aerial of radar Type 277

surface-search Type 271 series. With that set in production it was possible to consider other areas of use. The most obvious was for gunnery control and in 1942 all work on 50cm was stopped and development begun on a new series of gunnery radars. The three basic requirements were the same as before:
1) a low-angle gunnery set with sufficient range and bearing accuracy to control the main armament of battleships and cruisers
2) a high-angle gunnery set for controlling secondary armament on heavy ships and main armament on destroyers, with the capability of blind tracking in azimuth and elevation
3) a short-range set capable of detecting aircraft at 7000yds and then following the target automatically.
Whilst the basic requirements were the same as for Types 282–285 the increase in frequency was to be used to give greater accuracy and, in the case of 2) and 3), the hitherto unobtainable radar elevation tracking. That this required a

interrogations with coded pulses that could be displayed on the main radar's display or on a separate IFF display. The system was known as MkIII IFF and operated on a reserved frequency band of 157–187MHz. In common with RAF and US Forces it was adopted by the RN in 1942.

Because the system relied on active responses from other ships and aircraft there were no problems with sea returns so it was possible to use vertical polarisation, giving better vertical cover. The transmitters also required much less power for a given range so fitting of IFF presented few space problems. Initially the interrogator aerials consisted of directional Yagi arrays mounted on a horizontal 'U' bar but by 1944 the most commonly used aerial system consisted of four broad-band cage dipoles in a rectangular configuration and capable of power rotation. Various interrogators were used, numbered in the 240 series, the most common one being Type 242.

Transponders were given 250 series numbers and their aerial systems consisted of omni-directional dipoles, either broadband cage or eggtimer types, sometimes just a vertical dipole with an associated ground-plane. Both interrogators and transponders were fitted in convenient mast positions, although the interrogators associated with Types 279/281 and 271–273 were fitted 'in-line' with the radar aerials and rotated in synchronism.

Reports on the value of IFF vary according to the theatre of operations and range from 'worked very well' to 'never saw it used once in two years at sea'.

CENTIMETRIC RADAR 1943–44

Due to the urgent need for a radar to detect U-boats the initial emphasis in centimetric development had been on the

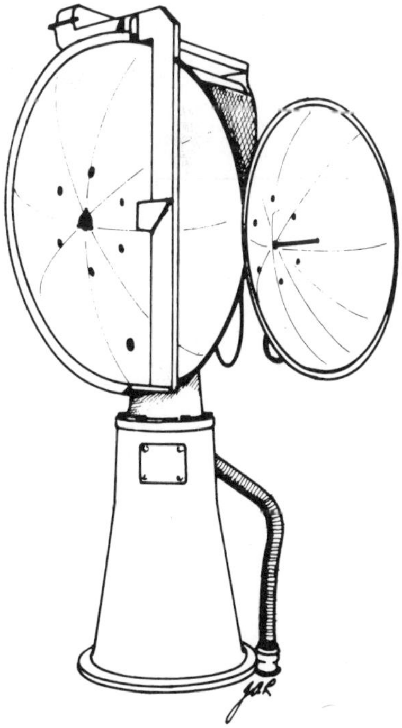

The twin aerial of radar Type 273Q

The cruiser *Liverpool,* returning to the Mediterranean Fleet in April 1948 with her radar outfit unchanged since 1945, was one of several cruisers that retained Type 279B (aerial on mainmast) throughout her postwar career. She still carries four Type 282 sets but they were removed shortly after this date. She also carried Types 274 (aerial on main DCT), 285 (aerial on after HA/DCT), 277 (aerial on tower abaft main DCT) and 293 (aerial at fore-topmast head)

Wright and Logan

The cruiser *Kenya* leaving for the Far East in October 1949 with a standard postwar radar fit of Types 277 and 293 on the foremast and 281B air-warning on the mainmast, all with associated Mk III, IFF. the main armament director carries the aerial of the 10cm Type 274 gunnery radar but the secondary armament still relies on the old 50cm Type 285 while the Type 282s of the close range armament have been removed completely.

Wright and Logan

The *Dido* class cruiser *Sirius* in July 1948 with a very dated outfit of Type 281 (still with separate transmit and receive aerials) and 50cm gunnery control Types 284 and 285. What appears to be a Type 268 is fitted on the foremast for navigation. The wartime fit of Type 282s for the pompom directors has been removed, without any replacement. Of the *Dido* class only *'Royalist',* transferred to the RNZN, was given a radar update after 1945.

Wright and Logan

considerable jump in technology and hence a long development time is evidenced by the fact that only the comparatively simple first requirement was met before the end of hostilities. The Type 274 set was designed in 1942 and was ready for shore trials early in 1943. It operated on a wavelength of 9.1cm (3296MHz) with a transmitted power of 400kW. To give good range resolution, 0.5 microsecond pulses were used at a repetition rate of 500 pps. With the short sharply rising pulses an accuracy of plus and minus 25yds was achieved and 14in shell splashes were displayed at ranges of 34,000yds. The aerial system consisted of a double 'cheese' with a 14ft aperture mounted on top of the main armament director. The beamwidth was 2° in the horizontal and 14° in the vertical plane, giving good resoltuion and stability. Beam switching was applied to the receiver cheese (the top one) by having two waveguides, one either side of the focus of the mirror, the receiver being switched to these guides alternately by means of a rotating shutter. Bearing accuracies of plus and minus 3 minutes of arc were achieved which was about the maximum possible.

Fitting of the set began in 1943-44 to the *King George V* class battleships and modern cruisers as they came in for major refits. Of the older battleships only *Nelson* was fitted, the remainder retaining Type 284/285 until withdrawn from service.

In parallel with Type 274, Type 275 was developed to meet requirement 2). To allow beamswitching in elevation and azimuth, conical scanning was introduced. As TR switching development would not allow the use of a conical scanning system that could handle a peak power of 400kW, separate aerials were again adopted. These consisted of two 4ft diameter dish aerials mounted in nacelles, the receiving dish being rotated to give the conical scan. Detection range was 36,000yds and it was hoped to give continuous tracking below 30,000yds. Bearing and elevation accuracy was 15 minutes of arc above a 3° angle.

With such a complex set development took a long time and the first sea trials did not take place until early 1945. The radar was mounted in the new Mk VI HA/LA director and went to sea in the first of the Battle class destroyers. Of the older ships only the battleship *Anson* was fitted, in 1944-45, and of the few ships fitted only the destroyer *Barfleur* entered the operational fighting zone, off Japan, before hostilities ended.

Even more complex was the third requirement and indeed the resultant Type 262 system with its associated 'Staag' mounting did not reach the fleet until 1946. Type 262 operated on 3cm and also employed conical scanning but with a common aerial. The system will be fully described in a future article on post-war radar.

TYPE 277 SURVEILLANCE

Having introduced the Type 271-3 into service the next step was to improve upon its performance, both in range and bearing resolution. The set that resulted, Type 277, was probably the finest of the radar products of ASE during the Second World War. It incorporated in its design most of the techniques used by modern radars today and was used during the war by all three Services. The transmitter employed a newly designed magnetron giving a peak output of 500kW with either an 0.7 microsecond or 2 microsecond pulse, depending on the range requirements, transmitted 500 times a second. A waveguide fed the energy, via a TR switch, to a single parabolic dish aerial 4ft 6in in diameter. The aerial had continuous power-rotation and was stabilised in the direction of radiation. PPI displays were incorporated in the design for the first time with facilities for additional displays as required. Initially the set was intended as a surface-warning set for frigates and corvettes but facilities were incorporated for elevating the mesh paraboloid and it was fitted to large ships for height-finding and low-angle cover. Detection ranges of up to 50 miles were obtained on low-flying aircraft; surface ranges, depending largely on the height of the aerial, were about 15–25 miles. Beamwidth was 4.5° giving good discrimination and a sharp PPI picture.

The use of Type 277 for low-angle cover filled the gap below the Type 279/281 but the narrow beamwidth led to difficulty in getting the Type 277 on to the target for height-finding. To overcome this another set was fitted, this time with a cheese aerial for target indicating. Designated Type 293 the set was identical to the Type 277 apart from the aerial which, being wider in the horizontal, had a beamwidth of 2°.

With Type 281 modified for PPI operation the method then used was to track with the Type 281 and 277 until the target appeared on the Type 293 PPI. The Type 293 PPI had an additional bearing cursor driven by the Type 277 aerial so the Type 277 could be rotated by hand until its bearing rested on the Type 293 target echo. It could then search in elevation for the aircraft, the information being displayed on an HPI (Height Position Indicator).

The radar information was therefore in a form for fighter direction and control and all the fleet carriers and larger ships were fitted with the system in 1944 especially those proceeding to the Far East.

In its original form Type 277 was fitted to the newest escort vessels, the Loch class frigates and Castle class corvettes, where it was very successful apart from problems with the aerial stabilisation. Its sister set, Type 293, was fitted to all new destroyers and AA frigates where its TI (target indication) facilities were used to bring the gunnery director on to the target. The TI system worked on the principle of an overhead optical system projecting bearing lines on the PPI. Operators could align them with targets and, in large ships, targets on four different bearings could be ranged-on, using sector display panels, without losing the all-round search capability. This system was later replaced by display outfit JL, in which any selected bearing was displayed on an 'A' scan which had an expanded trace, strobe tracing and IFF response facilities.

When the British carrier force joined the USN off Okinawa in March 1945 it had come a long way from the *Illustrious* with its Type 279 sailing into the Mediterranean in September 1940. Equipped with a comprehensive air-surveillance system and full aircraft-direction facilities they represented a tremendous four year technical advance on the part of the RN and if the various equipments were not as modern or numerous as those of their US counterparts they were certainly not outclassed.

Although Type 277 was fitted to corvettes it was too large for the

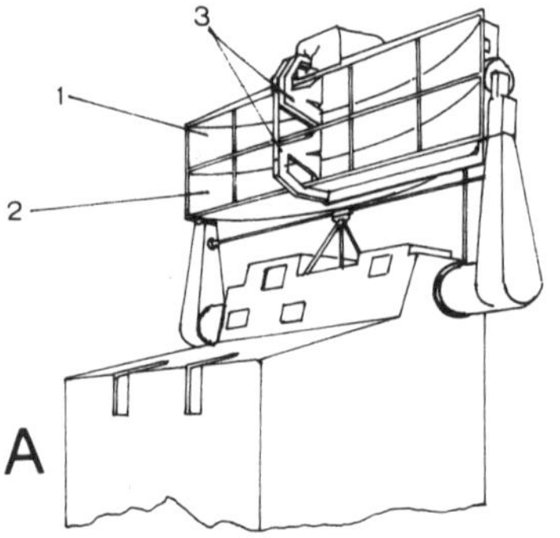

The equipment constituting the 10cm gunnery radar Type 274 as developed up to 1946.
A – Director control tower – 1. Receiving aerial, 2. Transmitting aerial, 3. Waveguides.
B – Radar panels 3AW and L30 – 4. Local indicator, 5. Panel 3AW control unit, 6. Wave monitor, 7. Trigger unit, 8. Modulator
C – Radar panel L31 – 9. Ranging tube, 10. Spotting tube, 11. Video filter unit, 12. Spotting box, 13. RTU (Range transmission unit) MkVII, 14. Wave monitor, 15. Oscillator.
D – Radar Panel L32 – 16. Coarse bearing tube, 17. Fine bearing meter.
Note: The L31 and L32 panels were located in the Transmitting station along with the AFCT (Admiralty fire control table), L32 being hung from the deckhead while L31 was deck mounted. The panels could also be employed with the surface warning radar.

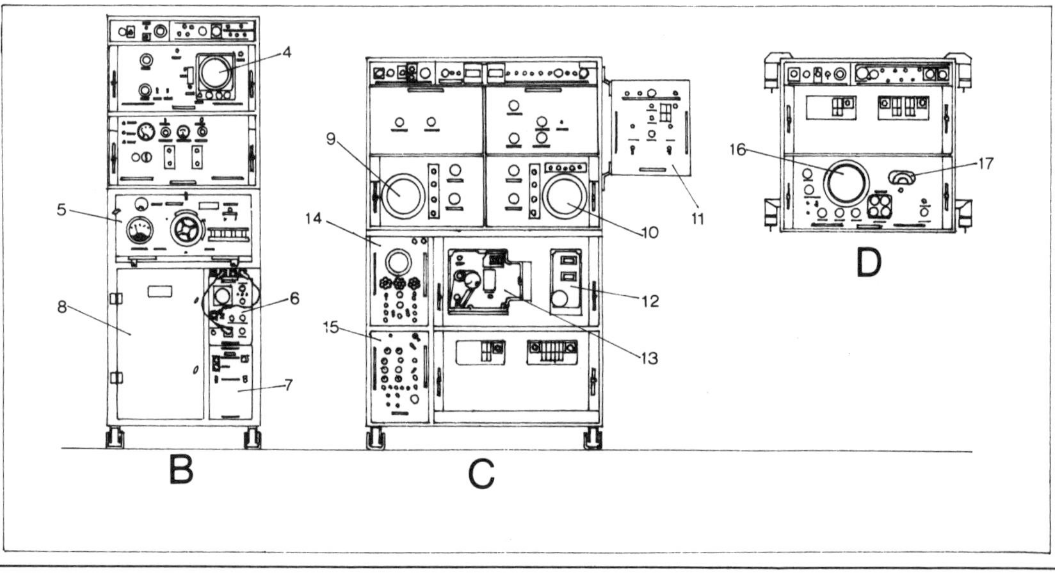

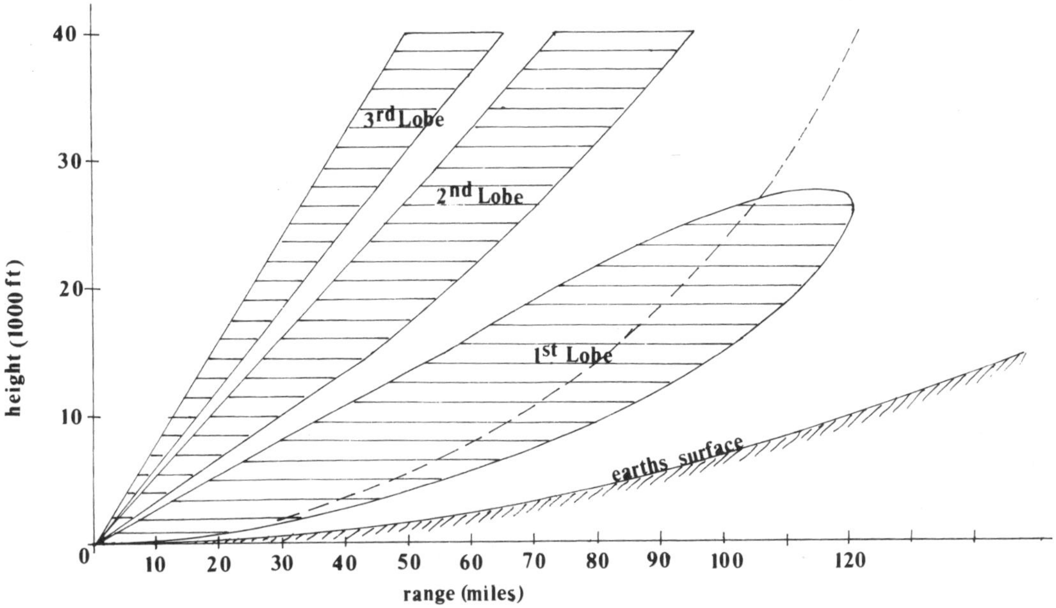

A vertical cover diagram for Type 281 showing the lobe structure. The broken line indicates the lower edge of the bottom lobe for Type 79.

The after section of the famous escort sloop *Starling* viewed from the foremast. In the centre of the picture is the Type 271 radar cabin with the perspex radome fitted above it. An early IFF aerial, employing horizontal polarisation, is mounted on the short pole mast extending from the roof of the radome – it is not clearly visible but is fitted about one quarter of the mast height from the mast base *IWM*

As modified for operations in the East Indies in 1945, the LSF *Boxer* carries a mixed bag of RN, RAF and US radars. Visible are the large aerials of a US – SK long range set on the quarterdeck, two SM fighter direction dishes, on the aftermost deckhouse and the foremast, and an RAF IFF aerial above the bridge. Type 281B and 79B aerials are mounted at the heads of the two after masts with IFF aerials above them. The other masts carry various aerials for RT, DF and homing beacons. In addition the large dome structure before the foremast is believed to have housed an RAF Type 11, 50cm, long-range radar aerial

IWM

The foremast of the *Algerine* class minesweeper *Espiegle* viewed from amidships. At the masthead is the aerial for Type 291 air-warning radar showing clearly the four dipoles mounted on an 'X' frame. The vertical dipoles at the ends of the upper yard are TBS aerials

IWM, courtesy D Sambrook

many smaller craft requiring a centrimetric replacement for the obsolescent Type 286. The requirement was met by using a small 3cm set developed in Canada and numbered Type 268. Introduced in 1944 it employed a small 'cheese' aerial with a 4° beamwidth rotated at 22rpm. The transmitter had a power output of 30kW giving a range of 500yds to 10 miles. A tremendous improvement on Type 286, it was widely fitted on many types from 'Hunt' destroyers to MTBs. It became the standard submarine set, combining with the air-search Type 291 to become Type 267. A ranging outfit, RTD, was also fitted in some ships for controlling the main armament.

Another 3cm set fitted in 1944 was Type 971 which was a naval adaptation of the RAF airborne H2S system. This set was widely fitted on landing craft as a navigational aid when approaching beaches but was also used by some for surface search when operating with light forces as in the campaign in the Adriatic. Apart from a problem with high sidelobes giving multiple echoes, the equipment worked well and proved reliable in service.

for low-angle warning and the Ground Controlled Interception (GCI) Type 7 that operated on 200MHz and whose aerial systems bore a striking resemblance to the US Navy's SK type. This set could handle four interceptions, gave cover to 90 miles and, with lobe switching, gave height information from $2\frac{1}{2}°$ to 20°.

The only other radars used by the RN during the war were the surveillance types fitted on the Lend-Lease ships from the USA. Largest of these was the long-range air warning SK fitted on the later escort carriers. Developed from the original XAF the set operated on

SECONDARY TYPES

Although the RN had adopted the airborne ASV and the later H2S radars, they made little attempt to use or adapt any of the ground radars for naval use. Initially, this was due to the large size of the arrays employed ashore and later of course they had a perfectly adequate set, in the Type 277 series, which indeed saw wide service with the RAF as Type 14. However, for the Normandy landings several landing ships were fitted as fighter direction ships to control RAF aircraft over the beachhead and as such were equipped with RAF radars including the 50cm Type 11

The light fleet carrier *Glory* leaving Portsmouth for the Mediterranean in November 1949 and still carrying the long-range mix of Types 79B (foremast) and 281 (mainmast) fitted in 1945. Level with the lower yard on the foremast is the aerial for Type 293 and at the fore-topmast head, abaft the 79B aerial, is the YE homing beacon. The main postwar alteration to the radar fit is the provision of a Type 277Q (aerial above forebridge), in place of the Type 277P, which gave improved height-finding capability and range. The hinged mast abaft the foremast sponson carries the aerial system for the VHF/DF set FV5

Wright and Logan

The cruiser *Glasgow* moored at Kings Wharf, Port of Spain, Trinidad, in March 1950. She carries radar Types 277 (aerial on tower abaft DCT), 293P (aerial at fore-topmast head), 274 (aerial on roof of DCT), 285 (aerial on HA/LA director) and 281 (aerial on main topmast head). Also visible, between the HA director and pompom mounting, is the aerial of the YE aircraft direction beacon.
USN

200MHz with an output of 200kW which gave a range of 120 miles at 10,000ft, roughly equal to the Type 281B. It, however, employed a much bigger array than Type 281 consisting of 36 dipoles in a 6 x 6 configuration backed by a 17ft x 18ft reflector mesh to give 20° horizontal/vertical beamwidths compared to the 35°/45° of the British set.

Also on 200MHz was the SA type fitted to the Captain and Colony class frigates. A much smaller set, it had a range of 20–30 miles on aircraft, its 3 x 2 dipole array giving it a beamwidth of 30°. Again it was on a par with Type 291 but with a slightly better beamwidth due to the use of 3 dipoles instead of 2.

The one air-warning set that really interested the RN was the fighter direction set SM-1. By employing their superior technical know-how in TR switching the Americans developed this 10cm set to drive an 8ft dish with conical scanning. This resulted in a set with a range of up to 70 miles, a bearing accuracy of plus and minus ½° and an elevation accuracy of plus and minus 500ft at 30 miles, which was much superior to the Type 277. The RN ordered 26 of them for fitting on carriers and FD (Fighter direction) ships but in the event only *Indomitable* was equipped. The reasons for sticking with Type 277 are not known but SM-1 was very heavy, complex and probably could only be fitted during a major refit. As it was not available until 1944, and ships were even having difficulties with the simpler Type 277, plans for widespread fitting were dropped although the LSP *Boxer* was given two in her 1945 conversion.

For surface warning the two US sets most widely fitted in British ships were SL and SG. Both operated on 10cm and came into service in 1944. The most common was SL with a power output of 120kW in a 1.5 microsecond pulse that gave a range of 10-13 miles with an accuracy of plus and minus 100yds. The aerial was a 42in paraboloid, with the top and bottom cut-off, giving a horizontal beamwidth of 6° which allowed a bearing accuracy of plus and minus ½°. The set was widely used by Captain class frigates in the closing stages of the European war to control light forces engaging marauding German S-boats.

Summing up wartime radars, it is evident that in the main, despite the phenomenal speed of development and with the honourable exception of the centimetric Type 271 series, that the crucial phase of the war to the end of 1943 was fought using the radars developed during 1937–40 or no radar at all! It says a lot for the design staffs, the technical development teams and the maintenance personnel in the ships, that something that didn't exist in 1935 was as successful as it was five years later.

USS WORCESTER

by Norman Friedman

USS *Worcester* using her 'washdown' system, installed to remove fall-out in a nuclear attack, 7 July 1954. From top down, her foremast carries an array of UHF ship-to-air radio antennaes, then a zenith search radar (SG-6), an air search radar (SPS-6B) and abaft the mast, a TDY jammer. The radar dish on her mainmast is for an SP height-finder. Her two Mk13 surface fire control antennas are visible fore and aft of her superstructure, as are two of her four Mk37 directors (with the dishes of Mk25 radars mounted on their roofs)

USN, courtesy N Polmar

Warships are usually designed to meet particular operational requirements, using weapons and other equipment either already available or in prospect. More rarely they are intended specifically as platforms for new weapons. As designed, the US *Worcester* class cruisers fell between these two categories. Ultimately, however, they were built largely to ensure that a new weapon, the 6in/47 dual-purpose gun, would actually get to sea, even though there was a preponderance of opinion in favour of terminating light cruiser production entirely so as to concentrate on the new rapid-fire 8in gun cruiser of what would become the *Des Moines* class.

EARLY STUDIES

The Bureau of Ordnance first proposed a dual-purpose (DP) 6in gun in about 1936; it appeared in Bureau of Construction and Repair sketch design, produced early in 1937, for small light cruisers in connection with the design of a new cruiser of under 8000 tons, as required under the new London Treaty. In fact a ship with four such

The *Worcester* (CL144) in August 1950, still has 20mm guns forward and has not yet been fitted with a twin 3in/50 in the forecastle position. The Mk 56 director is notable by its absence, its development not yet having been completed. The radar arrangement is unusual, with an SR-2 on the foremast and an SPS-6B on the second funnel
USN

twin mounts was actually chosen; only the sheepish admission of the Bureau of Ordnance that its new mount could not possibly be ready in time led to its ultimate rejection in favour of the all 5in gun design which became the *Atlantas*. Thus when more cruisers were to be ordered for Fiscal Year 1940, ie to be ordered late in 1939, the General Board favoured a design with the new heavy DP mounting and one with five twin 6in/47 was selected for development. It died only when it became obvious that emergency generators sufficient to operate the anti-aircraft battery (ie the 6in guns) in harbour, or at sea with a damaged main steam plant, would be impossible to fit within th 8000 ton limit. With such questions raised, the value of the all anti-aircraft battery was also raised. Perhaps more importantly, the beginning of the Second World Wa seemed to suspend all of the limitation treaties and a Construction and Repair proposal for a modified *Brooklyn* was soon accepted. The new 6in gun design now had no associated platform an work on it stopped in favour of weapons which seemed more likely candidates for installation, such as the new 5in/54 DP gun. For example, although early studies for what would become the *Montana* show the 6in/47 this weapon soon

gave way to the 5in/54.

The super-heavy anti-aircraft weapon was revived in response to war experience. Thus in June 1941 the General Board asked the Bureau of Ships, successor to Construction and Repair, for design studies of a new cruiser, with both twelve 6in guns and a very heavy armoured deck, about 6 to 7in thick; side armour was not required. Speed was to be the usual fleet speed of 33kts or more. Later General Board papers describing the responses to this request suggest a concern that no class of carrier or cruiser under construction would be able to survive or counter high-level bombing; it was not yet quite clear that such bombing very rarely found its targets.

The Bureau's sketch designs showed just how expensive that thick armoured deck would be. Early in July it presented two sketch Schemes, A and B, incorporating, respectively, four triple and six twin mounts. Neither, of course, had any secondary 5in guns, yet each was considerably larger than the contemporary *Cleveland* with its four triple 6in (capable of being modified to elevate to 60° but slow-firing at high-angles) and six twin 5in/38. In neither case could the full 6 to 7in be provided. It was far too heavy to place at main (weather) deck height, but on the other hand the weather deck would be the strength deck and therefore would have to be over an inch thick. It was therefore increased to 2in, with another 4in (3.5 over a course of 0.5, or 20lbs, STS) at the second deck level. This combination was considered equivalent to only a single thickness of about 4.8in, far short of the General Board's goal. The cost was a standard displacement of about 14,000 tons (*Cleveland:* 11,000). This was less than satisfactory to the Board so the designers tried to achieve a smaller, yet better-protected ship, trading off two 6in mounts for a 5in main armoured deck (on 20lb STS backing) plus a 1.5in weather deck, for a total thickness equivalent to 6.1in. This could keep out the 1000 and 1600lb AP bombs dropped by dive-bombers, whereas the previous arrangment would be penetrable by a 1000-pounder dropped above 7200ft, or a 1600-pounder dropped above 5000ft. All could be penetrated by the 2125lb AP bomb, although in the case of the new scheme, Scheme C, it would have to be dropped above 5800ft, which seemed less than practical. For the earlier schemes the limit was a far more practical 2800ft.

The preliminary designers pointed out that even though the heavy deck armour incorporated in these sketches would provide considerable immunity against direct hits, it would not save them from side damage due to near misses. For example, there was the recent case of the British carrier *Illustrious,* which had about 100ft of her side blown-in to a maximum depth of 5ft by a very heavy bomb exploding about 50ft away from her. There was also the issue of side armour: Scheme C had a 4in belt and barbettes but Schemes A and B would require an addition of about 200 tons to achieve the same level of barbette protection, plus about 600 for belt armour; together these would keep out 6in (105lb) shells at ranges beyond about 11,700yds (compared to 9000 in a *Cleveland*). Perhaps the only bright spot in a series of rather depressing design data was that the guns mounted on the 12,200 ton Scheme C were to fire about 1.5 times as fast as the earlier 6in so that her eight guns would be able to deliver the equivalent weight of fire of the twelve in a *Cleveland* during a surface action.

PRELIMINARY DESIGNS

Given these three sketch designs, the General Board prepared tentative Characteristics (ie Staff Requirements), which the Secretary of the Navy was to approve as a step towards actual construction. Dated 1 October, they called for a ship with the protection of Scheme C but the battery of Scheme B, ie, six twin 6in/47. The preliminary designers explained that this would require a ship of nearly 16,000 tons, with a 120,000 (rather than 100,000) SHP powerplant. Even cut to 100,000 SHP (about 32 rather than 33kts), this would mean a ship of 15,300 tons; a BuShips memo of 17 December observed that 'A standard displacement of over 15,000 tons seems more than should be concentrated in single anti-aircraft units of the type contemplated by the General Board. It is difficult to draw an arbitrary line and say that 15,000 tons is too much displacement but that 14,000 tons would be acceptable. On the other hand, if the major objectives of the design can be achieved on a ship of about 14,000 tons, then it appears questionable to allow the size to grow beyond this.'

Hence a new design, Scheme D, was prepared. Sacrifices included one 6in mount (at a direct savings of 600 tons) and a rearrangement of armour, with the protective deck over magazines dropped from the second deck level to the first platform and with the magazines aft protected by internal vertical armour. This was no more than standard practice in other US cruisers. In addition, the deck over the machinery was reduced by an inch to 4in. All of this cut the standard displacement to 14,200 tons; on 100,000 SHP such a ship, about the same size as A and B (664ft long on the waterline, beam 72ft rather than the 68.6 of the earlier studies, draft 22 rather than 23.3ft, 16,500 tons full load), would make about 32.5kts, which was perfectly acceptable.

Still, 'it appears, after considerable reflection, that the protection against direct bomb hits is over-emphasised and that the resistance to underwanter attack and to projectiles, even of moderate calibre, including those of light cruisers, is inadequate. The trend in air attack has certainly been to broaden the type of missiles and to increase the probability of torpedo attack or of large capacity bomb attack, retaining, however, as would be expected, a percentage of armour-piercing or semi-armour-piercing bombs. It appears, therefore, that a more conservative type, with a better balance of all qualities within some determined size which will permit

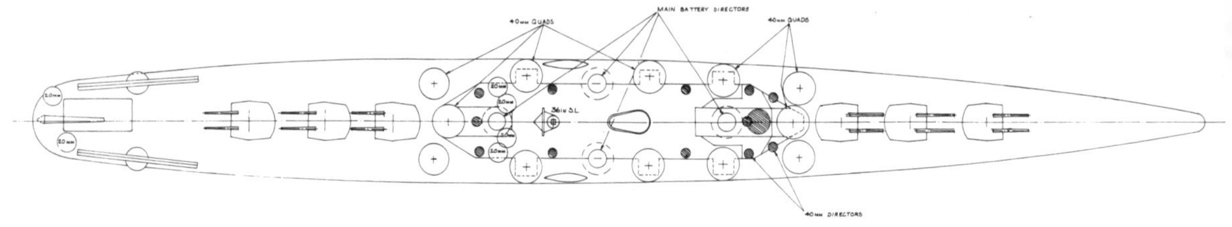

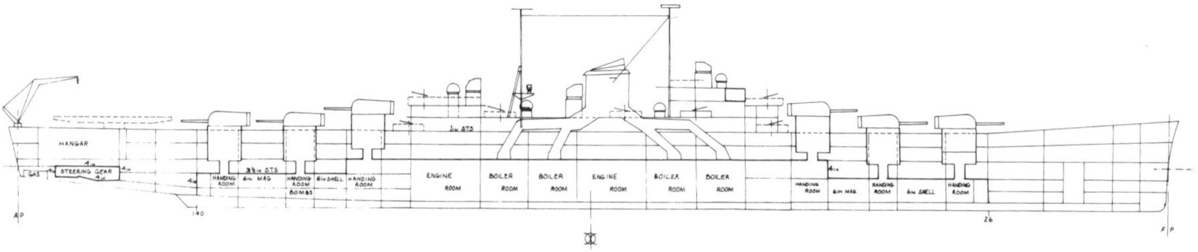

the number of 6in DP guns desired, will represent a more satisfactory final solution of the problem. The Bureau appreciates the thought to produce a special type design for a special purpose and the desirability of providing adequate anti-aircraft defence for a carrier striking force but, as outlined above, is inclined to believe that the defensive characteristics of such a ship should be more uniformly designed.'

In any case Scheme D seemed too large and provided too few guns for its size. The quickest way to cut that was to reduce deck armour which provided a large weight in itself and also required a large hull, particularly a large beam, to maintain stability. The result, Scheme E, carried five twin 6in on 11,500 tons, with only 3in decks over magazines and machinery. It was submitted in July 1942, just before American cruisers began to encounter Japanese cruisers around Guadalcanal. The Bureau of Ordnance objected to the reduced deck thicknesses and in a modified version the 1.5in weather deck was cut to an inch and the half inch transferred to the thicker protective deck where it would be more effective. Now, however, objections were raised that so large a ship should surely carry more guns and better side protection; by now it was painfully clear that cruisers might well be called upon to participate in surface actions.

The preliminary designers were caught between the general revulsion from very large cruisers and the series of requirements; moreover, the 6in/47 mount, as yet untested, was beginning to grow on the drawing boards. In January 1943 they submitted a new scheme, F, in which the sixth turret was restored and the side belt increased to 5in; barbettes were also increased from 4 to 5in. The immune zone now extended from 9000 (formerly 11,700) to 28,000yds, so that at its inner edge it matched that already standard in the *Clevelands.* The price for all of this seemed to be about 12,900 tons (15,450 full load), on roughtly a *Baltimore*-sized hull.

In submitting so large a design, the Bureau of Ships noted that 'The question now to be decided is whether 13,000 tons is too large a price to pay for a single modern 6in cruiser . . . the design study has several important advantages as compared with the CL 106 (*Fargo,* improved *Cleveland*) class which might well justify the increase in displacement. The most important of these are:

'(a) Improvement in horizontal protection against bombs.
'(b) Improved subdivision of machinery spaces (4 boiler rooms instead of 2)
'(c) Simplification of the design by omission of the 5in battery, which also permits a reduction of about 200 to 250 in complement.
'(d) Increased rate of fire of the main battery.
'(e) Improved fire control by virtu of having four main battery directors instead of two 5in and tw 6in directors as in the present 6in cruisers.'

Meanwhile the Bureau of Ordnance admitted that its new mount would not fire quite as rapidly as planned so the advantag over the conventional mount woul be less than a factor of 1.5. However, even if the new mount could fire only as fast as the conventional one in the surface rol it would still be far superior in anti-aircraft fire, in which the absence of power ramming in the older mount would prohibit very rapid fire. That was why the standard triple 6in mount was neve considered a DP weapon even though it was designed to permit elevation to 60°.

ISPLACEMENT PROBLEMS

cheme F was the result of a rather uick estimate using what the ureau of Ships later characterised s 'the usual short-cut methods mployed in preliminary design tudies, building up the estimate rom earlier studies which had been eveloped during the past year and half. After submitting Scheme F he design weights were thoroughly eviewed in comparison with recent veight reports of cruisers under onstruction and with particular ttention to the latest weight stimates of the component parts of he ship, such as the main nachinery plant and the ordnance nstallations . . . during the past year nd a half important growths have aken place in the weights entering nto this type of ship . . . a large part s due to the fact that special ight-weight alloys and light omponent parts are being replaced y heavier material, in order to onserve critical materials, or in ome cases, to facilitate onstruction. The net result of this tudy was to reveal that a larger hull han was proposed for Scheme F vould be necessary in order to etain adequate armoured reeboard, stability, and speed.' In act, the hull of a *Baltimore* class eavy cruiser would be needed, and better estimate showed Scheme F t 13,700 tons, a terrifying figure. ndeed, this result was impressive nough to lead the preliminary lesigners to try comparable four Scheme H, 12,550 tons) and five Scheme G, 13,150 tons) turret chemes. In both cases, many more ons per mount were necessary and n fact G and H seem to have been volved more as object lessons than s anything else: if the General Board wanted a 6in DP cruiser, it vould have to buy SCHEME F, or ome development thereof.

What the Board did want was ome indication of why it took so arge a hull to carry four, five, or six win 6in/47: after all, the five-turret lesign of 1939 had been within bout 200 tons of the 8000 ton limit n force at the time. In February 1943 the preliminary designers vere able to identify three major ources of increased displacement, as compared to the original design.

(i) Improved major characteristics, such as armour thickness (5in rather than 4in side, each over 20lb STS; 3.5in rather than 1.25in deck; increased machinery space subdivision) and the addition of a sixth twin 6in/47.

(ii) The effects of the normal development of hull, machinery, turrets. 'It must be remembered that the 8000 ton design was an attempt to obtain a ship of desired characteristics on a maximum standard displacement . . . as limited by the naval treaties. It was therefore necessary to skimp the design in some respects and to allow no margins for development. Light materials, such as aluminium alloys, were used extensively. There was no appreciable margin in beam to provide sufficient stability for the usual growth in topside weights. If this design were done over with proper margins and allowances for growth and under present conditions in which steel must be substituted for aluminium alloys, the displacement for the same characteristics would be much greater.' At the same time growth in the weight of the 6in twin mount amounted to 100 tons: it was now quoted at 360 rather than 260 tons per mount with, for example, a barbette diameter of 18ft 6in rather than the earlier 16ft 6in.

(iii) The effects of war experience, for example the vast increase in light AA battery, ammunition allowance and complement.

As usual, each direct increase required further increases in hull size, to provide strength, stability and speed; this might amount to as much as 150 per cent of the initial growth.

Nor was this all. The entire series, A to H, had been designed on the basis of practices common in about 1941, in particular aircraft and catapults amidships. The sole visible survivor of this generation of designs was the *Alaska* but in fact there had also been comparable heavy and light cruiser designs. Indeed these would have gone into production had not the exigencies of mobilisation prohibited departures from the basic designs (*Baltimore* and *Cleveland*) already in production. However, by the beginning of 1943 experience in the Solomons clearly showed the danger of fires due to shell hits on aircraft stowed amidships aboard cruisers. For example, gasoline stowed at the ends of the ship had to be piped about half the length and catapults amidships consumed valuable space. Moreover, in the 1941 series of cruiser studies it had been common to gain freeboard forward by breaking the hull abaft of amidships; this left a point of stress there, a discontinuity in the hull girder. Severe torpedo damage, a natural consequence of Japanese use of their 'Long Lance' torpedo, might well break a ship in half at such a point – as actually occurred in several ships, such as the *Helena,* with far stronger hulls. Moreover, the original justification for giving up the after aircraft location, that it left a large floodable compartment below the waterline, seemed far less convincing in a ship as large as Scheme F.

The preliminary designers therefore once again recast the design, with aircraft aft, as Scheme J. It was now necessary to adopt a flush deck, in order to provide volume aft for a hangar. This would raise the after turrets and with them the centre of gravity – thus demanding more beam and a larger hull. However, it proved possible to lower the main (weather) deck slightly abaft a point just forward of No 1 turret; this and other adjustments actually lowered the centre of gravity as compared to Scheme F and so avoided a great increase in size. In presenting Scheme J, the Bureau of Ships noted that 'if fire develops (amidships, due to aircraft) the smoke and heat from it are very apt to be drawn into air intakes from fire rooms and engine rooms.' Thus the alternative Scheme I, with aircraft retained amidships, was very soon drooped from consideration. The clearance of the amidships area in J permitted an increase in the number of quadruple 40mm mounts from 6 to 12, an additional advantage.

In presenting Scheme J in June

1943, the Bureau of Ships commented that it 'considers the design to be of sufficient importance to warrant further development into the usual contract plans and specifications, for possible use in case such cruisers are ordered in the future . . . the present 6in gun cruisers, including those under construction, have been developed by a long series of piecemeal modifications to the original *Brooklyn* class which was designed in 1933. It appears therefore that a completely new design is in order and the Bureau is proceeding on that basis.'

Meanwhile the Bureau of Ordnance was proceeding with the twin 6in/47. In fact that design seemed so successful that Ordnance wanted to try for similar results in a twin or triple 8in/55 design, which would soon be specified for a new heavy cruiser *(Des Moines).* Now Ordnance wondered whether it might not be better to substitute a 6in version of its new 8in triple rapid-fire mount; this would only throw out all of the design effort already expended by a hard-pressed Bureau of Ships design organisation. For its part the General Board wondered whether it might not be better to abandon light cruiser construction altogether in favour of the impressive new rapid-fire heavy cruiser.

DESIGN ANALYSIS

Admiral King favoured completing the 6in twin-mount design and by January 1944 it had progressed to the point where contract design could be completed within three months; it was time to decide whether to build. Seven light cruisers, tentatively numbered CL 143–149, remained in the authorised programme, and the Bureau of Ships suggested that procurement of hull material for them be deferred until the question of the new design was settled. At a General Board meeting on 24 January, the Chief of the Bureau argued that 'because of guided missiles the 5in/25 gun is entirely inadequate as of January 1944 as an AA gun and that we are accepting heavy, expensive and long delay, and a reduction of $\frac{1}{2}$ to $\frac{3}{4}$ of a knot in speed of the *Savannah* as typical of the early CL's to rearm with a 5in/38 AA battery . . . That we ourselves are working on the development of corresponding guided missiles which render even that battery of doubtful effectiveness . . . That in the *Cleveland* class there is serious interference between the battery and ship and fire control and considerable inter-battery interference . . . That while a number of ships of that class – *Denver, Birmingham, Montpelier* – have survived damaged in addition to *Boise, Savannah, St Louis,* and *Honolulu,* good fortune has been with each of them that the damage was not more severe . . . That the light construction necessarily adopted to keep the prototype *Brooklyn* class within the 10,000 ton Treaty limit has led to a greater number of lost bows than we expected . . . That we are discussing not ships to go into service against Japan in the spring of 1944 but ships which must look at least 2 to 3 years to the future and we hope to many more years.

'The contest between anti-aircraft guns and aircraft attack is exactly parallel to every other contest against a new offensive weapon . . . Torpedo plans and dive bombers have a tough time today getting in against an American ship. The controlled bombs struck from outside the range of the AA batteries of our 6in gun cruisers at Salerno. The present 5in/38 was the AA gun of 1930 developed to provide a DP gun for the *Farragut* class destroyers. The 5in/54 was selected for the 60,000 ton battleships in 1941 and confirmed in 1942 for the 45,000 ton CVB, before the advent of controlled missiles as an accomplished fact. Every chapter in past human experience promises further offensive developments of aircraft.

'Unless history fails fortunately to repeat itself in another respect, and remarks from Congressmen give no such promise, the ships we are now considering, ie the tail of the present programme of CLs, will be followed by few sisters in the immediate future; in fact, it will be very difficult to justify finishing them a *Clevelands* if the termination of hostilities should come before the are launched . . .

'It seems to me worth noting al in this connection that it will be fa more practicable, in the world policing job after this war, to control the building of prospective enemy cruisers, against which 8in gun cruisers will be needed, than t control aircraft construction and guided missile and long-range bor developments, against which effective gun armament will be needed if holding forces of surface ships and army ground forces are ever going to have to go into enen waters and territory again.'

The General Board felt that the 6in gun design would indeed 'produce a vessel with powerful main and close-in machine gun batteries, with other excellent military characteristics, despite thi however, such a large displacemn does not appear consistent with th battery carried, especially when compared with the 8in gun cruiser . . . The Board believes that the heavy cruisers carrying the 8in rapid-fire gun is the best design of large cruisers now available, embracing speed, protection, hea rapid firing main battery and dout purpose 5in and anti-aircraft machine guns. Therefore the Boa recommends that designation of CLs 143-149, inclusive, be change and that these vessels be laid dow and built of similar design to the heavy cruisers . . .'

This finished the 6in gun design unless, of course, Ordnance could not provide enough 8in guns and mounts in time. That proved the case; the Bureaus of Ships and Ordnance suggested that three of the ships be built as 8in gun cruise and four others as 6in DP. In that case all might be delivered during 1946, ie in line with the original

Another view of *Roanoke* late in her career but, in this case, with her SPS-8 radar aerial removed. Note the TDY on the foremast a the two small radomes of DBM (radar direction finder) on the mainmast

USN, courtesy N Polmar

USS *Roanoke* (CL145) late in her career, with the large aerial of SPS-8 height-finder radar on her mainmast and the forecastle 3in/50 mounting removed. The after Mk 13 surface fire-control radar is superimposed above her after Mk 37 dual-purpose director, opposite to the arrangement forward. The Mk 56 directors for her 3in/50 mountings are visible amidships, one being abeam of her after Mk 13 pedestal.

USN, courtesy N Polmar

schedule for the ships as originally proposed, as *Clevelands*. However, it would also have the very important advantage of producing a complete division of the new ships and thus a chance to test the unique qualities of their very long range anti-aircraft weapon. It was obvious that no further cruiser designs would be worked up during wartime and, therefore, that this was the last opportunity for the 6in DP gun which Ordnance had been developing since 1936. One cannot, therefore, avoid feeling that Ordnance was strongly inclined to assure the construction of the new light cruiser. CL 144–147 were selected as 6in gun ships.

FINAL DESIGN

The design was considerably developed in the next few months. For example, sketches of Scheme J show a single trunked funnel, as in the near-contemproary *Fargo* and *Oregon City* classes; she was completed, however, with two slender funnels. The original 100,000SHP powerplant was replaced by one of 120,000SHP (as in the *Des Moines)* after 1943 trials of the cruisers *Biloxi* and *Boston* showed that estimates of the speed performance of the two standard cruiser classes were too optimistic. One peculiarity of the design was the arrangement of its machinery. Originally that had been the usual alternate engine and boiler rooms. However, the six twin mounts consumed a great deal of centreline space. Therefore the two forward main turbines were mounted, battleship style, in separate compartments together with their boilers (unit system). Beam aft prohibited a similar arrangement, and a conventional boiler, boiler, engine room arrangement was used.

Of the four ships planned, only two, *Worcester* (CL 144) and *Roanoke* (CL 145) were completed; the other two were cancelled. They do not appear to have been entirely successful; they were considered wet, due perhaps to that attempt to avoid excessive hull growth through lowering freeboard amidships. The 6in automatic gun does not appear to have been nearly so successful as was the 8in developed from it and mounted in the *Des Moines* class; presumably this was the result of prototype development. It was certainly an impressive piece of machinery, with separate hoists for AP and AA projectiles for each barrel. It seems worth noting that it was far from the only postwar weapon of its type: the Royal Navy developed a similar gun for its *Tiger* class (among others) and Bofors developed a comparable weapon for the Swedish *Tre Kronor* and the Dutch *De Ruyter* classes. Like its US counterpart, the British gun was reportedly unreliable; it would be interesting to know how the Bofors product performed.

In any case, *Worcester,* completed in June 1948, was decommissioned in November 1958, the last US light cruiser in service. Her sister commissioned in April 1949 and decommissioned a few days earlier, in October 1958. Both then spent the next decade in mothballs (they were stricken 1 December 1970). Their large hulls inspired a variety of projects, including missile conversions and, about 1968, conversion to Major Fleet Flagships. However, by 1970 the entire reserve fleet seemed more a drain on maintenance funds than an opportunity for fleet expansion in an emergency and the two big cruisers were scrapped. They were handsome, impressive – if perhaps impractical – ships and they would probably have been far more impressive had not postwar air defence depended so heavily on carrier based aircraft and then guided missiles. However, that made the 6in anti-aircraft gun obsolescent at best, whereas the 8in, far from being primarily a ship-killer, was rather the ideal weapon for shore bombardment – which turned out to be a major postwar cruiser mission. Hence the survival of the heavy cruisers through the Vietnam War.

Apparently the General Board had been right after all!

ACKNOWLEDGEMENTS

Material for this article was taken overwhelmingly from the papers of the Preliminary Design Division of the Bureau of Ships at the Washington National Records Centre, Suitland. I am grateful to Mr Charles Wiseman, formerly at the Naval Ship Engineering Centre for making access possible. The joint Bureau of Ships/Bureau of Ordnance letter which made construction possible is taken from General Board papers at the Washington Navy Yard; I am grateful to Dr Dean Allard, to Mrs Kathy Lloyd (formerly of the Operational Archives) and to Miss Nina Statem (Operational Archives) for assistance. The sketch plan is from General Board File 420-8, preserved at the Operational Archives.

WORCESTER CLASS PARTICULARS

Displacement: 14,700 tons (std)

Dimensions: 664(wl) 679½(oa) x 70¾ x 21½ft

Armament: 12–6in DP (6x2) 24 – 3in/50 (11x2, 2x1) 24 – 20mm (12x2').

'Planned-*Worcester* actually had 6 twin 20mm *Roanoke* 8 in April 1949. Between 1955 and 1958 the forward twin 3in mount was removed.

Protection: Main belt, 5in tapered to 3in; 1in main deck, with 3.5in over belt to cover magazines and machinery. 4in bulkheads. 5in barbettes; turrets: 6.5in face, 4in crown, 2-3in rear. Protection against 6in 105lb shell between 9300 and 28000yds; against 500 and 1000lb SAP bombs from any altitude; against 12in 1000lb AP from below 7500ft (horizontal bombing).

Machinery: 120,000SHP = 33kts.

Designed endurance: with split plant operation, 7000nm at 20kts.

Generating plant: four 1250kW turbo-generators and two 500kv emergency diesel generators. This sufficed to carry the battle load on half the steam plant plus the diesel plant. Two turbo-generators were located in the auxiliary machinery room, the other two in each of the two firerooms. Emergency diesels were in separate compartments towards the ends of the armoured box.

Book Reviews

he Fleet Submarine in the US Navy
y Commander John D Alden
'S Naval Institute Press, Annapolis
Id, 1979 (Price $30.00)

is widely recognised that the Inited States Navy has the unique istinction of winning a big ıbmarine campaign outright. It is ırther recognised that one of the ıost important ingredients of that ɔlossal victory over the Japanese avy and their mercantile marine as the magnificent quality of the g fleet submarines, particularly ıe *Gato* and *Balao* classes.

Commander Alden has ccomplished the daunting task of riting a comprehensive technical istory of these boats, starting with *A1* in 1914 and ending with the *ench* class. Each class is described full, with technical details and board profiles and plans as well as hotographs. There are also ppendices with outboard profiles, uilding times from the various ırds, differences in constructional ıethods and types of machinery.

The story does not end in 1945, ıd all the subsequent ıodifications, both experimental ɔnversions and the 'Guppy' rogrammes are dealt with. The hole complicated Guppy I, II and I chronology is unravelled and ıere are many fascinating details of ıe 'Loon' and 'Regulus' missile ɔnversions, among others. A great eal of background information bout postwar submarine chnology is contained in the latter art of the book.

What the author has done articularly well is to plot the ɔrtuous course which led from the efficient 'T' class of 1920 and the ımbersome 'V-Boats' to the ıtstanding *Gato* design. It is a ory of trial and error, with many sappointments in the search for liable diesels. The attempt by the lectric Boat company to retain a ıonopoly of submarine expertise arallels the arguments between ickers and the British Admiralty efore 1914, while the otherwise ighly efficient Portsmouth Navy ard proved remarkably reactionary when the time came in the early 1930s to change from rivetting to welding.

The photographs are quite outstanding: large, well-printed and with informative captions. With the line drawings by John Humphrey, they complement the text admirably. Although each class is described in detail this does not rob the narrative of continuity and one can follow the main threads of American submarine development from the First World War up to modern times. Technical writers accustomed to dealing with the British authorities will envy the freedom of access to recent information. Would the Royal Navy permit an 'As Fitted' set of an *Oberon* to be used to illustrate a technical history of British submarines? Certainly nothing in this sort of depth has yet been written about any other navy's submarines. It is a handsome tribute to a remarkably effective generation of submarines and sets a new standard for the Naval Institute's books on technological history.

If I have a criticism it is the lack of information on torpedoes. These were after all the chief weapons of these submarines throughout but, apart from a brief reference to the disastrous performance of American torpedoes in the Pacific, there are virtually no particulars of performance. I suspect that this reflects a dearth of information rather than any blind spot in the author's perception of his subject, given his exhaustive treatment of every other relevant topic. Apart from the list published in the current *Weapons and Warfare* partwork I do not know of any comprehensive list of marks of torpedoes used by the US Navy.

Antony Preston

Conway's All the World's Fighting Ships 1860-1906, *Conway Maritime Press, London, 1979* (Price £24.00)

If ever a book needed to be compiled it was this one. The nineteenth century has never attracted the attention that the eighteenth century has from shiplovers and enthusiasts and I am sure that one of the overriding reasons is the lack of sources. Nothing puts off a researcher faster than that, and for most of the nineteenth century it is a matter of virtually no references, let alone untrustworthy ones.

The scope of the book is broad but specific. Only seagoing warships are included and, although there is a general lower limit of 400 tons, torpedo-boats are not excluded. The convenient distribution of Great Powers, Coast Defence Navies and Minor Navies, so beloved of nineteenth century commentators, has been followed, making comparisons between like and like much easier. All the major navies are given an introductory essay describing their background and then the various ship types are given their own introduction. These essays are gems of erudition in their own right and add considerably to the authority of the whole.

The book deserved to be read widely for it is clear evidence to refute the standard view that nothing much happened in the nineteenth century between Nelson and Fisher. This is particularly so of the Royal Navy, and hardly a book is published without scathing denunciation of reactionary admirals who thought iron wouldn't float and steam wasn't reliable. If this work does nothing else it will show just what a frenzy of technical change gripped the Royal Navy from the Crimean War to the turn of the century. Mistakes were made, as in other navies, but apart from a decade of financial parsimony imposed from outside, the RN coped remarkably well with the forces of technical progress.

Inevitably some of the later entries in the British section overlap with other published sources, notably Oscar Parkes *British Battleships* and Edgar March's British Destroyers, but this was necessary to complete the picture.

In the case of destroyers this version is a distinct improvement on March, for the destroyers are described in the way in which they were built, not according to the totally arbitrary 'A' to 'D' classification adopted in 1913. Thus, both destroyers and torpedo-boats are listed by batches from each builder, making it much easier to follow the quirks and foibles of each yard. For the United States' Navy the convenient distinction between the 'Old' Navy of 1860-1882 and the 'New' Navy which was created virtually from scratch has been followed.

What transforms this book from a weighty and valuable shelf-reference into something to delight the reader is the quality of the illustrations. The cream of the Conway Picture Library has been used and many more, while John Roberts' meticulous drawings fill in the gaps. Care has been taken wherever possible to use a drawing to illustrate the ship in a different rig to the one shown in the photograph, or to depict a sister-ship. The statistics speak for themselves: 471 photographs and 506 line drawings in only 448 pages. Above all the quality of paper and printing do justice to such a lavish spread.

Inevitably, the attempt to fill such a void has resulted in some minor omissions, but it would be churlish to make anything of this. We all have a pet ship, about which we know more than anyone else, and part of the pleasure of this book will be the fun of trying to find something to write to the editors about. But such is the quality of what has been done that I would warn all nit-pickers to be sure of their case before rushing into print.

Antony Preston

The British destroyer *Grenade* as completed. One of the vessels described in the reprint of *The British Destroyer* by T D Manning
MoD

British Battleships 1892-1957 *by Randolph Pears and* **The British Destroyer** *by T D Manning,* Godfrey Cave Associates, 1979 (Price £7.95 each)

These books first appeared in the 1950s, and have now been reprinted. Although both books have been overtaken by later, and more up-to-date, information both contain interesting sidelights on th old Navy. Unfortunately, the quality of photographs has not bee retained in the dot-for-dot copying and this was one of the best featur of the originals. 'Sailor' Manning's book in particular contained the cream of the Perkins and Oscar Parkes Society Collections. The discerning reader will, of course, spot the more obvious errors, and enjoy the books for the wealth of background detail which they contain.

Antony Prest

Editorial

Since the Second World War the nature of warships has undergone considerable change causing a certain amount of confusion with classifications. This has, in part, been due to an unwillingness to abandon the old classifications in favour of titles more closely describing a new ship's intended purpose. Some controversy was, for instance, caused by the classification 'guided missile destroyer' for the British 'County' class in the early 1960s. Some said that these vessels, at 5200 tons, were too large to be accurately described as destroyers and that they should be called cruisers, missing the point that warships should be classed by function and not size. Nevertheless, the term destroyer lost its original meaning very soon after being introduced (as 'torpedo boat destroyer' which has an obvious and accurate interpretation) since the resulting vessel both made the torpedo-boat obsolete and took over its function, becoming a super torpedo-boat capable of carrying out torpedo attacks and providing defence against them. It might well have been better if at this point the destroyer had reverted to the title torpedo-boat but the two world wars saw a substantial increase in their area of operations and in many ways they began to function as miniature cruising ships.

This brings us back to the 'County' class which were primarily designed as fleet escorts, with secondary roles as flotilla leaders, shore support vessels and, in peacetime, police duties and showing the flag. All these functions are traditionally those of cruising ships so perhaps they should have been designated cruisers but here one runs into a similar, but older, history of classification. The original term for such vessels was frigates but this was lost in the confusion of types developed in the early years of the Ironclad and the Victorians first began using 'cruising ships' to describe practically all vessels between the size of sloops and battleships and then as a specific classification for ships designated as fleet scouts, and trade protection vessels – both traditional frigate functions. Since then the term cruiser has been applied to a considerable variety of ships ranging from scout cruisers, which were little more than flotilla leaders, to the battlecruiser thus completely encompassing the warship types between destroyer and battleship which makes nonsense of regarding a cruiser as a specific type.

Oddly, the reintroduction of the term frigate for the 'River' class convoy escorts during the Second World War has given the modern Navy a reasonably accurate and traditional classification for new ships. Although an inaccurate term for ships as specialized as the 'Rivers' the postwar development of the frigate has produced general purpose ships with a wide variety of functions including fleet escort, trade protection and radar picket (the latter being a modern equivalent of fleet scout) – they even have the ability to act as communication links, a traditional role of the sailing frigate.

This however has been achieved more by luck than judgement and, with the odd exception, their has always been a tendency to hold on to traditional classifications whenever possible. The best definitions nearly always result from the introduction of a completely new type of vessel for which an accurate title is invariably found. A good example is the many types of specialised landing craft developed during the Second World War, some of which are described in this issue by Norman Friedman. The introduction of such a wide variety of craft, together with the need for a common US/British classification system, produced accurate definitions with a basic separation between landing craft and landing ships and then sub division into size and function – leaving little room for confusion.

John Roberts

THE DESIGN OF HMS INFLEXIBLE

by DK Brown

The *Inflexible* as completed in 1881 with brig rig

NMM

The *Inflexible* derived from the report of the 'Committee to Examine the Designs upon which Ships of War have recently been constructed' which was set up by the Board of Admiralty in January 1871, following the loss of HMS *Captain*. It was chaired by Lord Dufferin and, as well as some prominent Naval officers, included such famous men as Rankine, Rendel, R E Froude, Denny, Bidder, Rev Wooley and the almost inevitable Thomas Lloyd. The Board still did not understand the reasons for the loss of the *Captain* and wanted all other designs examined in case they too had unknown faults. It was an appalling confession of incompetence. The other designs had been prepared to meet the proper requirements of previous Boards by fully qualified Naval Architects and yet suspicion lingered as a result of this single disaster, when neither of these conditions had been satisfied.

The Committee, along with many other people, realised that impenetrable armour over the whole length of a ship was no longer possible. The 35 ton gun could penetrate 12ins of iron (weighing 480lbs/sq ft + weight of backing) at 600yds and much bigger guns were on the way. Thick armour therefore, had to be concentrated along the waterline amidships, where the ship was widest, in order to preserve stability.

A stern view of *Inflexible* after the removal of her sailing rig

CPL W/15/001

The design of the new battleship was the responsibility of Nathaniel Barnaby, Chief Constructor of the Navy, who was a brilliant naval architect but had rather strange religious views in which he saw the Royal Navy as the strong right hand of Jehovah. His attempts, lacking any real Naval Staff guidance, to produce a ship with thick armour and heavy guns, on a relatively small displacement and cost, produced a ship with many outstanding technical achievements but of little fighting value. Only in terms of the 1871 Committee concept of 'Centres of Naval Power', ie slightly mobile fortresses, does *Inflexible* make sense.

EARLY DESIGNS

Great ships are never conceived in the abstract. Sir Edward Reed had sketched out a range of studies in 1869, with the assistance of Mr John, a graduate of the second School of Naval Architecture, soon to become a leading Lloyd's surveyor. Though these studies range in size up to 10,938 tons (almost that of *Inflexible*), they were based on the 18 ton gun and the belt, of from 13in to 26in wrought iron, extended over a long citadel of about 200ft. No allowance was made, however, for the protection of guns or their gunners and the weight of any such armour was to be deducted from that of the belt. All these ships were intended to fight with the ends deliberately flooded to reduce freeboard, relying entirely on the armoured citadel for both buoyancy and stability.

Designs similar to these were considered by the 1871 Committee, as were many other ships with a central citadel and unarmoured ends. The Committee report was available to all and may well have influenced the Italian Naval Architect Benedetto Brin in his battleships *Duilio* and *Dandolo*, laid down in 1872. His ships were originally intended to have 4-35 ton guns but were soon adapted to employ the 60 ton gun. Following the introduction of the 80 ton gun in

A rare view of *Inflexible* with sails partially unfurled in 1881. The curved arm projecting from the bows is a trough which could be lowered and used to launch torpedoes directly into the sea – it was not a success

NMM

Inflexible, Brin changed his design yet again to carry 4-100 ton, 17.7in Armstrong guns (a similar weapon is still mounted at Gibraltar). The Italian ships had a 21½in belt of Creusot steel plate in a single thickness over a length of 169ft.

The Board of Admiralty had originally decided on an improved *Fury* for the 1873 programme and Barnaby produced a number of designs, similar to *Dreadnought* but with an increased armament. One of his assistants, Hounsom, produced a peculiar design with 10-35 tons guns in an open battery but none of these could match *Duilio*. What was never very clear was the role of the new battleship. She was to be the most powerful ship of the RN and, perhaps, in the world but how she was to be operated in company with older and less spectacular ships was not manifest. Her conception was that of an age obsessed with the might of individual ships as though, as in the age of chivalry, *Inflexible* the Champion of England would meet *Duilio* of Italy in a single joust. Though she was to be of such outstanding power, Barnaby was limited in size due to the Board's fear of losing their great ship to a single torpedo and to the need to keep the cost down.

THE CENTRAL CITADEL

Barnaby created *Inflexible* in three parts (Fig 1). First there was the central citadel amidships, an armoured raft, with sufficient buoyancy to float by itself and stability to ensure that it floated upright even with the ends destroyed. This citadel was protected with a double thicknesses of 12in wrought iron armour which, together with the 36in teak backing, weighed 1100lbs for each and every square foot. At such a weight, the armoured length had to be kept to a minimum. The citadel was just long enough to protect the machinery and guns. Originally, the armament was to have been 4-60 ton guns in two turrets but, in October 1874, it was decided to change the armament to 4-81 tons, 16in, MLR to match that of the new Italian ships. Including the heavier

Table 1

Ship	Calculated weight (tons)	Authorised changes	Revised Estimate	Weight on Completion	Error	Vertical CG (Kg, ft) Est	Revised Est	Actual
Alexandra	9508	–	9508	9492	− 16	25.97	25.85	25.77
Devastation	9062	+ 465	9527	9298	− 229	22.78	23.43	23.55
Shah	5700	209	5909	6250	+ 341	22.93		23.42
Shannon	5165	233	5398	5439	+ 41	21.73	21.98	21.77

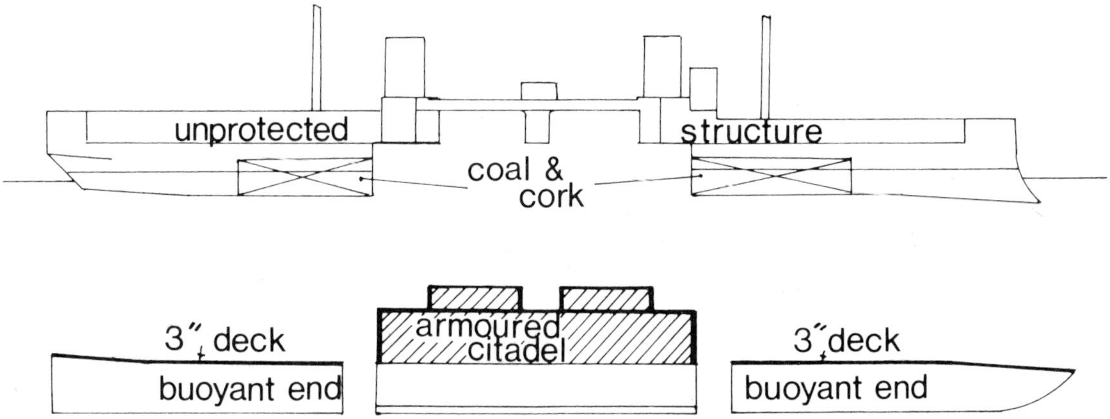

FIG 1 The main components of the *Inflexible* of 1876

mountings, etc, this involved an additional weight of over 200 tons above the upper deck without any change in the dimensions of the hull. These guns fired a 1700lbs studded shell (with a burster of 17½lbs) at a muzzle velocity of 1600ft/sec and could penetrate 22½ins of wrought iron at 1000yds. They were loaded from outside the turret, the guns being depressed into a glacis and the charge and shells rammed in, hydraulically, from below the upper deck. With this system the rate of fire did not exceed one round every 2½ minutes. (Two similar guns, in a slightly different turret, still exist near the train ferry terminal at Dover, and a studded 16in shell is exhibited at the Naval Armament Museum, Priddy's Hard, Gosport.) *Inflexible*'s turrets (but not the hull) were protected with the newly developed compound armour which had a hard steel face, to break up the attacking projectile, backed by a tough wrought iron layer, to support the face and to absorb the larger fragments. Even so, teak backing was still needed but such compound armour had about 25% more resistance to penetration than simple wrought iron and was as effective as the all-steel plates used by Italy and France.

The citadel also protected two sets of twin cylinder compound engines which, with a fuel consumption of 2½lbs/IHP/hour and stowage for 1300 tons of coal, gave her a good endurance. Even so, she was still fitted with a brig rig, weighing 100 tons, the equivalent of an extra 3ins of armour on the turrets. It was said that this rig was only for peace time crew training so that they could develop and maintain the skills still needed in the many RN ships which still used sail. It is easy to deride such vestigial features but, even now, there are those (not this author) who maintain that sail training is of real value to the sailor – as opposed to fun.

SOFT ENDS

The second part of *Inflexible* was her comparatively lightly built pointed ends. These had no side armour but Barnby had no intention of neglecting their buoyancy. 'We never intended to deprive the *Inflexible* of the value of these ends; on the contrary we devoted special attention to their construction, and to their lading with fuel and stores so as to delay the flooding and water-logging of these parts of the ship as much as possible; and we regarded the riddling and water-logging of these ends as the practical limit of the power of the enemy in an engagement.' The ends were protected by an armour deck, at about water-line level, which would preserve the buoyancy of the under-water portions against all but the most severe damage. Above the deck, there was a cellular arrangement of coal bunkers surrounded by cork which had been shown by full scale tests to be extremely resistant to shell fire. Damage would be localised and much of this waterline area would contribute to buoyancy and stability even after severe damage. The volume below the deck was closely subdivided. The third part of *Inflexible* was her lightly built superstructure giving freeboard for operation in bad weather, reasonably comfortable accommodation for her crew and stowage for boats, searchlights etc.

DESIGN CONSIDERATIONS

The design of such a ship was by no means easy. Some of the drawbacks of the low freeboard ships like *Devastation* were already apparent; in particular, they lost speed rapidly in head seas. Short ships were known to be more prone to such problems and yet Barnaby, driven by the need for a short citadel, was forced to design a short ship. On th other hand, William Froude's model tests had shown that the resistance of a relatively slow ship such as *Inflexible* (14kts) was mainly frictional and hence the smaller the wetted surface area the less would be the power required tc drive the ship, consequently *Inflexible*'s tubby hull required some 10% less power than her lean predecessors.

The stability of the ship with the ends flooded, 'riddled' by gunfire,

Inflexible with military rig in the late 1880s or 1890s
CPL W/15/002

Inflexible in 1887
NMM

required very careful calculation. The ship had to float upright in this condition but excess armoured length had to be paid for at 1100lbs per square foot. The detail design was in the hands of William White and Barnes; in their capable hands the accuracy of the calculations was no problem. Both weight and the position of the centre of gravity had to be estimated correctly and records show that this was usually achieved in that area (see Table 1).

The early mastless ironclads tended to have too much metacentric height due to the loss of the topweight of masts and sails. This led to heavy rolling, which reduced the accuracy of their gunnery, and *Devastation* and the other low freeboard ships relied on the sea washing over their decks to reduce the amplitude of rolling. For *Inflexible*, Froude introduced the much more effective bilge keel which allowed the freeboard to rise improving the speed in rough water. The stability calculations and the value of bilge keels were checked with a model weighing 1½ tons in Froude's tank at Torquay, both for the intact ship and with all the unarmoured structure flooded. The effect of forward speed on the

PARTICULARS OF INFLEXIBLE

Laid down: 24.2.1874
Launched: 27.4.1876
Completed: 18.10.1881
Builder: Portsmouth Dockyard
Fate: Sold for scrapping 1903
Load displacement: 11,880 tons
Length: 320ft pp, 344ft oa
Beam: 75ft
Load draught: 25ft 6in
Armament: 4-16in MLR (2x2) (each turret weighed 750 tons including guns), 6-20pdr BL (6x1), 2-14in torpedo tubes (submerged), 2-14in torpedo carriages (above water)
Armour: Citadel sides and forward bulkhead – 24in (12+12) at waterline, 20 in (12+8) at top, 16in (12+4) at bottom; Aft bulkhead – 22in (12+10) at waterline, 18in (12+6) at top, 14in (12+2) at bottom; Turrets – 16in (7in+9in); CT – 12in; Decks – 3in; Weight – 2205 tons (vertical) 905 tons (decks)

Machinery: 2 shafts 3 cyl Elder compound engines, 8 single ended and 4 double ended cylindrical boilers (working pressure 60lbs/sq in), trial performance 8407ihp = 14.75kts at 73rpm. 1200 tons coal.

Machinery weights: Engines 614 tons, propeller shafts and stern fittings 190 tons, boilers (dry) 522 tons, water for boilers 190 tons

Auxiliary engines: 2 surface condensers (16,000 sq ft cooling surface) 1 steam steering engine, 2 fire engines, 1 capstan engine, 1 small turning engine, 2 donkey engines for bilge pumps, 4 auxiliary feed engines, 4 ventilation fan engines, 2 condenser circulators, 2 reversing engines, 4 hydraulic engines (2 per turret, one being reserve) for shot hoists, rammers and turret training, 4 ash hoists, 2 40HP main-drain pumps, 2 shot hoists, 2 boat boom engines, 4-300 ton/hour ejectors, 8 air compressors for torpedo tubes, ?–800v generators

Boats: 12 (including 2 second class torpedo-boats)
Sail area: 18,500 sq ft
Complement: 440

waterline and wave profile with the model heavily trimmed after damage at one end only was also measured. Full scale trials were carried out with *Thunderer* to measure the angle of heel on the turn and the results were used to estimate the heel of *Inflexible*, intact and damaged. The diameter of *Thunderer*'s turning circle was about 1250ft with 31° helm. It took about 12 seconds to put the helm over and the steady speed on the turn was about two thirds that of the approach. Such figures must cast considerable doubt on the ability of ships of the era to ram a mobile enemy and, indeed, the rather frequent accidental collisions, while showing the danger of underwater damage, demonstrate their poor manoeuvrability. It is far more easy to avoid the impact of a friend than to achieve collision with an evasive enemy.

Inflexible in the 1890s; note the military rig with fighting tops and the 20pdr BL gun por on the flying deck forward

CPL W/15/003

THE INFLEXIBLE COMMITTEE

The accurate and refined calculations of Barnes and White, though confirmed by Froude's tests, were still greeted with incredulity by the innumerate. Even E J Reed attacked the design, following a visit to Italy during which he had inspected the *Duilio*. He claimed that this ship would be unstable if the unprotected ends were flooded and though this was denied by her designers they produced no evidence in support of their claim. Reed then launched similar attacks in the press and in Parliament on Barnaby (his brother-in-law!) and on the *Inflexible*. Parliament then set up yet another committee, with Wooley, Rendel and Froude as members, to investigage the likelihood of the unarmoured ends of *Inflexible* being riddled and the stability of the ship should she reach that condition. They found that total destruction of the ends was almost inconceivable but should it happen, the ship would just be stable. The reserve of stability would be very small, but since the free water would act as a very large roll damping tank, she should survive, though the guns would have to be worked with great caution and turning would be dangerous.

It was estimated that it would take some 300 hits to destroy the cellular ends and this figure received some practical confirmation when, in the Sino-Japanese war, two Chinese ironclads, designed on the same system as *Inflexible* were heavily damaged. It was said that they received some 200 hits apiece without serious effect, though the number of hits was probably exaggerated and most of them were of fairly small calibre. A diminutive of *Inflexible*, HMS *Edinburgh*, received even heavier damage during firing trials in 1910.

AUXILIARIES

Inflexible was full of novelties. She had two large tanks partially filled with water to damp the rapid roll expected in consequence of her 8ft GM. These tanks, invented by the young Phillip Watts, were too small to be effective and since the sloshing of water was very noisy their use was soon abandoned. However, during their trials Watts began a long friendship with the first captain of *Inflexible*, Fisher, which was to have important consequences many years later (see *Dreadnought*, *Warship* 13).

Inflexible completed in October 1881 and was later Fisher's command at the bombardment of Alexandria. She was the first ship to be given a full scale lighting installation. Power was supplied at 800 volts dc by generators manufactured (in 1877) by the British Electrical Co of America to a combined system of arc lamps in the engine and boiler rooms and Swan 'glow' incandescent lamps elsewhere. Distribution was complex, the Swan glow lamps being connected in circuits of 18 lamps in series between the main supply bus-bar and the arc lamps. Each glow lamp was fitted with an automatic cut-out bringing into circuit a substitute resistance approximately equal in value to tha of the lamp should the latter fail. This 800 volt installation was responsible for the first fatal accident due to electric shock (1882) in one of HM Ships. After this incident 80 volts dc (complete wire circuit) was adopted as standard RN practice. She also carried two small torpedo-boats, which could carry the new Whitehead automobile torpedo, and had launchers which could launch torpedoes from on board.

Inflexible was a technological triumph, packing enormous offensive and defensive power into a hull constrained to dimensions which were too small. She was the most fully investigated ship of the century and full of gadgets, but one still wonders what she was for. The naval staff did not exist and the Board gave no real guidance.

THE GERMAN TYPE VII SUBMARINE

by Erwin F Sieche

The Type VIIA submarine *U35* prewar. She
destroyers *Kingston, Kashmir* and *Icarus*
north west of Bergen. *E Sieche*
was sunk on 29 November 1939 by the

The conning tower of a Type VIIA submarine
USN, courtesy A D Baker III

A type VII approaches a submarine tender c the *Gustav Nachtigal* class across a frozen sea during work-up trials in the Baltic. Alongsid the tender is a Type VIIB *E Sieche*

The conning tower of *U48*, the most successful U-boat of the Second World War, pictured while returning from a sortie; the white pennants give the names and tonnage of merchant ships sunk *E Sieche*

Vhen German submarine onstruction resumed in 1935, two ypes of U-boat were ordered: the cean-going Type I and the coastal 'ype II. In 1934 design studies for ucceeding types had started, based n an enlarged Type I intended for pecial missions – the Type III, IV nd VI.

The original Type III carried 2 mall E-boats in a water-tight angar (similar to the later Japanese ircraft-carrying submarines of the : I, A 2, B 2 and STo types) while a ater version was designed as a ninelayer. The Type IV was a upply and repair submarine with a hree-month endurance and the 'ype VI had steam propulsion for oth surface and underwater ruising. At the end of 1934 it was lecided to cancel these projects ecause their tactical value was imited and, in case of the Type VI, team-propelled submarines were onsidered obsolete. It was next lecided to enlarge the Type II lesign from a 250 ton (300 ton ubmerged) submarine to an cean-going 550 ton boat with single-hull and external saddle tanks which subsequently became the Type VII.

The Type IA boats, *U25* and *U26*, had been commissioned on 6 April and 11 May 1936 respectively, while the first Type VIIA boat, *U33*, was commissioned on 25 July 1936. The two types were used in comparative trials when the Type VII easily proved superior to the smaller type and later became the preferred command of U-boat Captains, the favourite of Admiral Dönitz. Ten Type VIIA boats, *U27–U36*, were built, a significant feature of which was the single, stern, above water torpedo tube. In the follow-up design, the Type VIIB, a number of improvements were incorporated:

a. stern torpedo tube below water
b. length increased by 2m (6.56ft) allowing for longer saddle tanks containing an additional 40 cu metres (1412.7 cu ft) of diesel fuel – the displacement therefore increased from 500 tons to 517 tons
c. better manoeuverability provided by using 2 vertical rudders instead of one (this was also necessary because of the single submerged torpedo tube in the stern)
d. torpedo stowage raised from 11 to 14
e. Diesel engine output increased by 20% by use of turbo-chargers and, despite the greater displacement, the surface speed was raised by one knot

Twenty-four VIIB boats, *U45–U55*, *U73–U76*, *U83–U87* and *U99* to *U102*, were built and commissioned. The most famous of the U-boats belonged to this class: Endrass' *U46*, Prien's *U47*, *U48* the most successful submarine of the war (see *Warship* 12), Kretschmer's

The Type VIIB submarine *U45*, during prewar trials in the Baltic. The number painted on the tower and the telephone buoy on the foredeck confirms that this is a peacetime picture. The white grid painted on the bows served as a reference for recording the dimensions of the bow wave at different speeds. Note also, that the boat lacks the wave-deflector around the forward section of the conning tower. *U45* was one of the first of the class to be lost, being sunk by HM destroyers *Intrepid, Icarus, Ivanhoe* and *Inglefield* east of Ireland on 14 October 1939

PARTICULARS OF GERMAN TYPE VII SUBMARINES

Type	VIIA	VIIB	VIIC	VIIC/41	VIIC/42	VIIC/43	VIID	VIIE	VIIF
Length, oa (m/ft) Beam, hull (m/ft) draught, mean (m/ft)	64.5/211.6 5.18/19 4.4/14.4	66.5/218.2 6.2/20.3 4.7/15.4	67.1/220.1 6.2/20.3 4.8/15.7	67.2/220.5 6.2/20.3 4.8/15.7	68.7/225.4 6.85/22.5 5.1/16.7		76.9/252.3 6.4/21 5.0/16.4		77.7/255 7.3/24 4.9/16
Displacement, (tonnes) surface submarged	 626 745	 753 857	 761 865	 761 865	 999 1099	 Project only	 965 1080	 Design study	 1084 1181
Propulsion, surface	2 x Diesel 1160 HP	2 x Diesel 1400HP	as VIIB	as VIIB	2 x Diesel 2200 HP		as VIIC	2 x 12 cyl 2 stroke Diesels	as VIIC
submerged	2 x electric, 375 HP	2 x Electric 375HP	as VIIB	as VIIB	2 x Electric 375 HP		as VIIC	as VIIC	as VIIC
Sped (kts), surface submerged	 16.8 8.0	 17.2 8.0	 17.0 7.6	 as VIIC	 18.6 7.6		 as VIIC	 as VIIC	 as VIIC
Armament (original)	5 53.3cm TT (4 bow, 1 stern) 1-8.8cm 45 cal C/35 1-2cm AA C/30	as VIIA as VIIA	as VIIA as VIIA	as VIIA 4-2cm$_2$* 1-3.7cm	as VIIA 4-2cm$_4$* 4-2cm$_2$*	10 53.3cm TT (6 bow, 4 stern)	5 53.3cm TT (4 bow, 1 stern) 5 mine-tubes		5 53.3cm TT (4 bow, 1 stern) 27 torpedoes
Crew	44	44	44	44	45		44		46

Notes:
*2cm$_4$ = 2cm quadruple mounting
*2cm$_2$ = 2cm twin mounting
HP given is for each diesel/electric motor
For armament variations see text

Type VIIC41 showing appearance in 1944. Bow inset shows original 'net saw' which was removed at the same time as the 8.8cm gun.

KEY
1. After torpedo tubes
2. Pressure hull
3. Spare torpedoes, under floor
4. Motor room
5. After torpedo loading hatch
6. Diesel engine supports
7. Diesel room hatch
8. Diesel room
9. After battery room
10. Galley
11. Galley hatch
12. After battery hatch
13. NCO's compartment
14. 'Winter garden'
15. Diesel fuel
16. Attack periscope
17. FuMO radar aerial
18. 'Bali' antenna
19. Direction finder
20. Air-search periscope
21. Snorkel
22. Bridge
23. Tower hatch
24. Conning tower
25. Central hatch
26. Central control room
27. Diving cell
28. Commanders cabin
29. Forward battery room
30. Officers accommodation
31. Forward torpedo loading hatch
32. Ratings mess
33. Pressure proof containers for 5 man inflatable rubber boats
34. Capstan
35. Trimming cells
36. Forward torpedo tubes

All drawings by the author

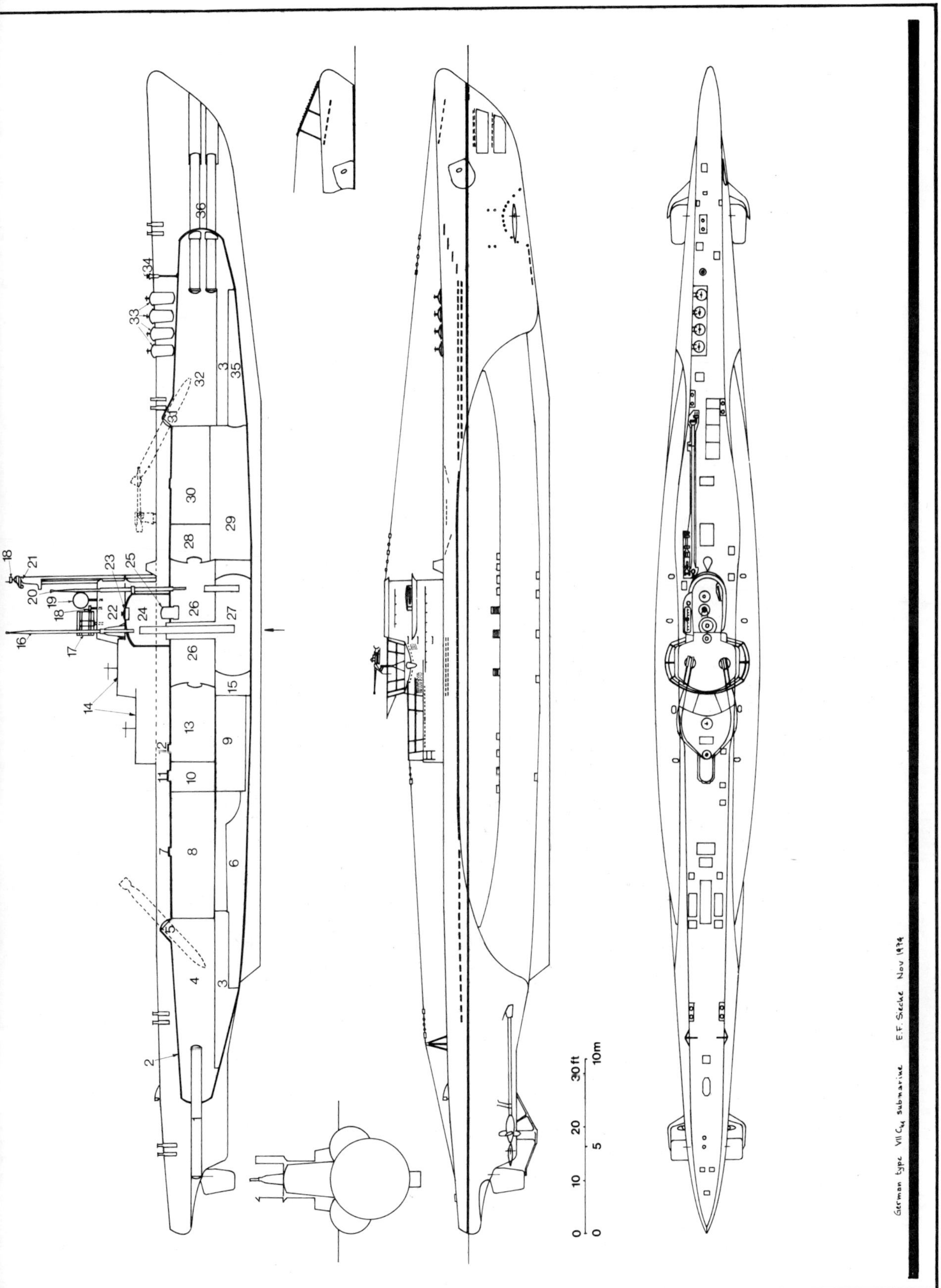
0 10 20 30ft
0 5 10m
German type VII C41 submarine E.F. Siecke Nov 1974

A wartime view of a Type VIIC, probably in the Baltic

USN, courtesy A D Baker III

U99 and Schepke's *U100*. Despite the success of the VIIB it was necessary to shift to another Type, the VIIC, to gain room for a new sound detection system, the Sondergeraet or 'pinger'. The hull had to be lengthened a further 0.6m (2ft) in the periscope section to accommodate this device but this also assisted in the provision of some further improvements:

a. installation of the 'S-Geraet'
b. enlarging the Type VIIB conning tower by 0.3m (1ft) in length and 6cm (2in) in width
c. installation of pressure-proof buoyancy regulating tanks, to port and starboard, to improve bad-weather seakeeping qualities
d. increasing the capacity of the inner oil-tank by 5.4 cu metres (190.71 cu ft).

The machinery was also improved, as follows:

e. engine-oil cleaning system fitted
f. replacement of the starboard turbo-charger by a Junkers compressor to save electric power
g. replacement of obsolete BBC switching-boards by modern AEG boards

Up to the end of the Second World War 572 units of the Type VIIC were built and commissioned; they formed the backbone of the German submarine war in the Atlantic.

Three further variations of the Type VII were produced. The Type VIID was a minelayer version of the standard Type VII design. It could carry 15 mines in an additional midship section containing 5 vertical mine tubes. The D version was 9.8m (32ft) longer than the VIIC; the saddle tanks also being longer meant that the boat could carry more fuel. Seven boats of this type, *U212–U218*, were built and commissioned.

The Type VIIE was an experimental design to study the development of a V-12 2-stroke Diesel.

The Type VIIF was an enlarged 'torpedo supplier' version of the standard design. The boat could carry 27 torpedoes. Four boats, *U1059–U1062*, of this type were built and commissioned.

WARTIME DEVELOPMENT

From October 1941 officials attempted to incorporate contemporary wartime lessons into the standard Type VIIC without fundamentally changing the design, which resulted in a sub-class known as the Type VIIC/41.

In February 1942 the chief engineer of *U157*, Olt Ing Kiesewalter, submitted a memo entitled 'Wartime lessons for the technical development of submarines' suggesting the following improvements:

a. increase surface speed to 22kts, as the Type VIIC was too slow to overtake a convoy
b. increase diving depth to 300m (984ft) to improve chances of escape when hunted
c. longer forecastle to improve seagoing qualities
d. demand for a high speed (50kt) torpedo

These were virtually the same requirements asked for by the above mentioned officials, the only exception being a total lack of notice of the first steps in Allied electronic warfare.

The Type VIIC/41 was characterised by simplified electrical plant and other weight saving devices to allow the pressure hull to be built of 21mm (0.83in) plates instead of 18.5mm (0.73in). The design also had a more streamlined bow and was 13cm (5in) longer. Of this type 70 units were built and commissioned, the first boats being delivered in August 1943.

The next step was to increase the diving depth to 500m (1640ft). As the yards could only work plates of 28mm (1.1in) maximum thickness, Krupp cemented armour steel was to be used. The additional weight of the pressure hull was balanced by increasing its diameter from 4.7m (15.4ft) to 5.0m (16.4ft). To reach a higher surface-speed, the output of the well-tried 6-cyl MAN Diesel (1050bhp at 470rpm) was boosted to 2200bhp at 530rpm by

Type VIIA

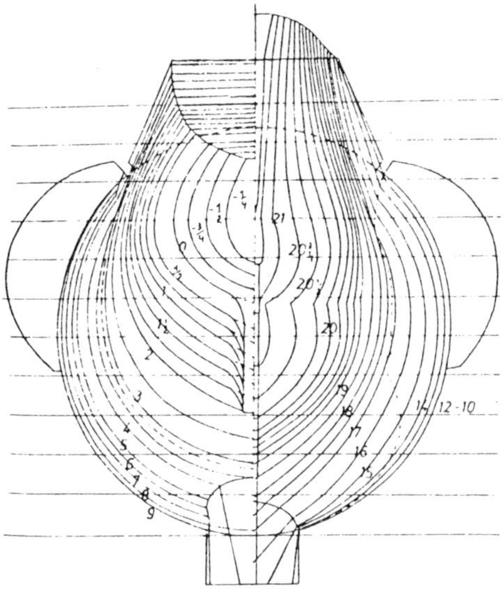

Type VIIB

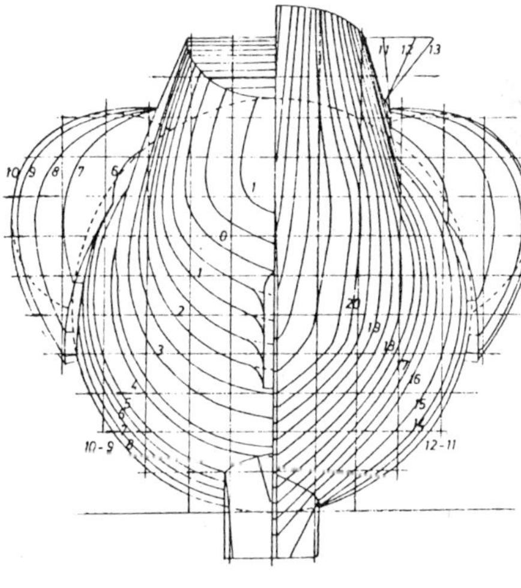

Type VIIC

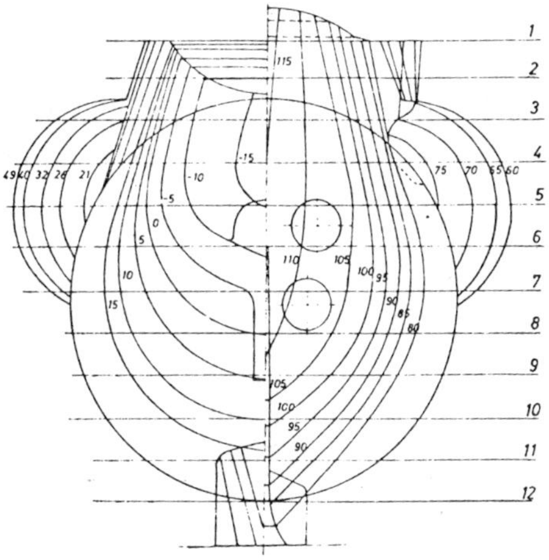

Type VIIC/42

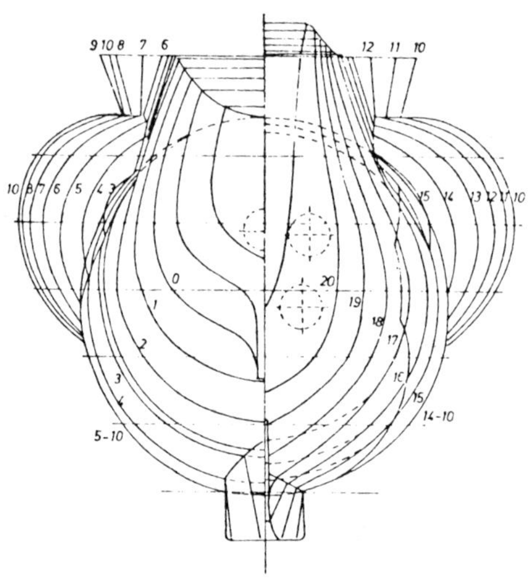

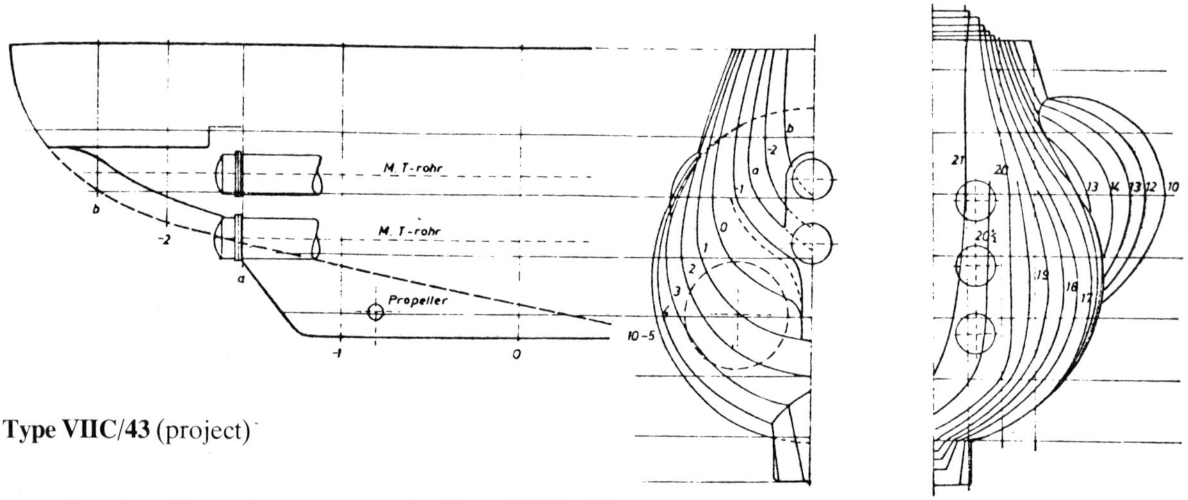

Type VIIC/43 (project)

supercharging. The surface speed was thereby increased from 17kts to 18.6kts. To provide space for the auxiliary engines, the Diesel room had to be enlarged by 0.8m (2.6ft) and the final design, Type VIIC/42, was 1.53m (5ft) longer than the standard design. The enlarged saddle tanks, meeting a demand for a greater endurance resulted in a final beam of 6.85m (22.5ft). This was the highest possible beam suitable for parallel slips, without changing their existing layout and rebuilding them.

In February 1942 the final design was adopted and a first series of 42 boats ordered but none were expected to be delivered before August 1944. A total of 176 units were ordered before the rejection of the old types in favour of the new Types XXI and XXIII but none of the Type VIIIC/42 boats were ever commissioned. It is worth noting that these boats would not have been battle-ready before 1945 by which time they would have been totally obsolete. It is dismaying to see how much industrial effort was expended in producing so great a number of submarines which were outdated from the moment of their keel laying, instead of giving top priority to the design, development and construction of a new generation of submarines.

During this same period front-line officers were demanding boats carrying as many torpedo tubes as possible. Due to improved Allied ASW techniques they had learned that there would, in general, be only one chance to attack a target and then with full salvoes. This lead to design-study VIIC/43 with 6 bow and 4 stern torpedo tubes but this entailed so many fundamental design alterations that the project was cancelled in August 1943.

The Types VIIC boats were built in steadily decreasing numbers until the end of 1944, when all construction work was shifted to the new Types XXI and XXIII.

WARTIME MODIFICATIONS

The first war modification was the provision of a slightly enlarged platform for the 2cm AA C/30 gun but in 1942 it was realised that this close-range weapon was insufficient to match Allied ASW aircraft. The German Submarine Command therefore ordered the adaption of the semi-automatic 3.7cm 83 cal AA C/30 to a new, C/39, submarine mounting. In the meantime Dönitz ordered the up-gunning of the boats with 4–151 MGs (4mm) in twin mountings and the fitting of a twin 2cm AA C/30 in a special sponson behind the tower, nicknamed the 'winter-garden'. As the 2cm twin mounting was not ready a number of interim measures were adopted resulting in the following proposed variations:

Modification I: consisted of one 2cm AA C/30 single on the 'winter garden' and 4–151 MGs in twin mountings. This was soon rejected, as the firepower of the 151 MG turned out to be ineffective. In November 1942 the following three alternatives were proposed:

Modification II: two 2cm AA C/38 in single mountings, one on the

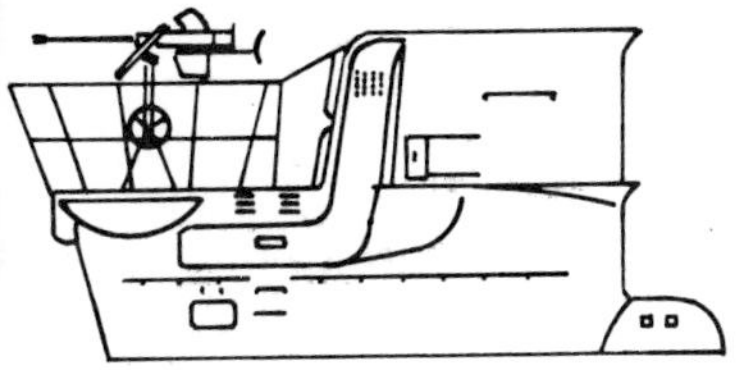

Type VIIB standard tower

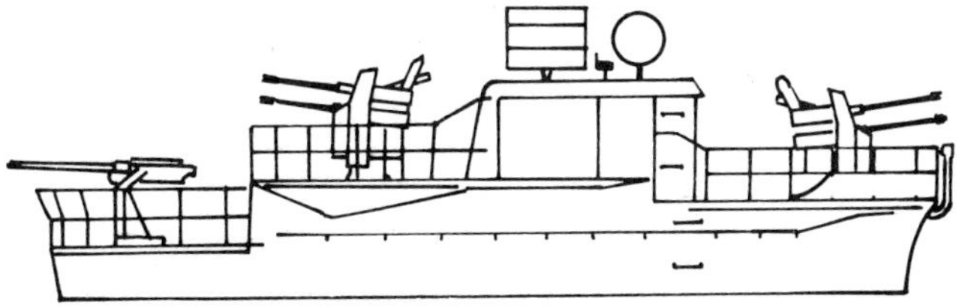

U Flak I, Two quadruple 2cm and one 3.7cm C30 U

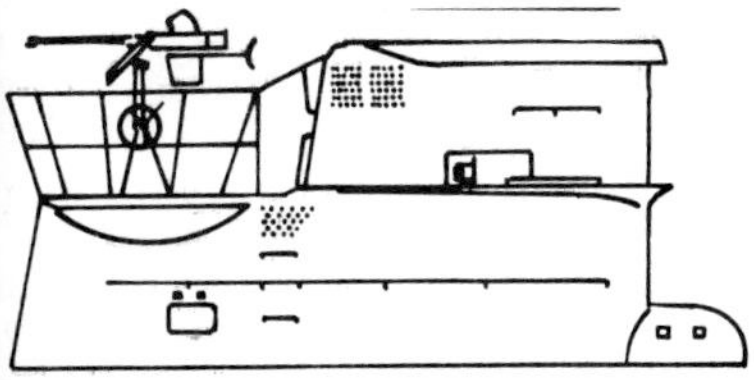

Type VIIC, standard tower 1940-43 with 1-2cm C30 AA gun

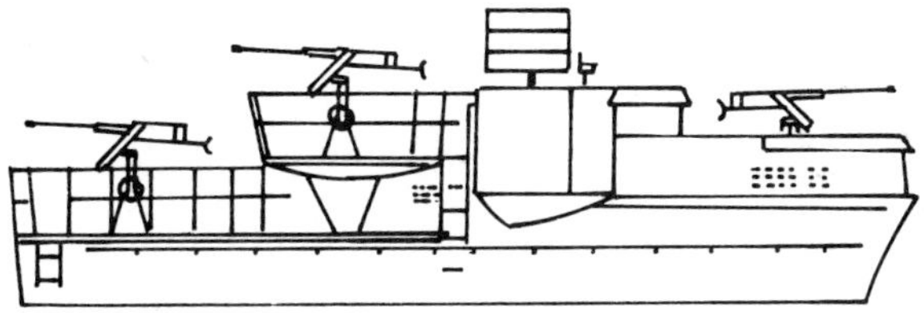

Modification V, *U363* with 3 twin 2cm 38M

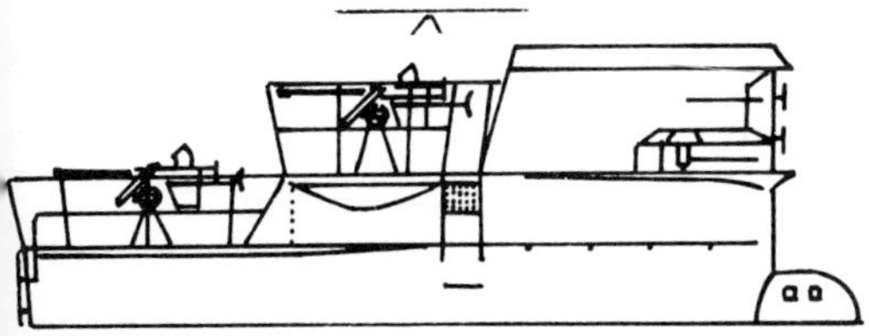

Modification II, enlarged 'winter garden' with 2 single 2cm C38 AA

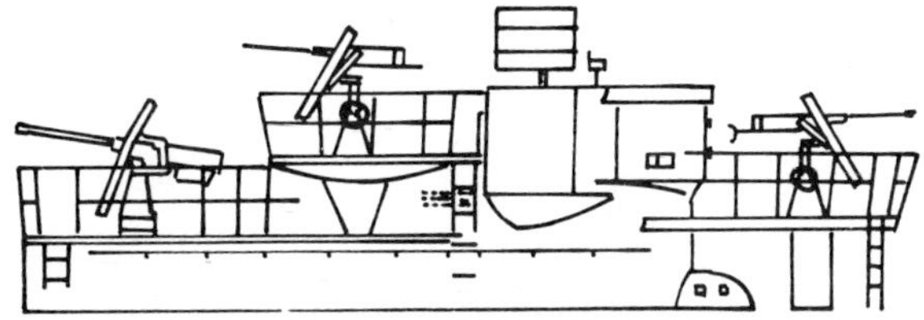

Modification VI, *U673* and *U973* with 2 twin 2cm 38M and one twin 3.7cm 42U

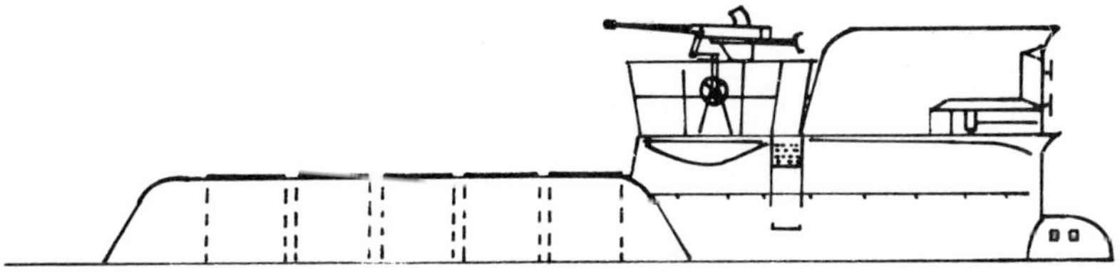

Modification III, for minelayers of Type VIID only – the raised section abaft the tower contained 5 vertical mine tubes extending down to the keel – armament was 2 single 2cm C38 AA.

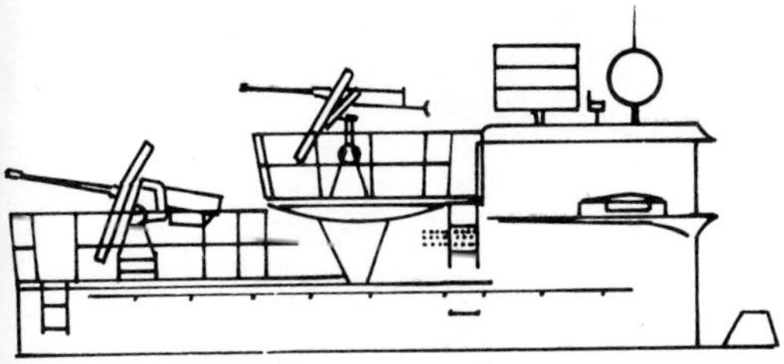

Modification IV, two twin 2cm 38M AA on upper platform and one 3.7cm, 42U AA on lower platform – sometimes this was varied to one quadruple 2cm 38/43U or one twin 3.7cm 42U in place of the single 3.7cm AA.

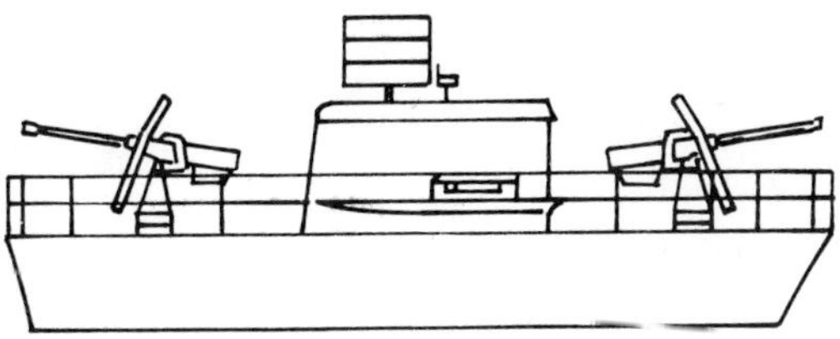

Modification VII, 2 twin 3.7cm 42U

bridge platform and one on the 'winter-garden'
Modification III: two 2cm AA C/38 in singles, both on the enlarged bridge platform. This was especially proposed for the Type VIID minelayers, as they had no place for an additional 'winter-garden' behind the bridge due to the raised mine-tube section
Modification IV: four 2cm AA C/38 in 38M twin submarine mounting on the bridge platform; one 3.7cm 83 cal AA M42U (M meaning 'Modell' in the sense of construction year, U meaning submarine version) on the 'winter-garden'.

Modification II was carried out until June 1943 but was only considered an interim solution until the 2cm twin, the 2cm quadruple and the 3.7cm single/twin became available on submarine mountings. In parallel with modification II all boats had their 8.8cm 45 cal C35 deck gun removed. On the one hand this weapon had become useless, as German submarines could no longer attack on the surface; on the other hand its removal was urgently required to compensate for the additional topweight of the new AA guns and radar equipment.

Until August 1943 all boats underwent modification IV as the equipment became available but a the 2cm quadruple was not sufficiently reliable it was decided replace it with the new fully automatic 3.7cm 83 cal, M42U. A complementary measure it was als decided to convert 6 boats into special AA 'Q' submarines. The first vessel of this type was *U Flak* (ex-*U441*); *U256* was to have become *U Flak 2* and four others were planned. As AA escorts they were intended to provide defence against Coastal Command aircraft while convoying their sisters

The stern of a Type VIIC showing the twin rudders

USN, courtesy A D Baker III

Ready for launch, a Type VIIC on the building slip

USN, courtesy A D Baker III

through the Bay of Biscay. Virtually the first mission of *U Flak 1* showed that a surfaced submarine had almost no chance against air attacks, especially as the aircraft could radio for quick support, so it was decided to reconvert the two AA Q subs to their original form.

In the summer of 1944 another increase in the AA armament was considered which led to Modifications V, VI and VII. Although similar to the 'U Flak' modification (both designs had all the space behind the tower occupied by the 'winter-garden', so additional weapons could only be placed in front of the bridge) the purpose was self-protection and not escort. Versions V, VI and VII differed from each other only in the shape of the sponsons; *U363* received modification V, the *U673* and *U973* modification VI but the additional forward sponson seriously reduced the level of stability and consequently modification VII was never carried out and the test boats were reconverted to measure IV.

ELECTRONIC EQUIPMENT

One of the deciding factors in the failure of the German submarine offensive was a fundamental misinterpretation of the capabilities and tactical value of electronic detection. Believing in their superior optical instruments, the German Navy was afraid of having its own – top secret – radar emissions detected and therefore mainly concentrated on the development of radar detection devices believing that the British were doing the same. The British in contrast developed more and more compact radar systems and the

A Type VIID submarine minelayer, the distinctive structure abaft the conning tower concealing the tops of 5 vertical mine tubes. This structure meant that the Type VIID could not carry an enlarged 'winter garden' as it was necessary to keep the tube hatches clear. The photograph shows *U213* immediately after launching at the Germaniawerft, Kiel; she was lost west of the Azores on 31 July 1942 when depth charged by the escorts *Erne, Rochester* and *Sandwich*

German Navy was trapped in a deadly circle of being forced to develop more and more detecting sets and antennas to fulfil her doctrine, while falling back in the development of active radar. The result of this fundamental error was that British commanders had full battlefield information provided by active radar while their German opponents had only passive electronic information and their eyesight. (Although it has nothing to do with submarine warfare, the sinking of the *Scharnhorst* is one of the best illustrations of the results of this German 'blindness'.)

RADAR WARNING EQUIPMENT

The increasing intensity of the 'microwave war' caused a steady re-equipment of all German submarines with detection sets which became more and more sophisticated. To detect the British ASV (air to surface vessel) radar all German submarines were equipped, from the summer of 1942, with 'Metox' detectors manufactured by the French firms Grandin and Sadir. The antenna was a simple wooden cross, nicknamed 'Biscay cross', which had to be turned by hand and withdrawn into the boat when diving. As the British switched to microwave radar in the spring of 1943, German submarines were attacked so often, usually without a chance to take countermeasures and sometimes on totally dark nights, that losses became catastrophic. German industry was, therefore, ordered to develop new radar emission detection receivers which resulted in the production of several sets ('Naxos', 'Cypern I', 'Cypern II', etc) which employed two interchangeable antenna systems serving two different purposes:

a. the pressure-proof round-dipole 'Bali' for improved omni-directional reception

b. 'Palau' antenna for direction finding, whose distinctive butterfly shaped antennas were situated at the rear side of the 'Hohentwiel' frame.

RADAR

Although Germany was leading in this field in 1939 certain factors meant that German submarines could not use radar extensively for observation and detection purpose Firstly, German scientists believed strongly, that it was impossible to bring down radar wavelengths to the centimetre (VHF) band so all German sets, in the first half of the war, suffered from the bearing inaccuracies inherent with metric radar and served merely as navigation aids. Only after the Germans had examined a British H_2S set, containing the revolutionary cavity magnetron, were they able, by copying it, to cross the VHF barrier. By then it was too late to regain all the lost technological and scientific development of the previous years

Secondly, the German Navy having overestimated the value of passive versus active detection, created a philosophy of 'radar silence' to avoid passive detection i the same way as 'radio silence' was intended to prevent signal

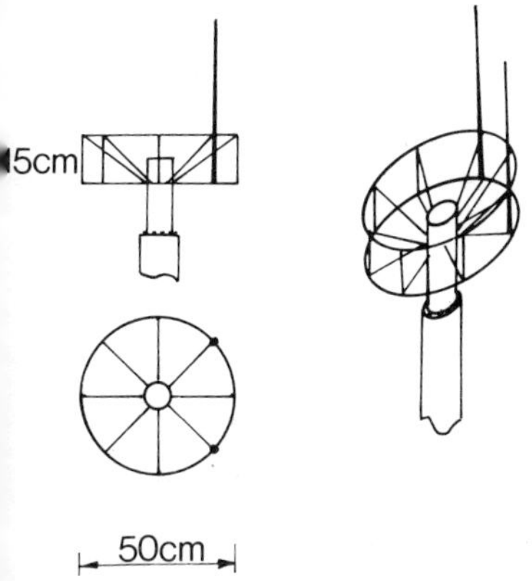

'he 'Bali' antenna for all-round (360°)
:ception of radar emissions

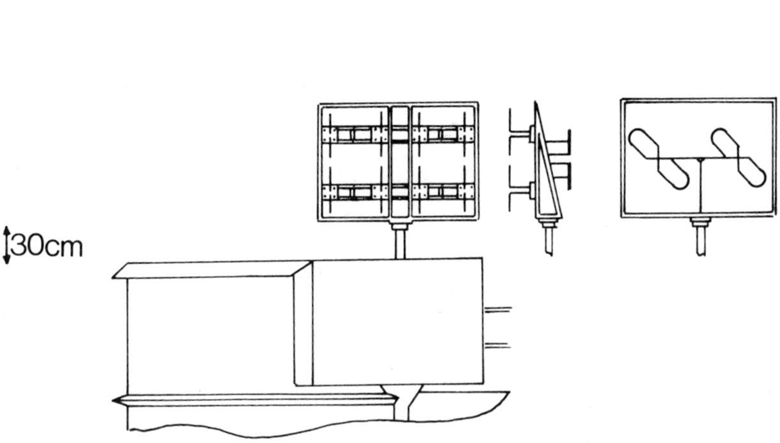

FuMO30 aerial, the second type provided for the 'Seetakt' radar, showing from left to right front, side and rear views. The frame carried 4 receiving dipoles in the upper row and 4 transmitting dipoles in the lower row. The frame could be retracted into the armoured van, shown here immediately below the aerial. The back of the frame carried the two 'Palau' broadband antenna for the reception of enemy radar emissions

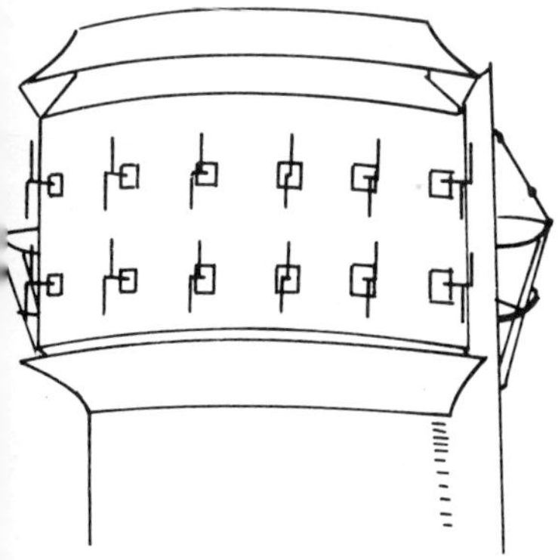

uMO29, first submarine version of the
eetakt' radar aerial, consisted of fixed
poles fitted on the forward side of the
›nning tower – the upper row contained the
receiving dipoles and the lower 6
ansmitting dipoles

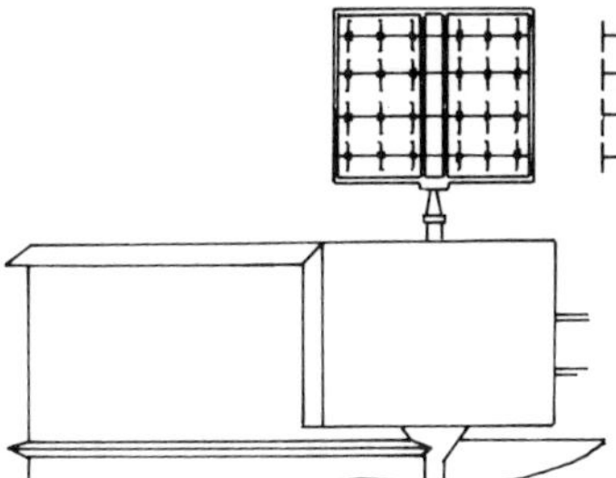

FuMO61 aerial for the 'Hohentwiel U' radar with 4 rows of 6 dipoles

ıterception and direction finding. 'his might also explain why the ʲavy had radar with vertically olarized transmission throughout ıe war; an arrangement which esulted in water surface isturbance of the radar signal 'grass'). In contrast, even the first ˌuftwaffe ASV detection sets had orizontal polarization to avoid this roblem. Hence, nearly all German vartime naval radars had aerials vith vertical dipoles. Lastly, the adar antennas were fitted in such a vay that they could not be used vhile snorkelling so, although the J-boats had radar sets, they were outdated, largely silenced and, for nost of the time, useless.

The first Radar employed on U-boats was a submarine version of the standard German 'Seetakt' (82cm wavelength, 368 megacycles). Set FuMO29 had a 2x6 fixed dipole array fitted on the forward face of the conning tower while FuMO30 had a small turnable frame aerial with only 2 x 4 dipoles. Results with these sets where unsatisfactory but the Luftwaffe, which held more advanced views on radar than the Navy and was now the leading service in this field, had a better type, the FuMG200 'Hohentwiel' (56cm wavelength, 556 megacycles). This set was 'navalized' and became the FuMO61 'Hohentwiel U' which employed a rectangular frame aerial (1m x 1.4m) carrying 4 x 6 dipoles.

THE SNORKEL

The idea of a breathing tube is as old as diving itself and has often attracted the attention of submarine designers. Probably both navies, British and German, saw their first snorkels when Dutch *O* class submarines were taken over by both after the fall of the Netherlands. As there was no vital necessity for this item in the first stage of the war, neither tried to develop or adopt it. When, however, German submarines were forced to dive during almost their entire missions, the snorkel aroused new interest.

In the spring of 1943 Helmut Walter added a new idea to the Dutch snorkel concept: the entire interior of the boat should serve as an air reserve in case of undercutting of the snorkel head. Against heavy opposition from serving officers, based on arguments of the impossibility of holding exact depths for long periods and the dangers of poisoning entire crews, German submarines were subsequently equipped with

The *U995*, preserved as a memorial at Laboe near Kiel, seen here in 1975. Her bridge carries the two twin 2cm and single 3.7cm AA guns of Modification IV

S Terzibaschitsch, courtesy A D Baker III

snorkels. This opposition did not of course vanish when the first snorkel boat, *U264*, was lost during its first mission on 19 February 1944. But, step by step, the German submarine crews learned to handle this life saving device and managed to run their boats at exact depth for long periods. However, the placing of the folding snorkel was very inconvenient in the Type VII submarines as it hindered the starboard forward periscope angle and sometimes the periscope was shrouded by exhaust smoke.

DECEPTION AND CAMOUFLAGE

The Bold: This was a small perforated tube containing calcium hydrite. It was ejected through a small lock in the stern section after which it floated at a depth of 100ft and produced a swell of Hydrogen bubbles making a dummy target for enemy Asdics.

Alberich: To prevent the hull from reflecting Asdic-impulses the entire wet surface was covered with a resonance-absorbing coat. This was codenamed 'Alberich' after the mystic dwarf who, with his cloak of invisibility, guarded the treasure of the Nibelungs. It consisted of a double layer of thin rubber, the first having a honeycomb structure, the second layer covering the first.

Chimney sweeper: In the later stages of the war the snorkel head was virtually the only radar target. To reduce its radar-reflection it was covered with a special rubber coat.

Aphrodite: This was a simple Hydrogen filled balloon carrying tinfoil strips* which created a dummy radar target.

Thetis: This was an ejectable buo which carried a radar-reflector on pole; it was intended to create a dummy target with echoes similar that of a conning tower.

*German codename 'Düppel', British codename 'window', US codename 'chaff'

THE JASKOLKA CLASS MINESWEEPERS

by Marek Twardowski

IRTH OF THE CLASS

ie story of *Jaskolka* class inesweepers begins in 1932 when e Polish Navy decided to build ur minesweepers/minelayers to place four ex-German FM class ineswcepers purchased in 1921. itially the ships were to be built road but it was later decided to nploy Polish shipyards which at that time were short of work. Three yards were chosen – the Gdynia Shipyard, the Naval Works at Gdynia and Modlin Shipyard (near Warsaw). The last one was a river yard but the most experienced naval engineers were there. All these yards were very small, and faced many problems, but the decision to build the ships in Poland was correct,

The *Zuraw* or *Czapla* under construction in the Naval Works at Gdynia. They were mainly of rivetted construction but also employed some welding.
Library of Gdynia

Fitting out: the *Zuraw* or *Czapla* having one of its diesel engines lowered into position. The lining-up of engine and propeller shaft required very careful attention, if vibration and shaft deformation were to be avoided

Library of Gdynia

especially in those years of the world-wide economic crisis.

The building contract was signed on 26 January 1933; the first ship was to be completed 20 months from that date and following vessels at 3 months intervals. The Modlin yard prepared the building drawings of the ships (including workshop drawings) which were later sent to the Naval Works and then the Gdynia Shipyard.

The minesweepers were to be built of steel, contrary to normal practice at that time, when small ships were usually constructed of wood. Plates and bars were ordered from Polish steel works to meet a ruling that as much material as possible should be ordered from home manufacturers. This decision, although right, delayed the completion of the ships because the steel works were $4\frac{1}{2}$ months late with their deliveries, while the cast steel stem for *Jaskolka* was not ready until after the hull was built!

These troubles did not end with the arrival of the plates and castings as the shipyards were not properly equipped for shipbuilding. For example, the Naval Works' main occupation was ship repairs and there was no proper slipway so one had to be constructed. Unfortunately, the new slip was supported by wooden beams instead of a concrete floor and, af every period of rain or after a hea piece was added to the constructic the hull sank into the ground and had to be laboriously straightened with spirit level, hammers and wedges. Things did go better in th other yards but at Modlin a flood caused work to be delayed for a fe weeks.

These troubles and difficulties were overcome, however, and the first to be launched was *Jaskolka* 11 September 1934 at the Naval Works in Gdynia. She was follow by *Mewa* from the Gdynia Shipyard, and in April 1935 by th

ŻURAW in 1948

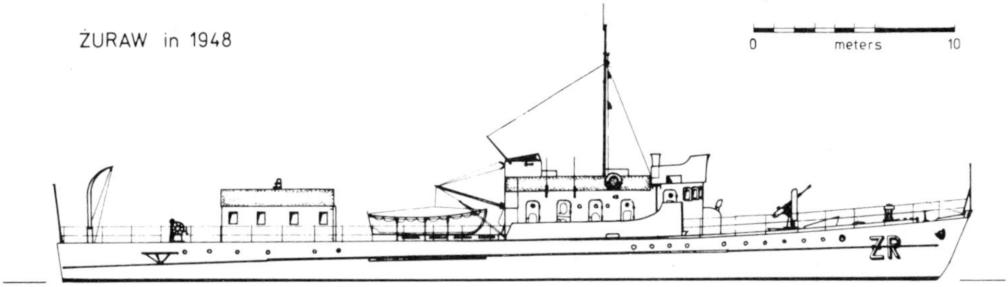

KOMPAS, ex ŻURAW
as rebuilt 1959 - 1963

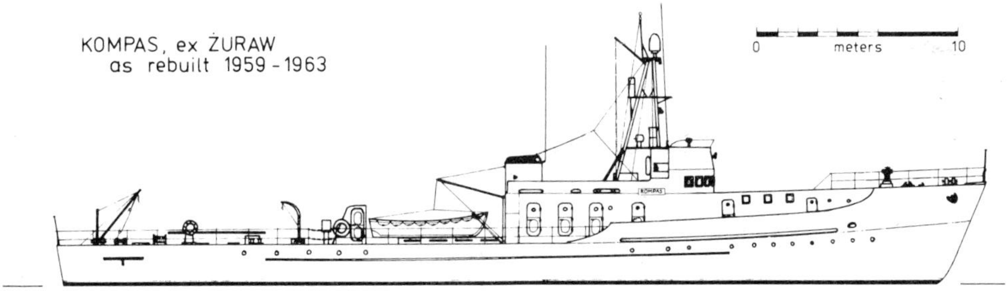

The hull of *Zuraw* or *Czapla* complete and ready for launch with the aid of a 100 ton crane borrowed from the Gdanzk Shipyard

Library of Gdynia

CONSTRUCTION DATA

Name	Launched	Commissioned	Builder
Jaskolka	11.9.1934	27.8.1935	Naval Works, Gdynia
Mewa	10.1.1935	25.10.1935	Gdynia Shipyard
Rybitwa	26.4.1935	21.12.1935	Modlin Shipyard
Czajka	10.4.1935	10.2.1936	Modlin Shipyard
Czapla	22.8.1938	1.9.1939	Naval Works, Gdynia
Zuraw	22.8.1938	1.9.1939	Naval Works, Gdynia

The exact dates of the commissioning of *Zuraw* and *Czapla* are not known and the above day and month are estimated. The engines were built by the 'Ursus' Works in Warsaw except for *Czajka,* which had Swedish-built engines.

last two – *Rybitwa* and *Czajka*, launched at Modlin. Fitting out took almost a year and *Jaskolka* finally completed on 27 August 1935 – 31 instead of 20 months after the contract was signed. The last vessel *Czajka*, was completed 6 months later.

TECHNICAL DETAILS

The ships' main engines were two 8-cylinder Diesel engines of the Nohab type built in Poland under licence from the Swedish firm Nyquist-Holm. Each of these engines developed 525bhp at 350rpm; they proved to be excellent, serving for over 30 years. Electrical power was supplied by three generator sets, of 40hp each, made by the National Gas & Oil Engine Co Ltd, the high power being necessary because all the auxiliaries were electrically powered.

Armament consisted of one 75mm dual purpose gun, 2 Maxim machine guns and 20 depth charges. When acting as minesweepers they were equipped with two sets of minesweeping gear, while as minelayers they could stow 20 mines on their deck rails. The Maxim guns were their sole close range AA armament and, as these and the use of speed and the ship's manoeuvreability proved to be insufficient protection from aircraft they were later augmented by one 13.2mm machine gun.

The cost of the four ships was 5,284,684 zlotys ($1 = about 5 zlotys), excluding armament, ammunition, minesweeping gear, W/T sets etc, while their contract price was 5,280,000 zlotys. The estimated total cost was a little less than 8,000,000 zlotys. Only 20.8% of the materials for the ships came from abroad, namely the minesweeping winches, steering gear, engine room telegraphs, engine revolution counters and generator sets.

Jaskolka running trials on 16 July 1935. Her minesweeping gear has been fitted but she carries no guns

Polish Navy Museum

Four of the *Jaskolka* class in harbour prior to 1937. At this time they were virtually identical apart from the bow recognition letter

Authors collection

EARLY SERVICE

Thus, in 1935, the Polish Navy received four small and handy 'mine vessels', as the new ships were officially called. Together with two gunboats, *Komendant Pilsudski* and *General Haller,* they were formed on 20 January 1936, into the 'First Division of Mine Vessels', a title unused since the old FM boats had been deleted.

The new minesweepers proved to be a success – small, fast and seaworthy in Baltic conditions. Little wonder that, when a Roumanian admiral visited Gdynia in the spring of 1939, he asked for the general arrangement details and cost estimates for the building of such vessels for the Roumanian Navy. These were probably sent but war was very close and nothing materialised.

Although the *Jaskolka* class minesweepers were ideally suited for coastal operations, the Navy needed something bigger. In 1937 it was proposed to build two 300 ton minesweepers but the necessary funds were not granted and the Navy had to be satisfied with two repeat ships of the existing design. The *Zuraw* and *Czapla*, as they were subsequently named, were virtually repeats of the *Jaskolka* with minor improvements, the most important being increasing the AA armament by the additions of two Hotchkiss 13.2mm machine guns. The Naval Works were chosen as the builders and the ships were launched on 22 August 1938. At the outbreak of the war they were nearing completion and on 1 September 1939 they were hurriedly commissioned into the Navy without official trials being conducted.

WARTIME ACTIVITIES

As already mentioned the minesweepers formed the main body of the First Division of Mine Vessels. Their main occupation was training and for this purpose the *Czajka* and *Rybitwa* were cruising along the Latvian and Estonian coast at the beginning of August 1939. As war seemed inevitable, however, they returned to Gdynia on about 20 August 1939. Nine days later the Polish Navy was ordered to close Gdynia harbour with a mine barrage, provided that no foreign ship was in the harbour, but as there was a Greek steamer in port, the barrage was not laid, although the existence of the minefield was announced on the radio.

On the morning of 1 September the First Division left Gdynia and joined the minelayer *Gryf*, steaming in the northern part of Gdansk Bay. Shortly after noon the Division's Commander Zdzislaw Boczkowski was ordered to dispatch two minesweepers to escort two ships through the fictional minefield. The

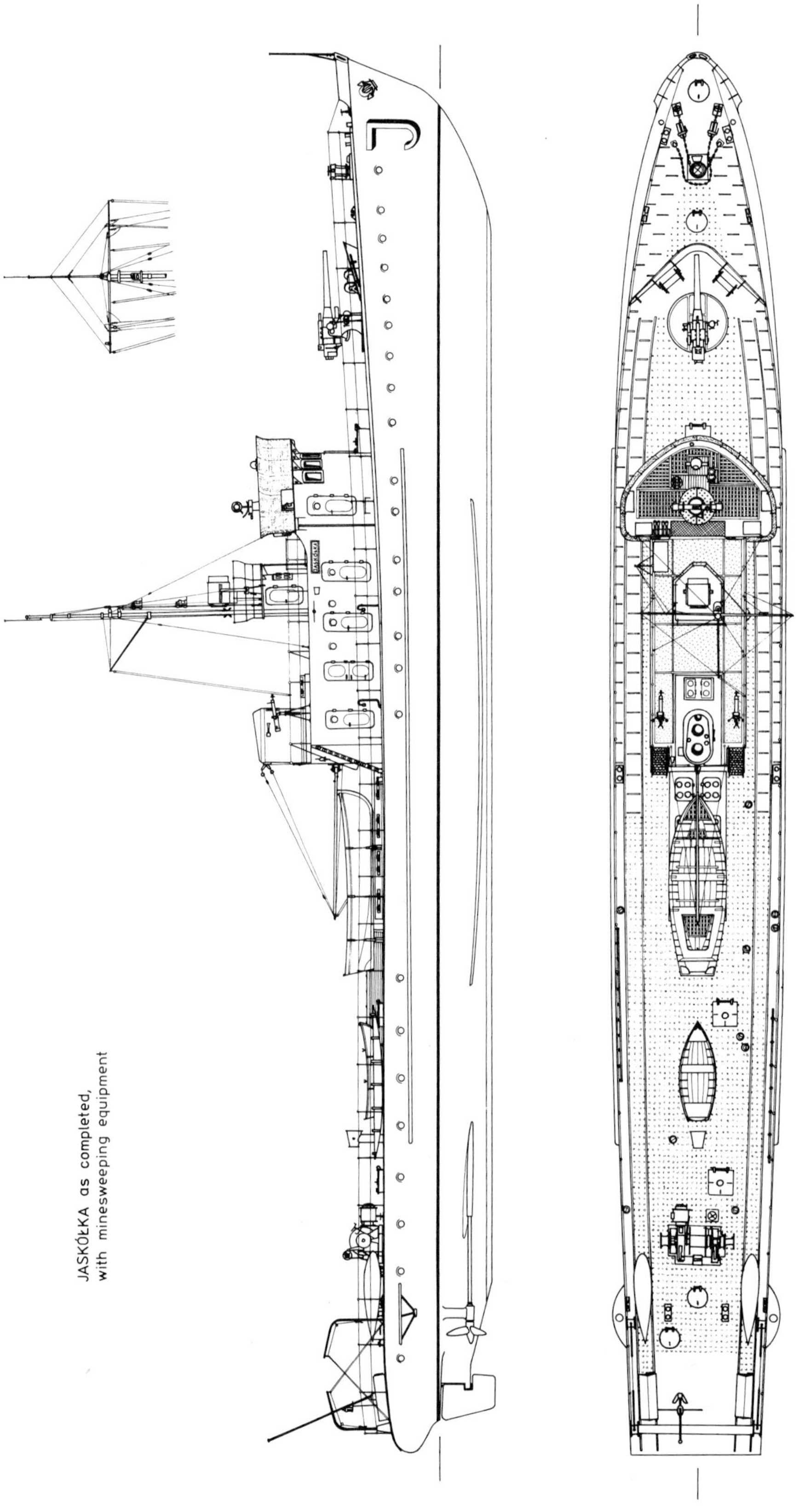

Jaskołka as built with minesweeping gear
All drawings by the author

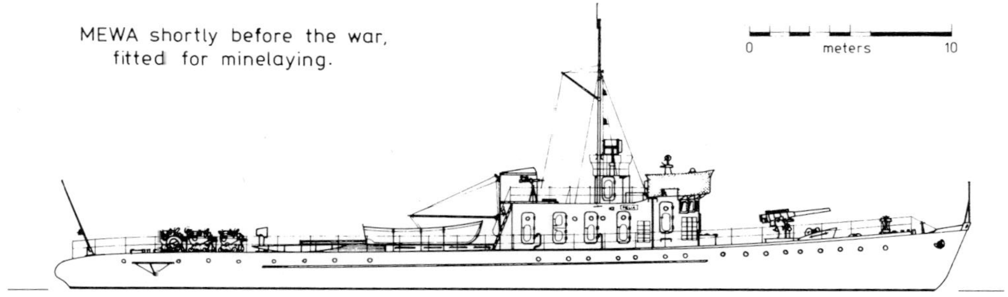

Mewa as a minelayer shortly before the war. Each ship could carry 20 mines which weighed about 700kg each. To reduce topweight, and clear the mine rails, the minesweeping gear, as well as a small boat, had to be landed.

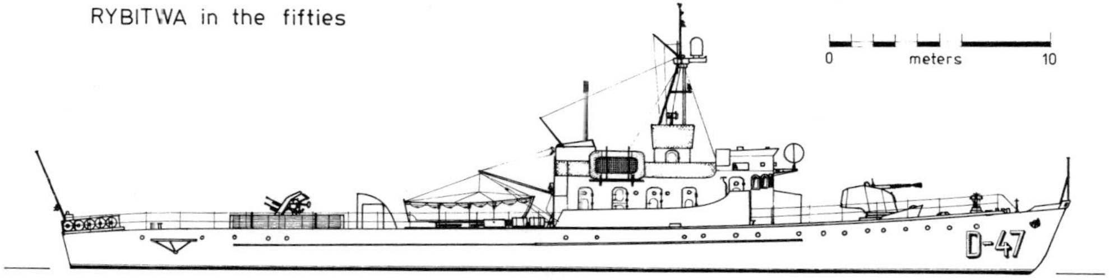

Rybitwa re-equipped and reclassified as guard (patrol) ships with new mast and 'Skin Head' radar. The new armament included the addition of depth charges.

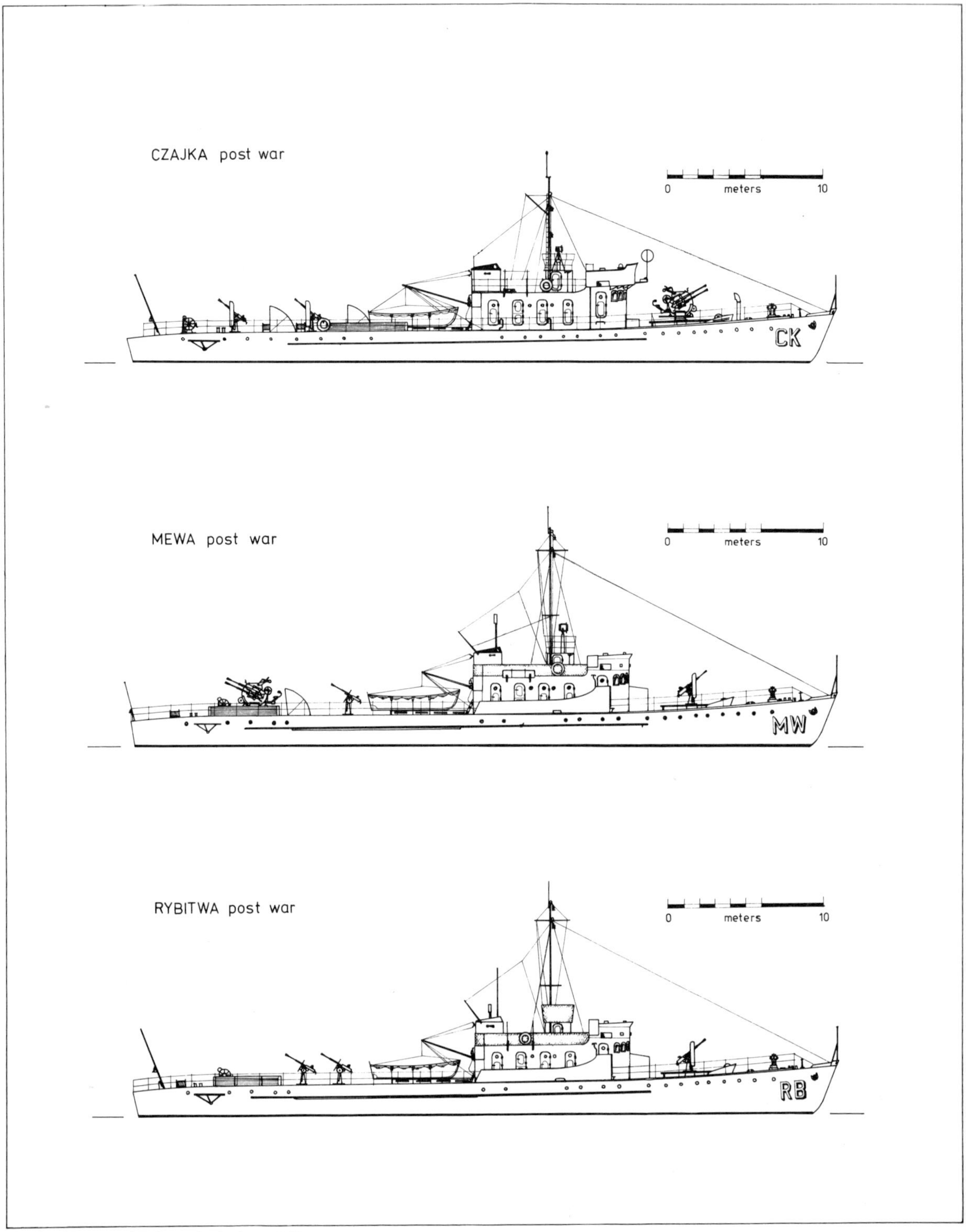

Czajka, Mewa and *Rybitwa* in the late 1940s showing their postwar variations in appearance. The *Zuraw*, not shown here, had two twin 20mm guns, one forward and one aft, and one single 20mm gun amidships. Note that only *Czajka* carried a direction finder.

TRANSLATION OF SHIP NAMES

Czajka	lapwing
Czapla	heron
Jaskolka	swallow
Mewa	gull
Rybitwa	tern
Zuraw	crane

Westerplatte, Rixhoft, Putzig, Oxhoft were German names of geographical places on the Polish coast. All ships seized by the Germans in 1939 were renamed in this manner.

The operational area of the *Jaskolka* class in September 1939.

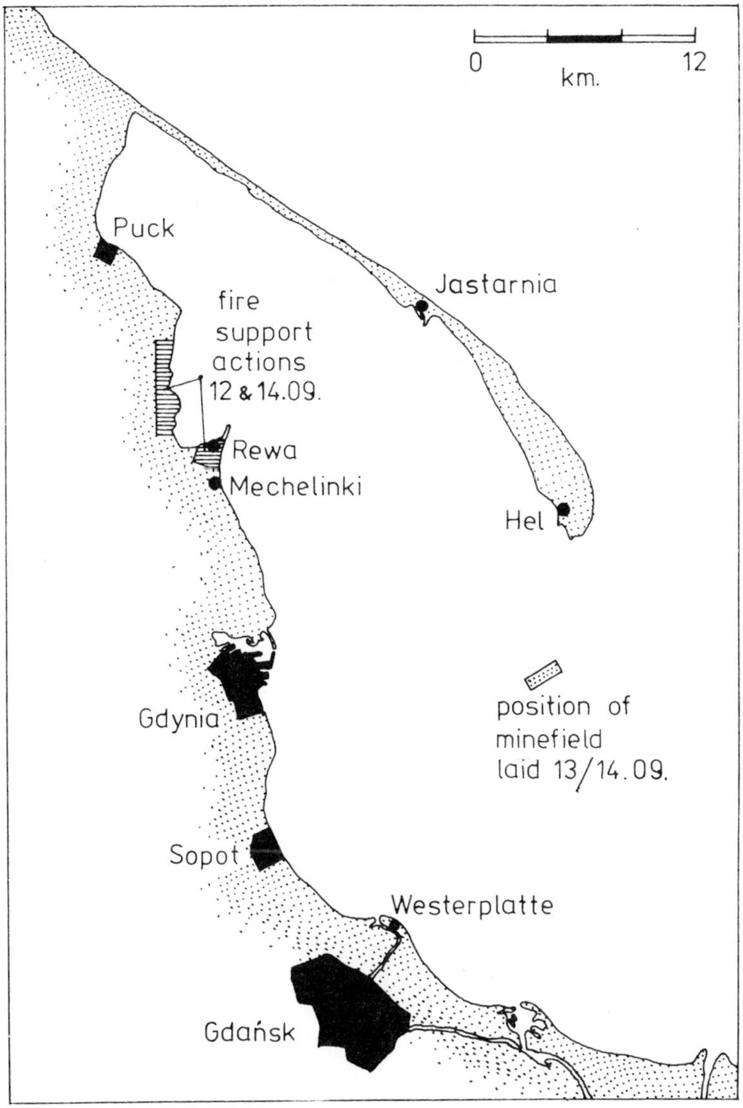

Czajka took care of the coastal passenger vessel *Wanda*, while *Rybitwa* looked after the previously mentioned Greek vessel. Everything went well until *Stuka* dive-bombers appeared in the sky. The Greek captain altered course, ordered 'full ahead' and trusting to luck sped right through the imaginary mines. After this the minesweepers rejoined the Division and at about 1600 the whole formation (minesweepers, minelayer *Gryf* and destroyer *Wicher*) proceeded to Hel, from where operation 'Rurka' (the mining of Gdansk Bay) was to begin. Two hours later the ships were attacked by 32 Stukas and for half an hour the minesweepers evaded bombs while trying to shoot down the bombers with their machine guns (the 75mm DP guns proved to be too slow in training and rate of fire to be of any practical use) but not a single plane was destroyed. Instead the Stukas managed to seriously damage *Mewa* when a bomb exploded in the sea

TECHNICAL DATA

a) as built
Displacement: 185 tons
Length: 45m oa
Beam: 5.5m
Draught: 1.55m
Armament: one 75mm DP gun, one 13.2mm and two 7.6mm machine guns (*Zuraw* and *Czapla* – one 75mm low angle gun, two 13.2mm and two 7.6mm machine guns), 20 depth charges; according to their role – either 20 mines of '08' type or 2 sets of minesweeping gear
Machinery: 2 shafts, Nohab Diesels, 1050bhp = 18kts at 350rpm
Crew: 2 or 3 officers and 26 or 27 men

b) after October 1939, according to Groener, they were armed with two 20mm guns.

c) when returned to Poland, 1946
Czajka eight 20mm AA guns (2 x 2, 1 x 4)
Mewa seven 20mm AA guns (1 x 1, 1 x 2, 1 x 4)
Zuraw five 20mm AA guns (2 x 2, 1 x 1)
Rybitwa four 20mm AA guns (1 x 2, 2 x 1)
This armament was retained while the ships served as minesweepers, until July 1949.

d) after 1949, when reclassed as patrol boats (guard ships) the armament was changed as follows: two 37mm AA guns (1 x 2), two 12.7mm machine guns (1 x 2), two depth charge racks.

e) Zuraw later Kompas
When employed on hydrographic duties only the forward twin mounting was retained. After her 1959-63 reconstruction details were as follows:
Displacement: 300 tons
Dimensions: 45.7 (oa) x 5.5 x 2.2m
Speed: 14.3kts
Crews: 38 officers and men
No armament

The *Czajka* in 1936 or 1937 during a naval review. The sailing ship in the background is the merchant training ship *Dar Pomorza*

Jerzy Micinski collection

close to her starboard side. Everybody on board was either killed or injured and *Mewa* had to be towed to Hel by *Rybitwa.* Although no ship was sunk, the Germans succeeded in preventing the mining operation as *Gryf*'s Commanding Officer had jettisoned his mines due to the danger of a direct hit on the mine deck.

One day later, on 3 September, the Luftwaffe succeeded in sinking *Gryf, Wicher* and *Mewa*, leaving the five surviving minesweepers as the only effective Polish surface vessels. Two of them, *Zuraw* and *Czapla,* were employed as AA guardships in Jastarnia harbour, as their steering gear was out of order and could only be hand operated.

Now began the days of action for the minesweepers. On 3 September *Jaskolka* and *Rybitwa* brought Stefan Frankowski (commander of naval section of coast defence) from Gdynia and supplies for the Hel garrison. Four days later the same minesweepers went to sea to find German pilots shot down near Mechelinki. None was found but *Jaskolka* downed one Stuka which attacked the ships. On 11 or 12 September the three vessels provided fire-support for Polish troops near Rewa and on the following day *Czajka* had to flush German airmen from their hide-out on the Mechelinki beach with her artillery. The last fire-support action took place on 14 September, about 0400, near Rewa.

The ships were to lay mines only twice; the first occasion was when Frankowski decided to lay 60 mines to prevent the old battleship *Schleswig Holstein* from shelling the Hel peninsula. The operation was scheduled to take place during the night of 9/10 September but, due to the full moon, had to be postponed until 12/13 September. Everything went well and a second batch of 60 mines was to be laid a night later but fire-support duties caused the operation to be postponed for 24 hours. However, the Germans, for whom the minelayers were a constant nuisance, decided to destroy them and at 10.00 on 14 September eleven *Stukas* appeared over the Jastarnia harbour. By the time they had left the Division had ceased to exist. *Jaskolka* was destroyed when a bomb hit the ammunition stores and started a fire in her oil fuel. *Czapla* had a hole in her starboard side and was listing. *Rybitwa* was hit by a bomb that did not explode and was repaired with wooded pegs wedged into the holes in her deck and bottom. *Czajka* and *Zuraw* escaped with minor damage.

After the air raid the crews and armament were landed but not before *Rybitwa, Czajka* and *Zuraw* had transferred to Hel harbour. When on 2 October 1939 Hel surrendered, these three ships were still there, together with the half sunk *Mewa* and the crews determined to destroy their ships. The kingston valves were opened and the three minesweepers, the gunboat *Komendant Pilsudski* and three mine barges were sunk. The waters were shallow however, and on 3 October the Germans raised *Zuraw* and *Czajka*, as they were in the best condition, having only shot holes in their superstructures. Incorporated into 7th Minesweeper Flotilla and renamed respectively *Oxhoft* and *Westernplatte,* they were transferred to Gdansk. Afterwards, *Oxhoft* served as Vermesungsboot (survey vessel) while *Westerplatte* became a 'Bojenboot' and was used

The gunboat *General Haller* with her sister *Komendant Pilsudski*, was built in Finland for the Russian Navy but never delivered. In 1921 both ships were purchased for the Polish Navy. By September 1939 they were too old and slow to be of fighting value and were laid up at Hel where *Haller* wa sunk by aircraft. *Pilsudski* was captured and, as the German *Heisternest*, was sunk at Nantes on 16 September 1944

Library of Gdynia

BOW INITIALS AND PENNANT NUMBERS

a) pre-war

Czajka	C
Jaskolka	J
Mewa	M
Rybitwa	R

When *Zuraw* and *Czapla* were ordered their initials were to have been Z and CP, *Czajka*'s was later changed to CY but after 1937 bow initials were no longer painted up.

b) when in German hands

Czajka	renamed *Westerplatte* 10.1939 and *TFA11* 20.8.1943
Mewa	renamed *Putzig* 10.1939 and *TFA7* 1940
Rybitwa	renamed *Rixhoft* 10.1939 and *TFA8* 1940
Zuraw	renamed *Oxhoft* 10.1939

c) 1946-1949

Czajka	CJ, then CK
Mewa	MW
Rybitwa	RB
Zuraw	ZW, then ZR; after 1948 – HG-11, the letters HG standing for 'okret HydroGraficzny' (hydrographic ship)

d) after 1949

Czajka	D-45, then 325
Mewa	D-46, then 326
Rybitwa	D-47, then 327

The letter D stood for 'Dozorowiec' (guard ship)

Zuraw	*renamed Kompas* 1951 (no pennant number), from 31.12.1971 renamed *BK-4*, BK standing for 'Barka Koszarowa' (barrack hulk)

The 'Birds' back in Poland in the late 1940s, from left to right *Mewa, Czajka* and *Rybitwa* no longer show their original uniformity in appearance. In the background is the destroyer *Blyskawica*

Jerzy Micinski collection

for laying navigational markings. In 1940 she became a 'Versuchsboot' (experimental boat) and finally on 20 August 1943 she became a 'Torpedofangboot' (torpedo recovery vessel) and was renamed *TFA11.*

Meanwhile, *Rybitwa* and *Mewa* were being repaired and rebuilt. The shape of the stern was changed

The *Rybitwa* as a minesweeper in the late 1940s

Jerzy Micinski collection

2

3

1 The *Zuraw* as a survey ship with the after guns replaced by a deckhouse

Jerzy Micinsky collection

2 With new armament but otherwise little altered, the *Mewa* as a patrol vessel in the early 1950s

Jerzy Micinski collection

3 The *Rybitwa* following her last modernisation with her tall pole mast replaced by a short tripod

Jerzy Micinski collection

1

and steam, instead of the original electric, heating was employed. Renamed respectively *Rixhoft* and *Putzig* they served from 1940 as 'Torpedofangboote' *TFA8* and *TFA9*. The burnt out *Jaskolka* and structurally deformed *Czapla* were not considered worth repair and were scrapped.

POSTWAR FATES

When the war ended, and the German Minesweeping Administration was created, the four ex-Polish minesweepers were among the ships operating in Danish waters. They were discovered by the Poles in Travemunde and brought back to Gdynia where they arrived on 12 March 1946. Before they left they were equipped with German 20mm guns and minesweeping gear and at Gdynia they were repaired, having been damaged by rough use in German hands.

The minesweepers were initially intended to operate from Gdynia together with nine ex-Soviet minesweepers but it was later decided to base them at Szczecin. The first to go was *Zuraw* on 29 November 1946 and two weeks later she was temporarily detached for hydrographic duties. At the time the Navy was trying to obtain a Merchant vessel of 500-1000 tons for use as a survey ship but nothing came of it and *Zuraw* officially became a permanent hydrographic vessel on 15 August 1948. Meanwhile, in June 1947 the remaining three had completed repairs. They served as training ships for the Naval Officers School until July 1947, when training duties were taken over by *Blyskawica* and then joined *Zuraw* at Szczecin for service as minesweepers.

By the middle of 1949 the area over which Poland was reponsible for minesweeping was considered safe for international navigation and at the end of July all the minesweepers except *Zuraw* were rearmed as patrol boats. They were employed for training seamen in anti-submarine work until 1960 when they became barrack ships and finally, in February 1970, it was officially stated that all three had been deleted. Two of them were probably scrapped, while one is still used for training seamen.

As a survey vessel the *Zuraw* received a new deckhouse near the stern. In 1951 she was renamed *Kompas* and in 1959-63 Gdynia Naval Shipyard rebuilt her with raised forecastle and tripod mast. On 20 February 1971 the flag was raised on ORP *Kopernik*, a new hydrographic ship and *Kompas* was paid off on 31 December 1971. Afterwards, she also became a barrack ship and this will probably be her last employment, although it may prove to be a long one.

The *Jaskolka* class minesweepers, when built, were a success. Despite a long building time, they proved that the infant Polish shipbuilding industry was capable of building warships. They also considerably strengthened the small Polish navy and played an important role during September 1939. These ships will long be remembered by all who served in them and those whose paths they crossed.

HMS GROVE

In October 1938, the Director of Plans minuted 'The recent crisis has once again drawn attention to our shortage of small craft available to carry out the many escort and screening duties.' The need for two main types of destroyer – one for fleet work and one for escort duties – was not new in principle, and had been applied for many years by allocating old destroyers to escort work as new construction replaced them in the fleet. However, when new construction dropped to one flotilla of nine vessels per year, too few old destroyers were made available for escort duties. The only alternative was to build special escort destroyers which, lacking the speed and torpedo armament of the fleet unit and armed with dual-pupose guns, could be reduced in size and cost to speed their construction. The sloop *Bittern*, completed in 1938, was taken as a model possessing most of the desirable qualities. A vessel of 1190 tons, she had dimensions of 262ft pp, 282ft oa x 37ft x 8¼ft, was powered by geared turbines of 3300shp for a speed of 18¾kts and was armed with six 4in dual-purpose guns controlled by a HA/DCT on the bridge.

The attempt to install this armament, plus a quadruple 2pdr AA mounting, in the Hunt Type I, produced a vessel of similar length but, with beam reduced by 8ft and an increase in installed power to 19,000shp, it was so short of stability that it proved necessary to drastically reduce topweight by the removal of one twin 4in mounting. This mounting was replaced with the multiple 2pdr, moved from its position abaft the funnel, to maintain a satisfactory trim. This lack of stability was largely corrected in the Hunt Type II vessels, to which *Grove* – shown here – belonged, by increasing the beam by 2½ft, which allowed for the reinstatement of the third 4in mounting and the moving of the multiple 2pdr back to its original location.

Whilst too short for ocean escort work and lacking in endurance, they enjoyed a high reputation in escorting East Coast and Mediterranean convoys where AA fire power was in great demand in the face of heavy air attack.

Few additions could be made to the armament of the Hunt class due to their small displacement but nearly all vessels of Types I and II had 2 single 20mm guns added in the bridge wings. All units were fitted with Denny Brown retractable fin stabilisers positioned well forward, which were subsequently removed from some units to increase the bunker capacity by 53 tons. In layout and construction the Hunt class followed normal destroyer practice except that they were the first vessels of this type where the officers were berthed forward, under the bridge, instead of in the traditional position aft. They naturally adopted the two boiler arrangement, while the engine room spaces were divided into separate turbine and gearing compartments.

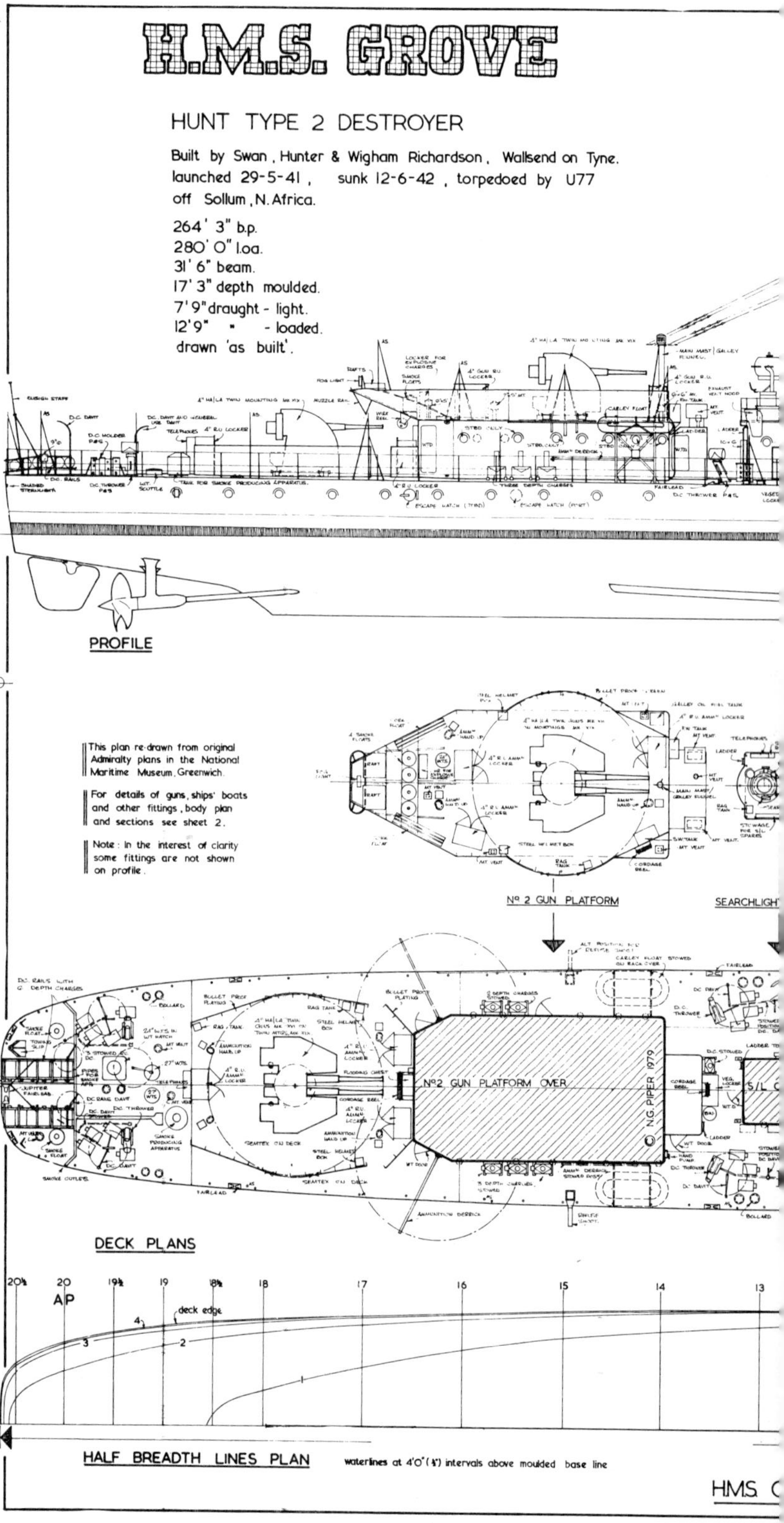

Dyeline copies of the accompanying drawing, at a scale of ⅛in = 1ft, are available from Sambrook Marine Plans, price £3.00 + 40p post and packing. A second sheet giving details of guns, boats, etc is in preparation.

Drawing by N.Piper

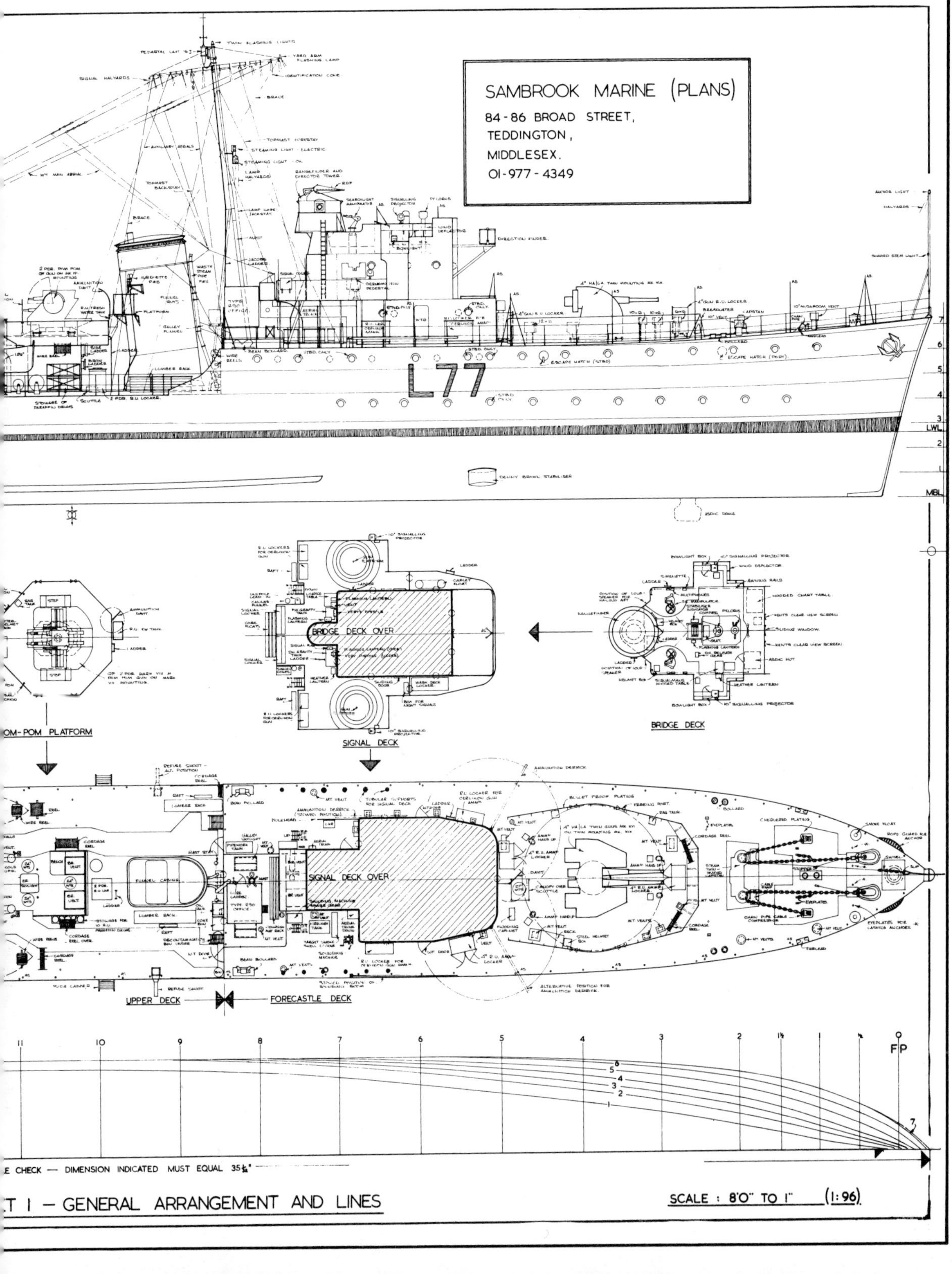

THE ROYAL SOVEREIGN CLASS OF 1889 by RA Burt

PART 2

A port quarter view of *Hood's* stern in 1894
MoD

The *Hood* in overall grey, 1904, with W/T gaff on main topmast and shields removed from QF guns in tops

NMM

HOOD

Apart from the turret mounting of her main guns and her low freeboard fore and aft the *Hood* was of basically the same design as the *Royal Sovereign* class. There were a few other minor differences, including thicker longitudinal bulkheads which, together with her 17in turret armour, gave her marginally better protection but against this was her poor performance as a seaboat and reduced accommodation, due to having one less deck than her half sisters. Particulars of the ship are given in the accompanying tables.

MODIFICATIONS

As completed the seven *Royal Sovereigns* were very similar in appearance and almost impossible to tell apart. There were, however, a number of small differences in funnel detail, positions of hawsepipes, etc and the *Empress of India*, *Royal Oak* and *Royal Sovereign* were the only ships completed with a bow scroll (see accompanying drawings). Alterations and additions made to the ships during their careers included the following:

Royal Sovereign class

1900: Bow scrolls removed and wireless transmitters fitted.

1902-1904: Upper deck 6in guns provided with armoured casemates (5in faces and 2in rears of KNC), above water torpedo tubes removed.

1903-1904: Victorian paint scheme replaced by overall grey.

1904-1905: Net shelf lowered to main deck level, early funnel bands painted up (*Empress of India* 2 dark, *Royal Sovereigh* 1 dark – painted out 1907). Note: funnel bands of this period cannot always be relied upon for identification purposes as they were often changed; the standard system did not come into use until 1909.

PARTICULARS OF HOOD

Laid down: 12 August 1899
Launched: 30 July 1891
Completed: 1 June 1893
Builder: Chatham Dockyard
Displacement: 14,150 tons legend, 14,780 tons load, 15,588 tons deep
Length: 380ft pp, 410ft 6in oa
Beam: 75ft
Draught: 27ft 6in mean, 28ft 6in max
Freeboard: 11ft 3in forward, 17ft 3in amidships, 11ft 3in aft
Armament: 4-13.5in BL, 10-6in QF, 10-6pdr QF, 12-3pdr QF, 7-18in torpedo tubes (2 submerged forward; 1 stern, 1 bow and 4 broadside above water)
Armour: belt 18in, 16in, 14in; upper belt 5in; fore bulkhead 16in; aft bulkhead 14in; turrets 17in-11in; redoubts 17in-13in; main deck casemates 6in; CT 14in; after CT 3in; communication tube 8in-3in; ammunition trunks to 6in guns, 2in; wood backing 8in-4in; main deck over citadel 3inch; lower deck fore and aft $2\frac{1}{2}$in
Machinery: 2 shaft Humphreys and Tennant 3 cyl vertical triple expansion engines, 9000ihp = 15.7kts (natural draught) 11,000ihp = 16.7kts (forced draught), 8 cylindrical boilers, working pressure 155lbs/sq in
Coal capacity: 900 tons legend, 1410 tons maximum
Endurance: 2780 miles at 14kts, 4720 miles at 10kts
Anchors: 3-115cwt Martins, 1-16cwt, 1-8cwt
Boats: 1-56ft vedette, 1-56 wood torpedo-boat, 1-40ft steam pinnace, 1-42ft steam launch, 1-36ft pinnace, 1-26ft cutter, 2-30ft cutters, 1-30ft gig, 1-28ft gig, 1-24ft gig, 1-25ft whaleboat, 1-18ft mail boat, 1-14ft balsa raft. Calculated life saving capacity of these boats was 682 but a note in the ships cover states that the boats could not take more than 431 which was quite inadequate for a complement of 690; there is no indication of any action on this point.
Cost: £926,396 actual, £820,000 estimated (£749,000 hull, machinery, fittings and equipment; £60,000 armament; £11,000 stores and first fittings)

1905-1909: Fire control system installed, originally in upper fore-top which was modified for the purpose.
1906-1908: Main deck QF guns on main deck and in fighting tops removed some of which were later remounted in the superstructure. Funnel bands reintroduced (*Ramillies* 1 white, *Repulse* 2 white on both funnels, *Resolution* 3 white – see accompanying drawings).
1909: *Revenge* was the only ship of the class to have stockless anchors and no torpedo nets at this time.
1910-1913: After bridge removed from all except *Revenge*. The 13.5in guns in *Revenge* were replaced by 10in guns for a short period but the original weapons were replaced in October 1912.
1914-1915: *Revenge* fitted at Portsmouth for bombardment duties. The 13.5in guns were relined to 12in calibre and elevation increased to $13\frac{1}{2}°$ giving a maximum range of 16,000yds. Fitted with bulges and a 'mine catcher' over the bows – most but not all of her small QF guns were removed.

In 1901-1902 it was proposed to modernise the class by fitting them with 12in guns, of the same type carried in the new ships of the *Formidable* class, mounted in armoured gunhouses. Total weight for the new weapons was 383 tons (four guns 200 tons, cordite and

Revenge in 1915 with the main topmast and torpedo net defence removed and the fore topmast reduced. Note the mine deflection gear, seen here in operation, over the bows
IWM

LEGEND WEIGHTS HMS HOOD

	(ton
Water for 10 days:	6
Provisions for 4 weeks:	4
Officers stores and slops:	2
Officers, men and effects:	8
Masts and rig:	7
Cables:	9
Anchors:	2
Boats:	5
WO stores:	5
Armament:	172
Main machinery:	105
Auxiliary machinery:	5
Engineers stores:	5
Coal:	90
Vertical armour on main belt:	135
Horizontal armour:	110
Protection to 6in casemates:	45
Teak backing:	8
Turrets:	149
(including 50 tons backing	
CT:	9
Hull:	475
Torpedo nets and booms:	4
Total	13,62
4% board margin:	53
Legend displacement:	14,15

Approved additions during construction:

Fighting tops (1889):	
Increased thickness boiler tubes:	2
Increased power for 18in torpedo tube air-compressors (1890):	
Increase in boatswains and carpenters stores (1890):	1
Raising funnels and steam pipes (1891)	1
Total	6

TABILITY FIGURES – HOOD

As inclined 24 March 1893
Legend condition with 900 tons coal on board and reserve feed tanks empty; mean draught 27ft, GM 4.1ft.
Deep condition, fully equippd with 1380 tons coal on board and reserve feed tanks full; mean draught; 28ft 6in, GM 4.4ft.
Light condition with water in boilers to working height; draught 25ft 10in, GM 3.95ft
Angle of maximum stability 34°, range 57°.
Folloing figures are based on a Metacentric diagramme dated 17 May 1893 which was produced from the above inclining experiment

Condition:	Load	Ordinary	Experimental	Light
Displacement (tons):	14,780	14,332	13,260	13,100
Tons per inch (tons);	53.34	53.28	52.65	52.6
GM (ft):	4.37	4	3.9	3.95
Mean draught (ft-ins):	28-3½	27-7	26-0¼	25-8

Empress of India in 1905
MoD

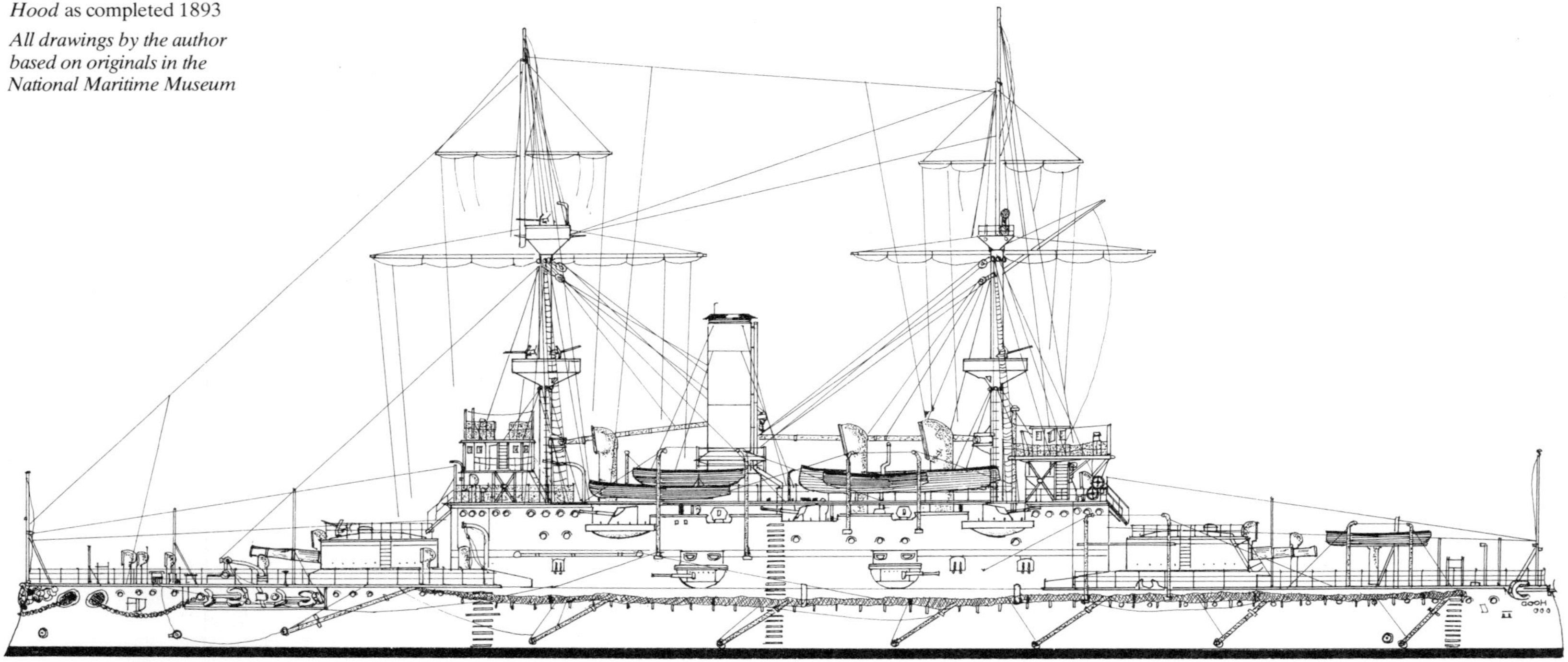

Hood as completed 1893

All drawings by the author based on originals in the National Maritime Museum

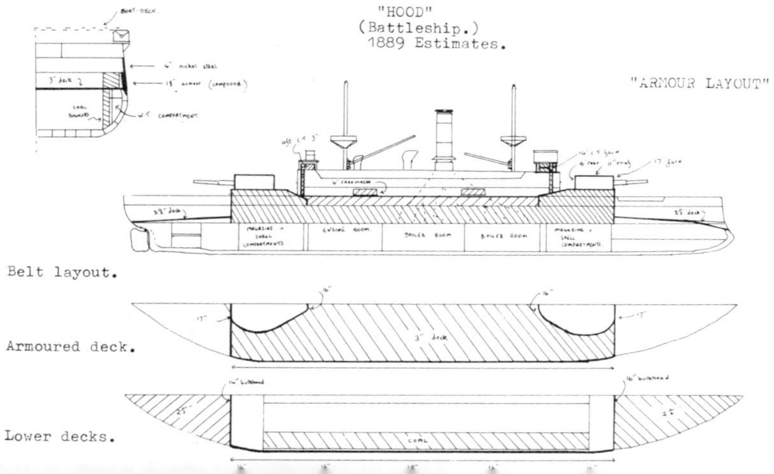

charges 59 tons, shells 124 tons) compared with 503 tons (four guns 268 tons, cordite and charges 53 tons, shells 182 tons) for the 13.5in. The turret weight would have been 180 tons and the idea was later dropped due to the adverse effect on stability and for reasons of cost (£200,000 per ship).

During the same period it was noticed that some high temperatures were being recorded in the 6in magazines, the worst case being 109°F in *Revenge*. The fitting of refrigeration units was recommended and *Empress of India* was equipped with an experimental magazine cooling plant which was successful but, for some unknown reason, no other ship was so fitted.

Hood
1899: net shelf lowered to main deck level.
1900: 3pdr QF guns in upper fore-top replaced by searchlights, above water torpedo tubes and bow scroll removed.
1902-1903: 6pdr QF guns on lower deck removed, wireless transmitter and WF gaff (on main mast) fitted.
1903-1904: Victorian paint scheme replaced by overall grey, main deck 6in guns removed.
1907-1909: 3pdr QF Guns

"IDENTIFICATION"

"ROYAL SOVEREIGN CLASS" (Battleships.) 1889 Estimates.

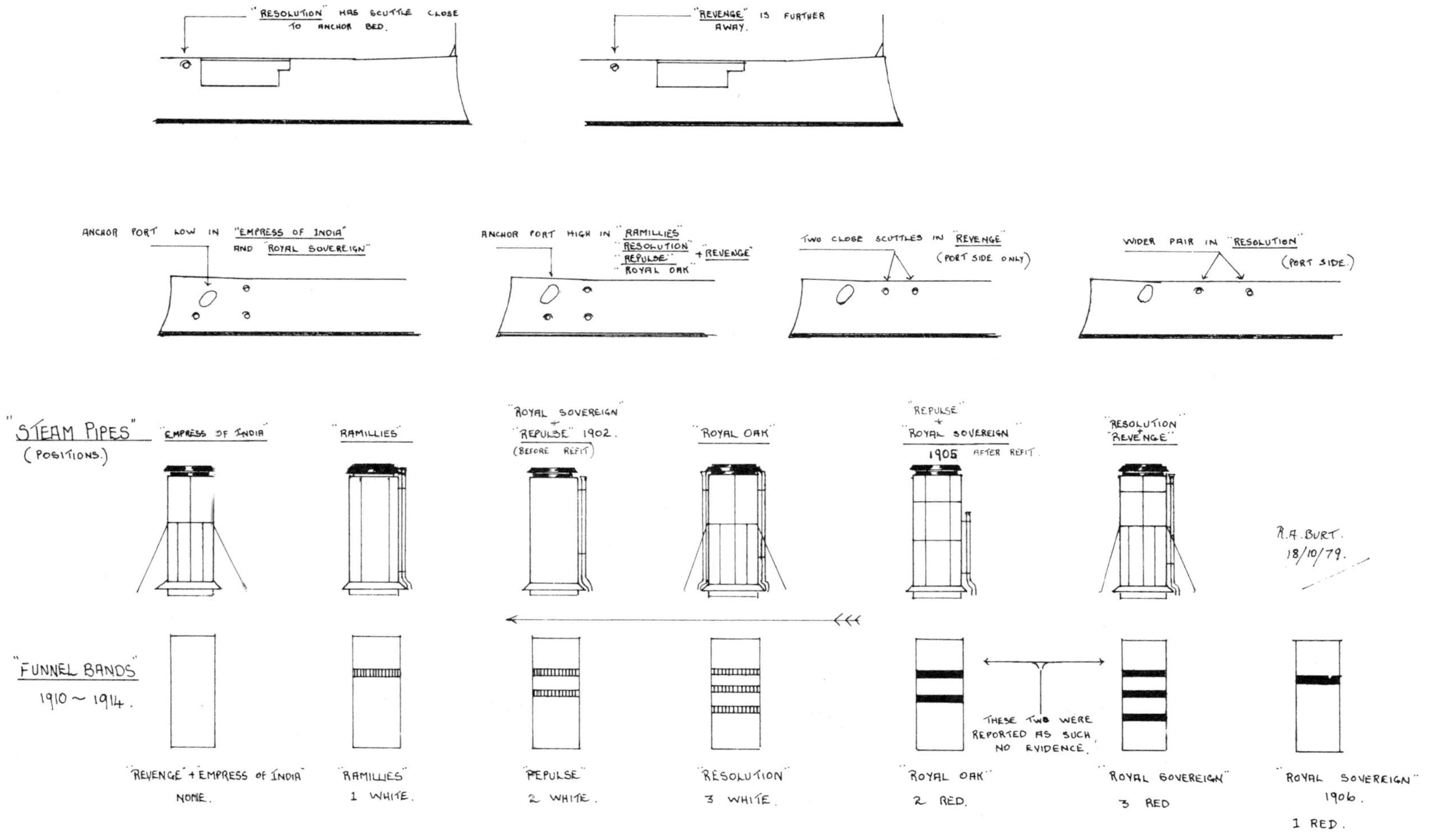

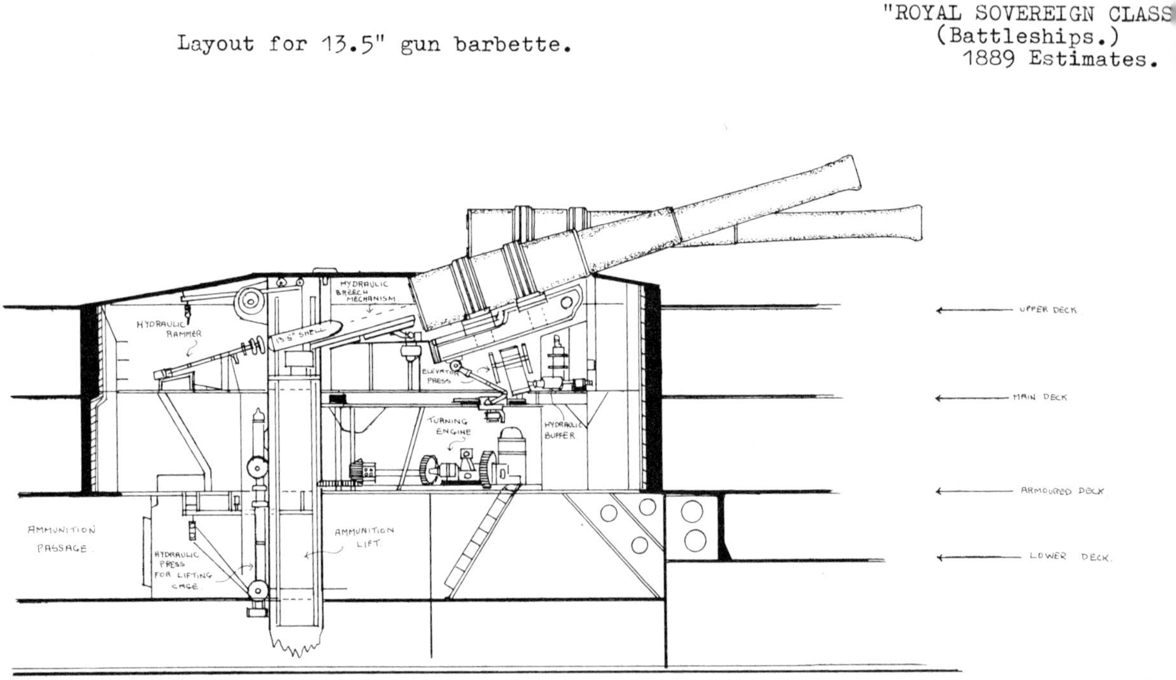

removed from lower tops.
1910: Partially dismantled for subsidiary service all 6in guns, searchlights, torpedo nets and wireless equipment being removed.

CAREERS

Royal Sovereign: Launched by Queen Victoria and the largest ship to be built for the Royal Navy to this date. Channel Fleet 1892-1897 and fleet flagship until December 1895, Kiel 1895, Mediterranean Fleet 1897-1902, Fleet flagship for coronation review at Spithead in August 1902, portguardship at Portsmouth 1902-1905 (under refit 1903-1904), reserve 1905-1911, placed on disposal list 1911, sold to Clarkson Co for £40,000 October 1913 but later resold to G B Berterelle and sailed to Italy for scrapping.

Royal Oak: Completed by Lairds at Birkenhead 12 June 1894 and served with Reserve fleet until 1896, flying squadron (see *Revenge*) January-December 1896, Reserve fleet 1896-1897, Mediterranean fleet 1897-1902, large refit 1902, Home fleet 1903-1905, reserve 1905-1911, placed on disposal list 1911, sold to T Ward January 1914 for £36,450 and scrapped.

Ramillies: Commissioned at Portsmouth 17 October 1893 for Mediterranean Fleet until 1903 (bilge keels fitted at Malta 1894) large refit at Chatham 1903-1904, Fleet manoeuvres July-September 1904, Reserve fleet 1904-1906, in collision with *Resolution* 15 June 1906, refit and repair 1906-March 1907, Home Fleet 1907-1911, placed on disposal list and sold to Cohen Shipbreakers Ltd for £42,300 on 7 October 1913, scrapped at Swansea.

Resolution: Commissioned at Portsmouth 5 December 1893, Channel Fleet 1893-October 1901, Reserve fleet November 1901-February 1902, 2nd flagship Reserve squadron 1902, large refit at Devonport 1902-1903, hull and steering gear damaged during passage from Holyhead to Plymouth February 1903, Fleet manoeuvres 1904, Reserve 1904-1906, in collision with *Ramillies* 15 June 1906, repair and refit at Chatham, 4th Division Home Fleet 1909-1911. Sold to Dutch shipbreakers 7 October 1913.

Repulse: Completed at Pembroke Dockyard 25 April 1894, Channel fleet 1894-1902, Mediterranean Fleet March 1902-1903, Large re[fit] April 1903-December 1904, Reserve 1905-1906, refit 1906-1907, 4th Division Home fleet 1907-1910, sold to T Ward f[or] £33,550 July 1911 and scrapped a[t] Morecambe.

Empress of India: Commissioned Devonport 11 September 1893, Channel fleet 1893-1897 (refitted with bilge keels 1894), Mediterranean fleet 1897-1901, large refit 1902, Home fleet 1902-1904, private ship 1904-190[5], refit at Devonport 1905, reserve May 1907-April 1909, 4th Divisi[on] Home Fleet 1909-1911, placed o[n] disposal list 1912 and laid up, use[d] as target and finally sunk 4 November 1913.

Revenge: Completed by Palmers 2[?] March 1894, reserve at Portsmou[th] 1894-1896, commissioned as flagship of special Flying Squadro[n] 14 January 1896 formed as a resu[lt] of growing European crisis which later subsided, Mediterranean flee[t] November 1896-April 1900, larg[e] refit June 1900-April 1901, coastguard ship at Portland April 1901-May 1902, Home Fleet Ma[y] 1902-August 1905, Portsmouth Reserve September 1905-June 1906, gunnery training ship June

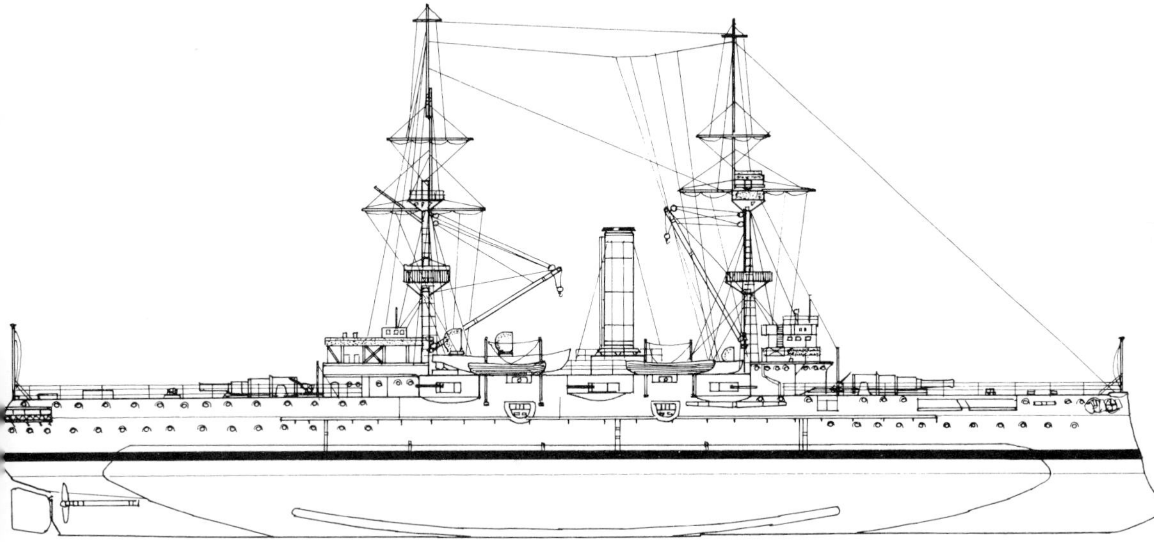

The *Revenge* as she appeared in the autumn of 1914 after her refit for bombardment duties. Her guns were relined to 12inch, bulges added, the foretop was enlarged and the rigging slightly altered

npress of India, 1905: note the lack of a /T gaff, the lowered torpedo net shelves, nission of shields from the 3pdr guns in the ps and the new 6in gun casemates on the per deck

MM

Royal Oak in 1896

MoD

The *Revenge* in 1914, as modified for bombardment duties off the Belgian coast. She has been given a deliberate list to port, a device to provide greater elevation and therefore greater range for the guns. Note the false bow wave and that she still has her torpedo nets and booms *IWM*

Revenge leaving Portsmouth in 1906 *IWM*

The *Hood* as completed, 1894 *MoD*

Hood, 1894 *NMM*

1906-May 1913, reserve 1913 and placed on disposal list but on outbreak of war refitted for service as bombardment ship. In this guise she was employed to bombard the northern end of the Western front and other targets on the Belgian coast. She was not however, completely suitable for these coastal operations because of her deep draught. Commissioned October 1914 for service with 6th Battle squadron, Channel fleet; Renamed *Redoubtable* May 1915 as her name was required for one of the ships of the new *Royal Sovereign* class; badly damaged by 8in gunfire from German shore batteries December 1915 but subsequently repaired; with completion of purpose-built 15in gun monitors withdrawn from active service and later used for firing trials and other experiments; accommodation ship Portsmouth January 1918; Sold to T Ward for £42,570 November 1919 and towed to Appledore for scrapping.
Hood: Mediterranean fleet 1893-1900, reserve and portguard ship at Pembroke 1900-1901, Mediterranean fleet 1901-1902, large refit 1902-1903, Home fleet 1903-1904, Reserve fleet 1904-1910, placed on disposal list 1913, employed in experiments with anti-torpedo bulges and hull construction, scuttled as blockship in southern entrance of Portland harbour November 1914 where she still lies.

Royal Oak in the Solent about 1900
Beken of Cowes

The *Revenge* with an early form of camouflage in 1914/15
IWM

Revenge in 1910 after the addition of a new, square, fire-control top on the foremast and re-equipping with stockless anchors
NMM

THE ORIGIN OF THE J CLASS DESTROYERS

by Alan Payne MINA

In 1936 the Board of Admiralty, after discussion with fleet officers of destroyer experience, decided that a design for a destroyer smaller than the 'Tribals' was required. Requirements for the new design included a powerful HA/LA gun armament, two torpedo tube mountings and minesweeping gear as in the pre-'Tribals'. The resulting design was the very successful *J* class which was repeated in the following *K* and *N* classes.

ORIGINS

The 'Tribal' class of the 1935 and 1936 Programmes were the largest and most heavily armed destroyers designed for the Navy up to that time, and were designed in reply to the Japanese *Fubuki* class which mounted six 5in guns. The designer of the 'Tribals' was Constructor A P Cole, who wrote, 'The Tribals, which were my first design were a breakaway from convention, but were exceptional ships to fulfill a particular function. When the staff requirements of the *J* and *K* classes were received, I had certain revolutionary ideas but knew that strong opposition would be encountered to their adoption'. The revolutionary ideas referred to included the introduction of longitudinal construction, adopting two boilers instead of the normal three and using twin 4.7in gun mountings as in the 'Tribals'.

The Naval Staff at first objected to twin mountings in a fleet

The *Javelin* under refit in dry dock in late 1944 or early 1945

IWM

The rear of one of *Javelin*'s twin 4.7in MkXIX gun mountings in 1944. The large frame extending behind the shield is the counter balance weight and its support and the items on which the men are working are the loading trays

IWM

lestroyer, although they had been ıpproved for the Tribals, and the Director of Naval Ordnance trongly opposed the adoption of he new mountings until the prototype had been thoroughly tested in the destroyer *Hereward*. Moreover he did not consider the supply situation would allow them to be produced in sufficient quantities to arm the *J*s and *K*s as well as the 'Tribals'.

The Engineer-in-Chief had very strong objections to the adoption of two boilers for a variety of reasons, one of which was its effect on boiler cleaning routines. The point about Cole's idea was that, if approved, it meant all destroyers would have only one funnel and consequently a smaller silhouette. Oddly enough the last British destroyer with a single funnel had been as far back as 1894! But, in view of the Engineer-in-Chief's strong objections, the only way to have a single funnel destroyer was for the Controller of the Navy to over-ride the engineers. This would normally have taken a miracle, as there was absolutely nothing any naval constructor could do about it, but it

The quadruple MkVII pompom mounting of the destroyer *Jackal*

IWM

was here that Lord Louis Mountbatten came into the picture. In 1936 he had been appointed to the Admiralty, after commanding a new destroyer in the Mediterranean. At Malta he had met Cole and they had become friends so it was quite natural for Mountbatten to look up his friend and, having served in command of destroyers, he became very interested in the new design on which Coles was working. 'Officially Mountbatten should have had nothing to do with destroyer design', wrote Cole, and 'I think he appreciated this, and his introduction to Kenneth Poolman's book cleverly deals with this point'.*

'I, of course, was wildly enthusiastic with his idea of doing away with the third boiler', wrote Mountbatten, 'but I pointed out that very special designs would be required to give at least 40,000 horse power out of two boilers.

'I knew that brilliant engineer, Rear Admiral Dight, and went and saw him with Cole and he undertook to design the necessary boilers to make this possible.

'Then of course we ran into opposition with the naval staff who objected to having less than three boilers in a destroyer as they felt that, if one boiler was out of action, the ship should not go into harbour without two boilers, for safety. I thought this right out of date and Dight entirely agreed. The naval staff would not, however, yield. I then went and saw the Controller of the Navy, Admiral Reggie Henderson, who lived near me in the country and had a long talk with him about it. I convinced him that this was a break-through in design and would lower the silhouette of destroyers and give space for other requirements through not having a third boiler room. He then said he would over-ride the naval staff and gave the necessary authority to go ahead with the construction of the *J* and *K* class destroyers.'

It was extremely fortunate that Admiral Henderson was a strong man, probably the best Controller the Navy ever had. If he had been a all weak, it seems unlikely that anything Mountbatten had said would have had the desired effect. While the idea was Cole's, the execution of the idea was entirely due to Mountbatten's very considerable powers of persuasion. The result was that all British

**The Kelly*, by Kenneth Poolman, Kimber 1954

The *Jervis* in 1945 showing her full outfit of war modifications; from forward she has twin 20mm Oerlikons in the bridge wings, the radar aerials of Type 285 on her range-finder/director, a tall lattice carrying the aerials for Type 276 radar and the FH4 HF/DF, single 20mm mountings abreast the searchlight platform (from which the searchlight has been removed) and a pole mast aft carrying the aerial of radar Type 29. In addition to these alterations she, in common with the rest of the class, had her after torpedo tube mounting replaced by a 4in AA gun in 1940 but the surviving ships had the tubes replaced during 1942/43

IWM

The *Javelin* in 1942

IWM

destroyers from the J's to the Battle class had only two boilers and one funnel.

Various reports in the press, after the new destroyers were completed, were grossly inaccurate and gave the wrong impression of exactly what Mountbatten had done. As a result of the press accounts, Cole had to report to the Director of Naval Construction to explain the true position, which naturally caused him considerable embarassement. But Cole was most explicit about Lord Mountbatten's contribution to the design. 'All destroyers subsequent to the *J* class (and they included the large *L*s and *M*s as well as the smaller emergency destroyers designed at the beginning of the war) incorporated the ideas, worked out by Mountbatten and myself in collaboration, in the *Javelin* and *Kelly*, so it can be said that his influence on destroyer design extended far beyond the *Kelly* and the Fifth Flotilla.'

Many famous designs have started on the back of an envelope and Cole's *J* class was no exception. Soon after Mountbatten and Cole met again, at the Admiralty, they arranged to visit the new high-pressure wind-tunnel at the National Physical Laboratory in Teddington. 'On our return', wrote Cole many years later, 'we halted for a while in Bushey Park and I mentioned to Mountbatten that I had just received the staff requirements for a new destroyer design, to be called the *Javelin* class. I had several revolutionary ideas for this design . . . I made a sketch on the back of an envelope showing how I thought the ship would look, and mentioned that I anticipated strong objections, particularly to the boiler arrangement.'

LONGITUDINAL CONSTRUCTION

The most important of Cole's revolutionary ideas was the adoption of the much stronger longitudinal system of construction, which was entirely his own idea and was to save a number of damaged destroyers in the coming war. There was nothing new about longitudinal construction, but it had been used for destroyer construction only once before; this was in Denny's *Ardent* in 1912. Instead of having closely spaced transverse frames with a few weak longitudinals, the longitudinal system has more widely spaced transverse frames with a large number of strong longitudinals which made the light hull of a destroyer much stronger. 'This was to prove an absolute winner,' wrote Mountbatten, 'and there is no doubt whatever that the *Kelly* owned her survival to this when she was torpedoed.'

Much to his surprise and delight, Captain Lord Mountbatten was appointed Captain (D) of the Fifth Destroyer Flotilla in 1939, in *Kelly*.

The 5th Destroyer Flotilla in the Mediterranean on 11 May 1941 following a night bombardment of Benghazi. Nearest the camera are *Jackal* and, on the right, *Kelly*; the three remaining ships (including that from which this photograph was taken) were *Kelvin, Kipling* and *Kashmir.* Within two weeks all except *Kelvin* had been sunk!

IWM

Seen from the forecastle of a battleship, four 'Tribal' class destroyers and the *Jackal* (nearest) leaving Scapa Flow in 1940

IWM

Survival depended very largely on the great designed strength and good workmanship of the destroyer and in particular with the damage control routine worked out by Mountbatten and Cole. After the *Kelly* was torpedoed the designer worked out some calculations and found that had Mountbatten not removed all the surviving crew, except 18 officers and men to man the guns, the ship would have turned over before she got back to harbour. 'When the King saw the photographs of the *Kelly*'s condition on being towed back to the Tyne', wrote Lord Mountbatten, 'and when I told him that her survival was entirely due to Cole's brilliant design and the help he given me in damage control, the King on his own, sent a message to the First Lord that he wished Cole to receive a knighthood for his outstanding contribution as a Naval Constructor, but the Civil Service Union were too strong for his Majesty, for they had the nerve to reply that Cole was not senior enough. I think he was about the most brilliant constructor we have ever had but he was much too forthright to bear fools gladly'.

BRIDGE DESIGN

The actual layout of a bridge in a new class of destroyer is always decided by a mock-up and this was done for the 'Tribals'. The shape had been decided by the designer, but the actual layout could be modified by various experts. However, before the mock-up was dismantled', wrote Cole, 'I invited Mountbatten to inspect it at Thornycrofts and give his views on the design of the *J* class bridge. Mountbatten's proposals were drastic, and as detailed in his letter. They were practical because of the redesign of the structure mentioned above, and were incorporated in the bridge design almost without modification. At the subsequent mock-up of the *J* class bridge, the revised layout was accepted and, I believe, resulted in the most successful 'open' type bridge ever fitted'. The redesign of the structure referred to was Cole's modification of the bridge structure of the *I* class destroyers in which the wheelhouse was fitted forward of, and below, the bridge enclosed in bullet proof plating and shaped to reduce the windage on the bridge. The *I* class type bridge was also fitted to the 'Tribals'.

The designed standard displacement was 1690 tons but this was increased by about 30 tons after the design was approved by various modifications and not surprisingly the weight of the hull and fittings, as well as the machinery weights, were slightly underestimated, so that the actual tonnage came out at 1760 tons. The designed weights and those as built are given in table 1.

The average speed on trials of the *J*s was about 33½kts, although *Jackal* averaged 37.5kts during a 6 hour trial, on a very light displacement of 1700 tons with 42,838shp, and *Javelin* averaged 35.1kts on 1927 tons with 40,182shp. The endurance for the class was 5500 miles at 15kts and 3700 miles at 20kts. The oil fuel capacity was 485 tons, which was actually less than in the 'Tribals'.

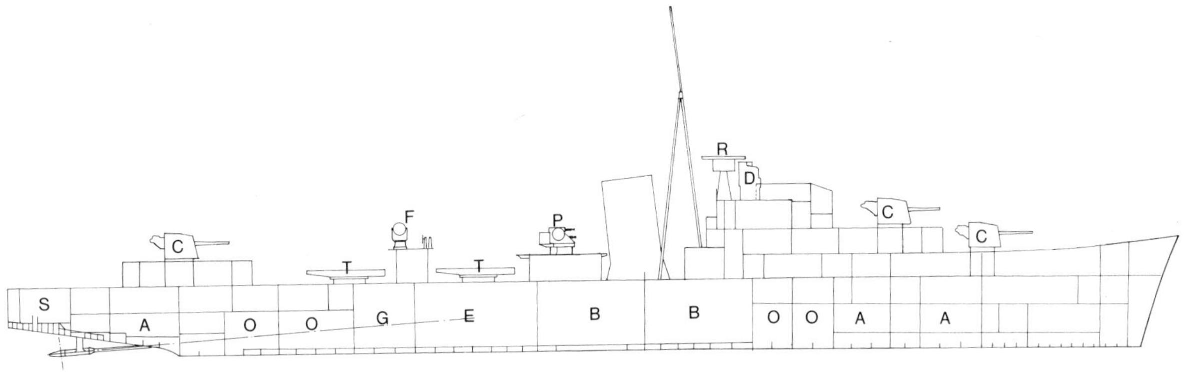

A magazine
B boiler room
C twin 4.7in mount
D director
E engine room
F searchlight
G gearing room
O oil fuel
P pompom
R rangefinder/director
S steering gear
T torpedo tubes

'J' CLASS-GENERAL ARRANGEMENT

CONCLUSION

There is no doubt that longitudinal construction saved the *Kelly* but it is very doubtful that it saved the *Javelin.* Both ships had been commanded by Lord Mountbatten and *Javelin* was struck by two torpedoes, one forward and one aft. The ship heeled considerably, but recovered with a list of 5° to port and then became upright when the stern dropped off. What was left of the ship was safely towed into harbour and the ship was subsequently repaired. Other destroyers had their bows or stern blown off but only *Javelin* had both and survived.

The Director of Naval Construction, Sir Stanley Goodall, showed a considerable interest in Cole's insistence on two boilers in lieu of three, as he had also proposed the idea when he was in charge of the Destroyer Section back in 1927. 'By a slight advance in the size of *Amazons* boilers, two boilers should suffice for 34,000shp.' He submitted a sketch design showing a saving of 10ft in length, the same speed with 2000 less horsepower and a considerable saving in money. In the case of the *J*s the boilers seem to have ended up a bit longer than hoped for, as only about 9ft was saved, compared with the 'Tribals' which had more powerful machinery, so not so much was gained as had been hoped for. The idea had one disadvantage in that, if the ship was hit in the forward boiler room, it was most likely that both boilers would be put out of action and this actually happened in the case of *Nestor*. It is most probably that the *Nestor* would have been saved if there had been a third boiler room. It must also be admitted that the ship might well have been saved if there had been a diesel generator available, as the main reason for her loss was a complete lack of power.

The 'Tribals' set the pattern for heavier gun armament and larger displacements and this trend was

TABLE 1. J CLASS WEIGHTS

	Weights as built (tons)	*Weights as designed (tons)*
Hull and fittings	920	860
Machinery	535	525
Armament	215	212
Equipment	90	90
Standard Displacement	1760	1690
Reserve Feed Water	30	
Oil Fuel	485	
Diesel Oil	55	
Deep Displacement	2330	

TABLE 2. PARTICULARS OF J CLASS

Length (oa):	356ft 6ins
Length (wl):	348ft
Length (pp):	339ft 6ins
Standard Displacement:	1760 tons
Deep Displacement:	2330 tons
Armament:	Six 4.7in QF MkXII, twin mountings One 2pdr pompom, 4 barrel Mk VII mountings Two 0.5in 4 barrel MG, Mk III mountings Two sets 21in Pentad torpedo tubes
Machinery:	Parsons single reduction geared turbines, 40,000shp = 32kts deep
Boilers:	Two Admiralty 3-drum (working pressure 300lbs/sq in)

Stern view of *Javelin* under refit

IWM

A postwar view of *Jervis* leaving Malta on 27 December 1945

A and J Pavia

followed in the *J*, *K* and *N* classes a[s] well as the larger *L* and *M* classes but none of these classes met the requirements of rapid wartime construction. It was obvious that there would not be enough twin 4.7in mountings for the war flotillas so all the war emergency destroyers were designed with single mountings and were simplified as much as possible. The first two emergency flotillas were cut down versions of the *J* class but the 3rd to 14th Emergency Flotillas were repeat *J*s except for the single mountings and minor modifications. Thus the majority were based on Cole's design which started out on the back of an envelope in Lord Mountbatten's car in Bushey Park.

By the end of the war all the British fleet destroyers of both the British Pacific Fleet and the East Indies Fleet were destroyers designed by Cole, who had retained a tight grip on destroyer design during the war first as the Chief Constructor in charge of the destroyer sections and later as Assistant Director of Naval Construction.

AMPHIBIOUS FIRE SUPPORT

by Norman Friedman

Among the new types of warships developed during the Second World War was a wide spectrum of amphibious fire-support craft. Although prewar exponents of amphibious assault had considered fire-support essential, they had thought in terms of shellfire delivered by conventional warships standing offshore. What they did not appreciate was the vulnerability of the landing craft, and of the troops themselves, as they approached and crossed the beach. At that point they would be engaged by small pillboxes and other beach defences not even visible from any great distance. Enemy troops, who had been forced to take cover by the pre-landing bombardment, would be able to emerge and counterattack invading troops still awkwardly emerging from their landing craft. The ships offshore would not be able to distinguish friend from enemy at such close quarters, and tanks embarked in the landing craft would not yet be on the beach in sufficient numbers to overwhelm local defences.

LANDING CRAFT DEVELOPMENT

It was necessary, then, to provide craft which might bring their weapons to the beach itself. US practice, which is described in this and a subsequent (postwar) article, evolved in wartime as the result of a combination of British thinking (largely applied to the problem of a cross-Channel assault) and the problems of Pacific warfare. The latter had shaped US prewar amphibious doctrine; it required that US landing craft be either seagoing or transportable aboard seagoing ships. At first this meant that most craft were limited to the 36ft length which could be accommodated in davits; later a

The newly-completed *LCS(L)3 50,* at the Albina Engine and Machine Works, Portland, Oregon, 19 September 1944. The two covered objects just abaft her 3in/50 gun are rocket launchers. Her radar is an SL surface search type, flanked by a pair of IFF transponders. The stern anchor and the large winch right aft are for retraction from a beach; she can follow her wave of boats all the way in.

USN, courtesy James C Fahey Collection, US Naval Institute

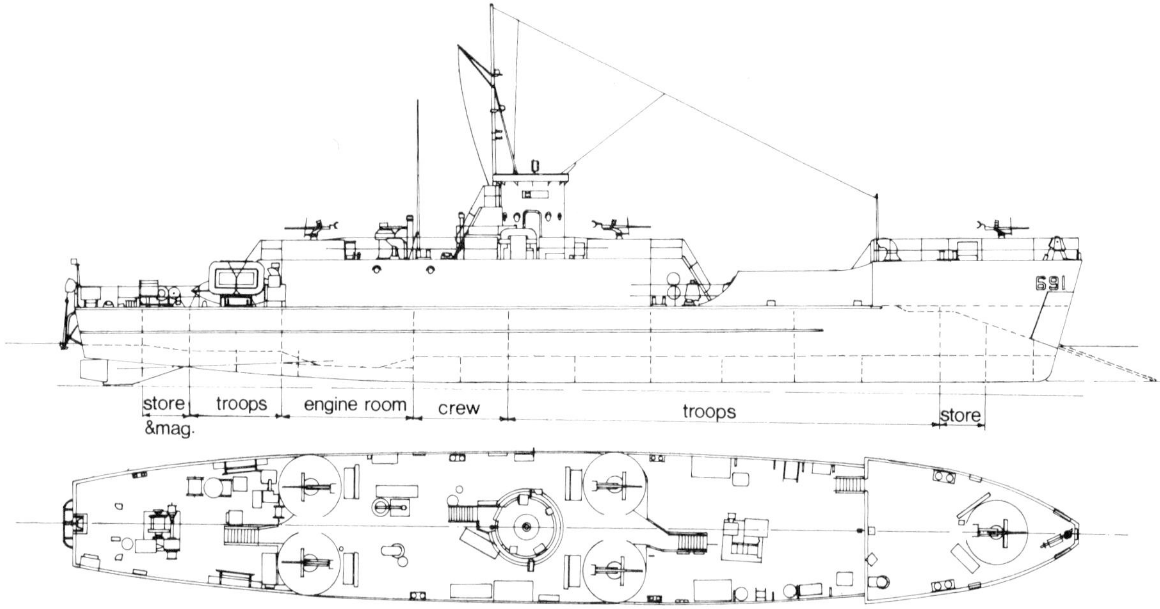

A STANDARD LCI(L)

50ft craft, the LCM(3), could be carried aboard attack transports and handled (empty) by special booms. In either case the craft had to be loaded after it had been launched and this made for a slow progression from ship to beach. The Royal Navy was able to produce much larger beaching craft, the LCT(1) to LCT(5), because great range and, for that matter, seakeeping, was not so important; they would only sail from Britain to France. However, as it became clear that the United States would built most of the craft for the cross-Channel invasion, the Royal Navy had to contend with its own problems of ocean passage from the USA. Its solution was the Atlantic Landing Craft, the LST, which could cross the ocean and then land tanks in numbers on the other side. Alternatively, there was the Landing Ship Dock, which could transport several LCTs filled with tanks.

The advent of specialised long-range landing caft meant that there could also be specialised long-range fire-support craft of comparable seagoing performance. Actually the LST was too slow and far too large to be an efficient gunboat, and the complementary LSD was in inefficient means of transporting a sizeable gunboat of its own. However, early in 1942 the Royal Navy proposed yet another new ocean-going type, the 'raiding craft', able to transport 200 troops for substantial distances at high speed (about 15kts) for Commando operations. It became, in US hands, the Landing Craft, Infantry (Large), or LCI(L), and turned out to be the ideal basis for a gunboat. The advent of a fast infantry beaching craft in turn suggested to the US Navy the value of a tank landing craft of similar endurance and sea speed, which became first the LCT(7) and then the Landing Ship, Medium, or LSM. The LSM in turn suggested further possibilities for a new order of support craft, which became the US standard postwar.

EARLY SUPPORT CRAFT

This development was not, of course, nearly so smooth in practice. In the Fall of 1941 the US Navy had 75 support landing boats (LBS) on order pending development, with another 75 planned for the FY43 budget, ie the budget for naval construction to be ordered from 1 July 1942 onwards. This programme was intended to suppo three Marine Divisions and two pools (Norfolk and San Diego); in November there were plans for boats for another Division, which would have included 33 more LBS This 37ft boat was modelled on British practice, in which a machine-gun armed armoured support boat beached with a wave of landing craft, driving back counter-attackers and attacking pillboxes. A British officer, Captai Hussey, suggested a standard ratio of one to every ten assault landing craft; he also wanted a joint US/British design. By the Fall of 1942 the US and Royal Navies had a unified system of landing craft designation, and the LBS had become the LC(S) or Landing Craf (Support), and then the LCS(S) or Landing Craft Support (Small), as the Royal Navy had graduated to much more powerful craft. Thus th LCS(S) was limited to 2-0.5in or 3-0.3in machine guns, some of which might be replaced by rockets to lay down a short barrage. Meanwhile the Royal Navy was working with converted LCTs, larg semi-oceangoing craft mounting ex-destroyer 4.7in guns (LCG, Landing Craft Gun) or with 2pdr

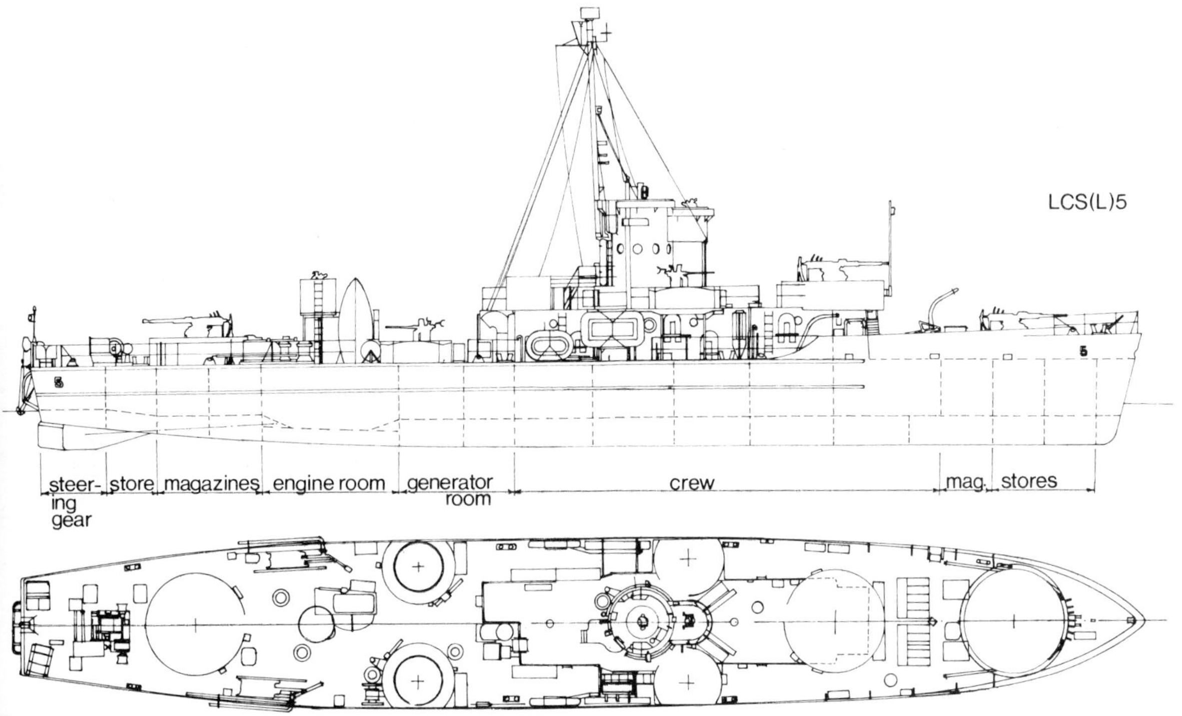

pompoms (LCF, Landing Craft Flak); it had LCS(M, Medium) and even LSC (L, Large), the latter a non-beaching type derived from the Fairmile landing craft, 105ft long and armed with a 6pdr in a turret.

On 1 May 1943 the C-in-C, US Fleet (and CNO), Admiral King, asked the Commanders of the fleet Amphibious Forces to recommend characteristics for new support craft; he noted that most existing types were exclusively British, although there was some possibility that the US LCT(5), which was considerably smaller than the British LCTs that had formed the basis for the LCF and LCG, might be converted. He thought in terms of a shipborne LCS, which in its large (L) form might mount an Army 37mm tank turret forward and a 20mm or twin 0.5in machine gun aft. The LCG, which would not be shipborne, was to mount at least two medium calibre guns for surface fire forward, with AA in addition if practicable; the LCF would have 'not less than 8 AA guns of 40mm or 20mm calibre. The use of dual-purpose guns may permit a combination of gun and flak craft.' As for size, the LCS would be limited to 50ft, although Admiral King was willing to accept about 100ft for LCG and LCF, which would also require an endurance of at least 1000nm. All three types would have to be able to beach 'in order to provide a stable gun platform'. A similar letter asked the Bureau of Ships to prepare sketch designs.

WAR EXPERIENCE

Fleet reactions were mixed. In May 1943 the United States had yet to conduct a seriously opposed landing. Guadal canal had not been opposed until after the Marines had been put ashore. North Africa had not been very much worse and the bloodbath of Tarawa was yet to come. Nor was it clear from the Cominch letter that wholly new types of US manufacture were to be provided. For example, Seventh Fleet (South Pacific) suggested that the LCS was preferable because of its similarity to other US 36ft landing craft whereas, although 'the LCS(M) and LCS(L) are desirable because of their greater fire-power; however, this advantage is believed to be off-set by the maintenance problem involved due to their being of British design and make.' The LCS(S) was indeed, so valuable that it should replace all LCM, ie all tank-carriers, assigned to attack transports. Tanks would, after all, be carried by LSTs and by the large LCTs, and there would be LCMs aboard the attack cargo ships. As for the LCT(5), it seemed unsuited to conversion to LCG/LCF 'due to low speed, poor sea keeping qualities, and lack of manoeuverability . . .'

The Atlantic Amphibious Force wanted a 20mm gun in place of the lighter machine guns in the LCS, and indeed wanted to replace the current beaching hull with the faster non-beaching picket boat hull. The LCT(5) would make a fine LCG or LCF. Admittedly it could not accompany a flotilla of LSTs at sea under its own power but unloaded LCTs could be carried to the combat area on the main decks of LSTs or in the well decks of LSDs. An armament of 2 to 4–5in/38 or 4–3in tank destroyer turrets was proposed; the LCF would have 6 to 8–40mm with a director. In addition, an LCT might be converted for rocket bombardment, as the British were already doing. A British LCT(U) could carry 700 to 900–4.5in rockets.

The Pacific Amphibious Force, on the other hand, was sceptical of the value of both LCF and the large version of the LCS; the flak function could in any case be performed by the LCG with a dual-purpose gun. 'In general, a destroyer can render more effective close support both against shore and anti-aircraft targets. Where a shoal area or coral reef extends far enough off shore to prevent such operation, it is believed that an LCG might be useful if of shallow draft.' For his part, Commander of Amphibious Forces South Pacific was sceptical of the value of any ship-carried craft, and wanted a seagoing combination LCF/LCG. He was already looking towards an island-hopping campaign in which distances of about a thousand miles would be encountered.

Commander of Amphibious Forces in North African Waters had experienced an actual assault but he had to admit that opposition had been light. He wanted all shipboard assault support craft combined into a single type, for economy of construction. 'Armament capable of neutralising pillbox installations and personnel at a range of about 2000yds is considered desirable . . . The speed requirement is governed by the expected speed of the wave the craft will support and should be at least 3kts greater than that of the wave. The endurance should be measured in hours and should not be less than six hours at maximum speed. The fuel used should be the same as that of the landing craft carried by the APA to which the LCS is attached. In most cases this will be diesel. The draught and design of the hull should allow it to beach and retract comparably with an LCM(3) . . . Support boats are considered essential in an amphibious operation. Two per APA should be provided. Two should not necessarily be carried in each APA as the facilities of the accompanying AKA can be utilized.' LCG 'should have a fair degree of seagoing ability and their characteristics should be such that they may be fuelled and serviced by the APAs and AKAs with which they will operate. Their armament should be such that they can neutralise fairly heavy types of beach fortifications and obstacles. Effective fire against gun batteries on reverse slopes should be possible . . . The speed of the craft must be governed by (1) the speed of the APAs with which they operate and (2) the speed of the waves the craft are to accompany in assault. Speed of LCGs should be 3kts more than the greater. As LCGs are to operat at sea with APAs and AKAs their seaworthy qualities must be considered in relation to their speed. Endurance should be not le than 1000nm. LCGs should be abl to beach and retract comparably with an LCI(L) and for that reason their draught should not exceed about 5ft . . . Two per AKA shoulc be provided in an opposed assault. Anti-submarine armament is desirable and should be installed, if possible, after provision of the othe characteristics noted above . . .' Th LCF was also to be seagoing, and indeed 5kts faster than APAs, with an endurance of not less than 3000nm, to beach and retract comparably with an LST, and 225 to 300ft long compared to the 100 to 200ft allowed for the LCG. Three were to be provided per APA and the LCF should have ASW equipment so that it could screen attack cargo ships and transports en route.

It fell to the Bureau of Ships to make sense of these conflicting requirements, of which the Commander in North African Waters had produced the stiffest. For example, his LCF came close t a destroyer escort with more anti-aircraft weapons and yet capable of beaching, given the 15 o

One of the original 12 LSM(R)s, firing her rockets. The 5in gun was mounted atop a self-contained armoured box, presumably the one the Bureau of Ships originally planned for its gunboat with 3 such weapons.
USN

16kts of some of the transports. At a conference of the Landing Force Development Board (18/20 May) it proposed a modified version of the one seagoing landing craft it had in production, the 153ft LCI(L), to mount, 2–3in/50 dual-purpose guns, 2 twin 40mm machine cannon and 4–20mm. The 2–3in were both close together, forward of the bridge, in response to the 1 May letter defining the LCF. The Marines and Admiral King were also represented; summarising the results of the conference, the Vice Chief of Naval Operations asked the Bureau to improve forward AA fire and install rockets and depth charges. Alternative LCG and LCF versions were to be developed.

CONVERSION PROPOSALS

The Bureau was also to consider conversions of existing hulls: the LCT(5), the new anti-submarine craft which were beginning to appear in numbers (the 110ft and 136ft sub-chasers and the 173ft PC), and even the PT boat. As for the ship-carried support craft, the new command and control boat (LCC, 56ft long) was to be considered as a new heavy assault-boat with, perhaps, 20mm guns fore and aft as well as rockets. Even the existing LCS(S) was to be modified, with either a 20mm gun or a twin 0.5in machine gun in place of the original battery.

The LCI(L) hull proved quite flexible. Although a twin 40mm weighed more than the 3in/50, it was possible to provide one such mount, superfiring forward, in place of the second 3in gun. Rockets, depth charges and sound gear could be provided at a cost of 2–20mm and, in June 1943, the Bureau of Ships estimated that the resulting small gunboat would be capable of 15kts, with an endurance of 4000nm at 14kts. Protection would consist of 10lb (1/4in) STS over gun bulwarks and pilot house and 10lb HTS (high tensile steel) over the forward hull above the waterline. The result would be a 329 ton dual-purpose gunboat quite capable of escorting amphibious convoys. As a single-purpose LCG it could mount 4–3in/50 dual-purpose guns, 2 forward and 2 aft, and 4–20mm; as an LCF it could mount 3 twin 40mm (2 forward) and the 4–20mm. The Bureau much preferred the single type as this

The 'production' rocket-firing LSM(R). *LSM(R)403*, in December 1959, is little changed from her 1945 configuration. The twin 5in rocket launchers occupy the decked-over well-deck space, with the single 5in abaft them. Note the large rangefinder for fire control and the numerous main deck vents. In contrast to the LCS(L), the LSM(R) has no unbeaching gear right aft. Electronics are more elaborate than in the LCS(L), comprising a series of short dipoles for communication with aircraft, as well as an SU surface search set, IFF and what appears to be a surviving wartime TBS tactical radio. The 'production' series comprised *LSM(R)401-412* and *501-536*.
USN

would not only simplify production but would also simplify their assignment operationally.

The other alternatives were far less attractive. For example, the 136ft and 173ft hulls would require some source of power for a twin 40mm mount and that would be limited by available space and weight. The 110ft sub-chaser, already overloaded, would in any case have very little capacity for more armament. In fact both 110ft and 173ft types would soon be converted into motor gunboats for the fire-support mission but only in small numbers. As for the PT boat, 2 single 40mm or one 40mm and one 20mm could replace the torpedo tubes but the Bureau noted that 'the space available for stowing ammunition on these vessels is so limited that it would probably be necessary to stow the 40mm boxes in various compartments throughout the craft.' The LCT(5) hull could support either 3in or 5in/38 guns but it would require modifications (sides plated up and the upper deck extended to full length) which precluded the usual practice of carrying the craft disassembled. Nor would the modifications solve the basic problems of low speed, low endurance and very poor seakeeping. A modified LCT could not be carried as LST deck cargo and therefore would have either to be assembled near the scene of action or else would have to be carried by an LSD, in which case it would be useless as an escort for the amphibious convoy.

It proved possible to modify the LCS(S) to provide a final battery of one twin lightweight 0.5in machine gun mounting as well as two projectors for 56 bombardment rockets. The existing craft were so close to overweight that the proposal for a 20mm gun was impractical. The resulting modified LCS(S) became its Mark 2 version and was built in quantity. As for the larger craft to be modified from an LCC, the proposal was practicable, at some cost in stability and at a loss of 2kts in speed. It was not, however, proceeded with.

On 27 September representatives of the Chief of Naval Operations and C-in-C US Fleet, the Bureaus of Ships and Ordnance, the Marine Corps and the Sixth Amphibious Force met to consider future action, based in part on the observations of the forces afloat. Clearly two classes of support craft were called for, one ship-borne and one seagoing; equally clearly the latter would have to be the modified LCI(L). Production would interfere with production of existing LCI(L); for example, if full interference were to be accepted, deliveries could begin in about 6 months and reach 15 per month in the following month. Alternatively, deliveries could begin in about 9 months, depending upon diesel production.

In addition, conversions of non-amphibious types were again considered, now including also the destroyer escort and the 180ft PCE. By this time the submarine menace was beginning to recede, leaving very large ASW construction programmes just reaching full strength. Conversions were, therefore, attractive and the large DE hull could accommodate a very heavy battery indeed. It was estimated that the new 5in type could take an additional (third) 5in/38, as well as 3 twin 40mm, 4–20mm, and limited ASW equipment. The existing torpedo tube was to be sacrificed. This possibility was so attractive that it was further developed in 1944; ultimately plans were made to convert 20 ships for shore bombardment upon the end of the U-boat war. By then, there was so much shore bombardment firepower in the Pacific that the project was no longer necessary. Moreover, conversions to anti-Kamikaze radar pickets were considered more urgent. The PCE was also attractive, although little improvement on its escort battery was to be expected.

Meanwhile pressure for LCI(L) conversions grew, particularly from the Amphibious Force in the Pacific. Experience in the invasion of Sicily also showed the value of fire-support craft, and the performance of British LCGs and LCFs excited comments. In fact the Service Squadron (South Pacific Force) began to produce its own conversions, which were designated LCI(G). These could not be nearly as efficient as the Bureau design, in which the former superstructure was almost entirely replaced by gun mounts; on the other hand, some troop-carrying facilities remained. As of October 1943 the Pacific scheme contemplated one 3in/50, a single 40mm, 4–20mm, and 6–0.5in machine guns. With this, much lighter battery, it would have roughly the same stability characteristics as the Bureau design and it was accepted on that basis. In practice alternative batteries were installed in many cases; examples included: 2–40mm, 3–20mm, 6–0.5in and 10 Mk 7 rocket launchers; 3–40mm, 2–20mm, 5–0.5in and 8 Mk 1 rocket launchers; 2–40mm, 4–20mm, 6–0.5in, 10 Mk 7 and 2 Mk 22 rocket launchers; 3–40mm, 4–20mm and 10 Mk 7 rocket launchers; 1–40mm, 3–20mm and 24 rocket launchers. About 86 were converted through 1944. In addition there were 42 LCI(M), with 3–4.2in mortars and 1–40mm and 4–20mm, as well as 36 LCI with 6–5in rocket launchers, 1–40mm single and 4–20mm.

Then there were the production LCS(L) Mk 3, the Bureau design, of which 130 were built. They had 10 Mk 7 rocket launchers forward but, in the end, no ASW gear. By the time the design had been completed, displacement (fully loaded) had risen to 387 tons and rated speed was dtwn to 14.5kts, with an endurance of 5500nm at 12kts. It is not clear whether, as produced, they had the 1/4in protection over the forward part of their hulls.

FINAL WAR DEVELOPMENTS

Even with the large production run of small gunboats, there were not nearly enough. In June 1945 the Chief of Naval Operations approved a request from the Pacific Fleet for further LCI(G) conversions to increase the existing total of 207 ships (82 LCI(G) plus 125 LCS) to a total of 400; conversions were to be completed at the rate of 20 per month, to provide in each ship 2 twin 40mm (director controlled), 3 twin 20mm, and 2 single 0.5in.

The LCI was relatively small, and it could not accommodate the largest US small-ship gun, the 5in/38. In the late Fall of 1943 the Bureau of Ordnance began to press for a new fire-support craft which could (at about the same time the Bureau of Ships was designing a new and much enlarged LCT which became the LSM – the seagoing Landing Ship, Medium). Ordnance asked Ships to reconsider an LCT(6) conversion, ie the conversion of a ship not much larger than the LCT(5) considered earlier in the year. The result, reported in January 1944, could mount one 5in/38 (enclosed), 2 single 40mm, and 2–20mm. Once more, low speed and poor seakeeping doomed the project. In August Ordnance proposed an LCI(L) conversion, but this was entirely impossible. However, early in September the Bureau of Ships commented that the LSM, now in production, would be an excellent platform; it might accommodate three 5in guns on the tank deck, a modification requiring only the removal of the bow twin 40mm mount and the reloctation of the 4–20mm. Indeed, the handling rooms and supporting structures of the 5in guns could be manufactured and shipped as separate units so that the ships might be converted at forward bases.

In fact by this time rockets were considered as important as guns for close support; they could smother beach defences, whereas the single 5in gun could deal with individual strong points. Thus the support LSM was redesigned as a combination gun and mass rocket ship, with only a single 5in/38aft. The first eight conversions, *LSM(R)188—LSM(R)195,* were fitted with aircraft-type rocket rails (75 four-rail Mk 36 and 30 six-rail Mk 30); the next four, *LSM(R)196—LSM(R)199,* with an automatic hopper type (85 Mk 51). In either case, an extremely heavy barrage could be laid down, although it could not be sustained for very long. These were extremely simple conversions, retaining almost the entire LSM structure. However, the Bureau of Ships continued to develop a more elaborate version, fixing upon 'Scheme F' within a few months. Thirty-six of the latter were built as a production model, with a fully decked-over hull, a new superstructure, with the 5in gun forward of the conning tower, and 20 continuous-loading 5in spin-stabilised rocket launchers in place of the hopper type of the earlier conversions.

The new LSM(R) first appeared in combat at Okinawa in January 1945. At the end of the war it alone remained, in its final (production) version. The ship-borne support craft were soon discarded, as were the converted or redesigned LCI(L). Some of the latter went to foreign navies as gunboats and so survived into the 1970s. In US service they were effective but retention in peacetime would have been expensive. The LCI(L) itself was an accident of wartime history, in that a pure troop-carrier was useful only because of the British commando concept. Otherwise the US idea of moving troops only by large ships carrying beaching craft would have been quite adequate. Without the LCI(L), its LCS(L) cousin was bereft of a source of spare parts, and in addition must have seemed rather expensive. At the same time those concerned with amphibious assault support could point to the existence of a new type of heavily armed amphibious tank, the LVT(A), which could furnish useful support at it moved through the water and then up onto the land.

Thus the future, if any, of close fire-support for amphibious troops would probably lie with rather large rocket- and gun-armed craft, perhaps something like the existing LSM(R).

Another view of *LCS50* on 19 September 1944

USN, courtesy James C Fahey Collection, US Naval Institute

LSM(R)412 in December 1959 shows an identical layout to her sister, 403

USN

USS TEXAS

The forward superstructure and funnels viewed from the starboard side showing the secondary battery of 5in guns. Originally 21 of these guns were carried but 5 were removed during the First World War and during her 1925-27 modernisation the 3 foremost on each side were raised one deck to the position shown here. The other 10 5in guns were removed in 1942.

J Chesneau

This pictorial features part of a collection of close-up and on-board photographs of the USS *Texas* taken by Joananne Chesneau in the summer of 1976. The battleship *Texas* (27,000 tons, 10-14in guns, 21kts) was completed in 1914 and saw service with the Grand Fleet in 1918. She was modernised during 1925-27 having her boilers converted from coal to oil firing, her two funnels reduced to one and her cagemasts replaced by tripods. She was active throughout the Second World War and, following initial employment as a convoy escort, saw extensive action as a bombardment ship. She was present at the Invasions of North Africa, Normandy and the South of France, in the European theatre, and then transferred to the Pacific where she provided support during the assaults on Iwo Jima and Okinawa. She was laid-up in 1946 but in 1948 was presented to the State of Texas for preservation as a floating museum. A special berth was built for her at San Jacinto, Texas, where she still lies. Her present condition is reasonable but the cost of maintenance is probably very high. Some of her wooden deck planking has rotted and repairs have been carried out with concrete which although not elegant will certainly prevent a repetition of the problem.

A quadruple Mk 2 1.1in MG mounting, designed for close-range AA defence and introduced in the 1930s, was not a success and was superseded by the 40mm Bofors during the early stages of the Second World War.

J Chesneau

The starboard crane which, with that to port, serviced the boats carried abreast the funnel and the aircraft catapult originally mounted on Q turret but no longer carried.

J Chesneau

Looking aft from the roof of the secondary battery, port side, Q, X and Y turrets and the short tripod mainmast are visible.

J Chesneau

A 3in/50 AA gun mounted on the roof of the secondary battery. During the Second World War *Texas* carried 10 of these weapons.

J Chesneau

The *Texas* off the US East coast on 9 December 1943.

USN, courtesy A D Baker III

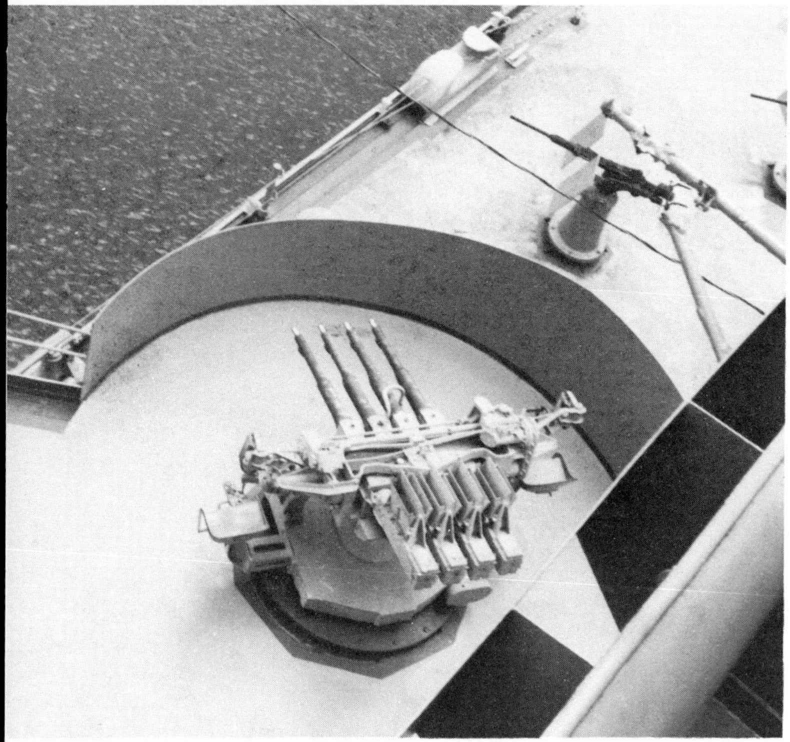

A quadruple 1.1in MG mounting, positioned to port of B turret, seen from the bridge.
J Chesneau

The port crane; it varies in detail from that to starboard having an additional platform (with handrails) around its central post. During the First World War 3in AA guns were mounted at the top of these posts but were removed after the war. The same position was utilised for single 20mn guns during 1942-43 after which it was used for Mk51 directors, for which the tubs shown in these photographs were designed.

J Chesneau

The funnel and mainmast viewed from the upper bridge. Note the second SG radar aerial on the main topmast head.

J Chesneau

The platforms and structure built around the tripod mainmast.

J Chesneau

A 20mm Oerlikon AA gun mounted on the quarterdeck. Several of these, and some 1.1in and other weapons, have been mounted for display purposes and, although weapons of this type were carried by *Texas*, they are not all original equipment and have been positioned where convenient.

J Chesneau

One of two searchlights mounted on the struts of the tripod foremast.

J Chesneau

The Bridge, control top and B turret with the conning tower abaft it. Note the SG navigational radar aerial above the control top.

J Chesneau

A's & A's

GV/BALTIMORE from Keith arton, Greenwich, London
have two points regarding the ›ries on the *King George V* class by obert Dumas. Firstly, *Prince of 'ales* had radar Type 281 air arning not 279; secondly I believe ıe pompom mountings on Y turret f both *King George V* and *Prince f Wales* were 4 barrel not 8. *Varship* 9, page 14).

Regarding the cover picture of *'arship* 11, and the photo on pages 50-51, the heavy cruiser in the ackground is the *Baltimore:* ›ason – two cranes at stern, ıerefore one of the first four *altimore* class ships, and hotographs taken August to)ctober 1944 at Mare Island – at ıis time *Quincy* was in the Atlantic, *'oston* and *Canberra* in the Vestern Pacific while *Baltimore* was nder overhaul on the US West oast so it can only be the latter essel. Information on this point omes from the *Dictionary of merican Naval Fighting Ships*.
)ne point of interest is that the *'altimore* was completed with her K radar on the mainmast and ıese photographs show that it has een moved to the foremast.

Ir Barton is correct in all points ıised, the 4 barrel pompom on 'GV is referred to correctly in Varship 11, page 113. The error ›ith regard to Baltimore was not the ıthor's — thanks are due to the ther readers who pointed out the ıtter mistake. Editor.

MTB *102* arrives at Portsmouth to take part in the 1979 Navy Days

Vosper Thornycroft Ltd

The *Alliance* leaving Southampton for Gosport in September 1979

Vosper Shiprepairers Ltd

'Molch' type midget submarines

A 'Neger' one man torpedo prepared for launch

CPL

VIRIBUS UNITIS

The following is a compilation of additional details and corrections received from various sources regarding the article in *Warship* 6.

1. *Tegetthof* was laid down on 24 September 1901, *Viribus Unitus* was launched on 24 June 1911 anc completed on 5 December 1912, *Prinz Eugen* was completed on 17 July 1914 and *Szent Istvan* on 13 December 1915. The disagreemer between various sources regarding the name ship of the class is apparently due to the fact that, although *Viribus Unitus* was the fi to complete, Emperor Franz Josef ordered them to be called the *Tegetthof* class because he did not wish his personal motto to be shortened into *Viribus* class.

2. The *Szent Istvan* was sunk on 1C June 1918 in position 44° 12.7′N, 14° 27.5′E.

3. The human torpedo employed i the attack on *Viribus Unitis* in 191 was a converted Italian torpedo an not a salved German torpedo. This is, apparently, a very old rumour – the unit expended was designated *S1* and a second human torpedo, *S2*, is still in existence, displayed at the Naval Museum in La Spezia.

4. There is no real historical evidence that *Viribus Unitis* was renamed *Yugoslavia* when taken over by the Yugoslav National council. At the time there was no well established naval authority and, with many other more pressin problems, there was little time for the discussion of a new name. Ther are several rumours regarding names which vary according to national origin, eg *Yugoslavia* for the Yugoslavs, *Frankopan* for the Czechs.

GERMAN MIDGET SUBMARINES

The accompanying photographs of German midget submarines arrive too late for inclusion in the article i *Warship* 14. One, from the author, shows British officers, guided by German naval personnel, inspectin 'Molch' type midgets after the German surrender. The remainder were discovered in the CPL files after '14' had gone to press.

The perspex dome of a 'Neger' being placed in position over the cockpit
CPL

ƽKI RAMP FOR HERMES
n February 1980 MoD(N) ınnounced that the anti-submarine :arrier *Hermes* is to be fitted with a ;ki-ramp to enable her to operate Sea Harriers more efficiently.

The main section of the ramp has ılready been constructed, at Portsmouth Dockyard, and will be itted at the forward end of the light deck during a forthcoming efit. Unlike *Invincible*'s ramp, which is set at 7½°, the *Hermes* ramp s set at 12° as a result of research ınd development work by the RAE ıt Bedford. The shape of the 230 on ramp was precisely defined by he RAE and its surface has been constructed to a higher degree of accuracy than is normal for ship structures. It is 45ft wide and 150ft long, the main 93ft section, which will rise 16ft above the present flight deck, being that part already built while the initial 57ft, which rises only 2ft, will be constructed on board the ship during her refit.

The first planning meeting to discuss the construction of the ramp was held early in December 1978 and, following the preparation of detailed design by the Dockyards Design Division, lines were laid out in the mould loft early in 1979. *Hermes* is due to begin her refit at Portsmouth in the summer, the ramp is scheduled for fitting at the end of the year and she should be back in service in the summer of 1981. In the meantime the ramp is to be broken down into units for painting and storage.

WARSHIPS PRESERVED
The 'A' class submarine *Alliance* completed an extensive refit by Vosper Shiprepairers at Southampton in September 1979 and then sailed to Gosport where she is to be put on display to the

public. She has been given a general exterior overhaul, 20 tons of cast iron ballast has been removed and, to prevent the hollow box section keel being crushed by her own weight when she is set up on concrete cradles, the ballast tanks have been filled with 90 tons of concrete. She is to go on show to the public after extensive interior work has been completed at Gosport.

MTB102, built by Vosper as a private venture in 1937 and then purchased by the Admiralty, was the forerunner of a long line of Vosper MTBs and MGBs. After the war she became a private motor cruiser, then a houseboat but in 1973 she was purchased by the Blofield and Brundall group of Sea Scouts who started to refurbish her. In 1976 Kelso Films completely restored her in return for her use in the film *The Eagle has Landed*. Since then she has been kept up by the Sea Scouts and used for training. Voluntary contributions for her upkeep are being sought by the Sea Scouts.

MISPRINTS

In *Warship* 13, photograph caption 2 on page 41 should be dated 1905 not 1906 and in the table on page 67, endurance should be 3000 miles at 15kts and oil fuel stowage 330 tonnes.

PRINZ EUGEN from E L Bacon, Connecticut

Regarding the article and photographs of the *Prinz Eugen* (*Warship* 9) I have just received word from the US Department of the Navy that the ship has not bee broken up, and remains exactly in the same capsized position. The date of the sinking was 21 December 1946.

Although reported as broken up during 1962-65, this was not carrie out due to fears of contaminating t steel market with radio active steel. Editor

A 'Neger' being recovered after its missior *CPL*

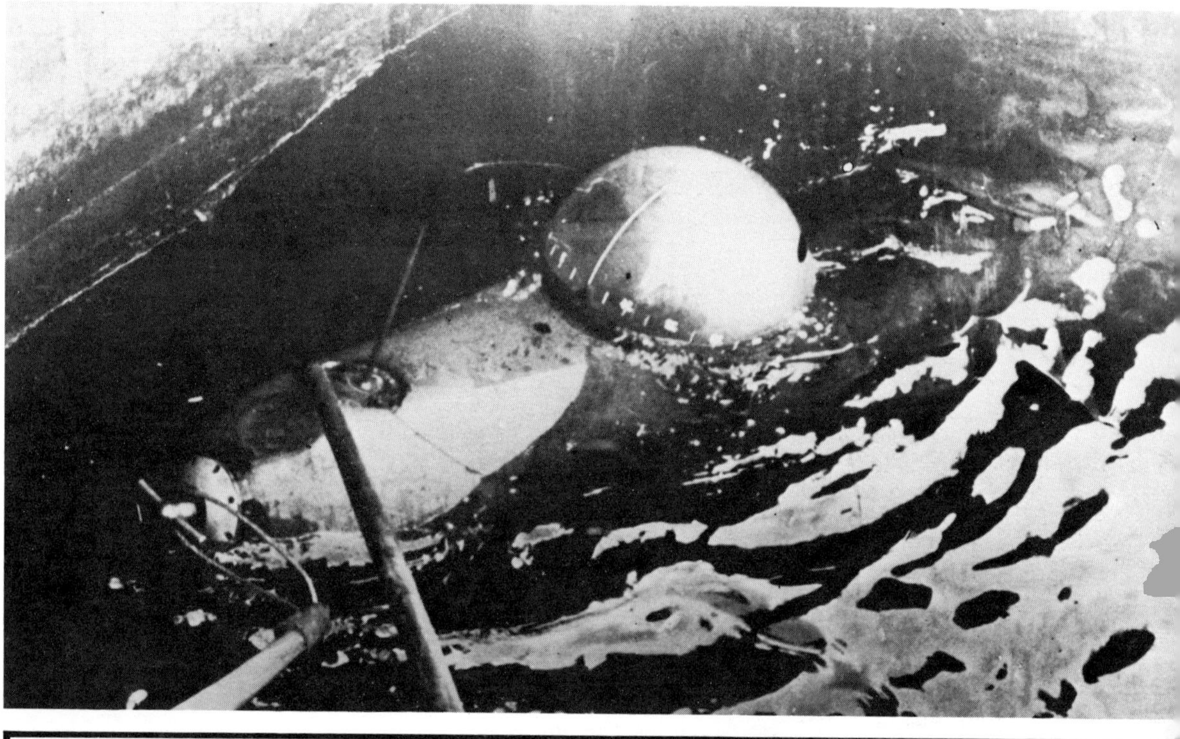

The Great Continental Bookshop For Naval Books

Naval books from Austria, Germany, Great Britain, Italy, Japan, the Netherlands, Spain, USA and other countries available. Moreover, I stock many ship plans, and a large collection of 1:1250 ship models.
Please ask for my catalogues: a) books + plans; b) scale ship models

CHRISTIAN SCHMIDT KG D 8000 MÜNCHEN 70
Sauerbruchstrasse 10 West Germany Tel: 01049-89-703227

Editorial

The amount of research which has gone into the many warship publications which have appeared over the years varies considerably, ranging from what publishers describe as 'deeply researched' to those which are stitched together from previously published information. To many the term 'research' implies some mystic function of the historian, whereas it is actually a comparatively easy task if one has a reasonably knowledge of the subject. The difficulties lie in locating the required information and finding the time to investigate it. Official records, particularly in Britain and the USA, are usually easy to locate although they are often scattered among several centres of research. In Britain for example the key documents to warship design, the Ships Covers and Admiralty drawings, are held by the National Maritime Museum, the majority of general Admiralty documents and papers by the Public Records Office and most declassified official Admiralty photographs by the Imperial war Museum, while much other Admiralty material is to be found both in these places and the various service museums, the Naval Library and so on. In addition, of course, there are the records of private companies (shipbuilders, armaments firms, etc), some of which are held by museums, and personal recollections of those people who were involved in the events being researched – who may have written memoirs or, perhaps, still be alive.

All historical writers have blind spots in their research (apart from the obvious impossibilities of covering all the avenues of investigation) my own being that I have never attempted to locate the personalities involved, although I do regard written memoirs as a prime research source as these often contain much information that would not otherwise be available. In contrast Alan Payne, who described the design history of the J class destroyers in the last issue, has gained a considerable insight into exactly who did what and when by locating and writing to members of the RCNC and the Naval Officers involved in the design of British Warships. Such information as this is not often available in official documents as they are written, for the most part, in detached terms which do not always give a complete picture of the origin of a design or the personalities which brought it into existence.

Many historians rely almost entirely on written documentary records which, although by no means an unworthy approach, can leave many easily available sources untapped. Photographs and official drawings can provide a wealth of information – details of modifications particularly (an area in which the quality of written evidence varies substantially) and, in the case of the latter, details of both internal and external layout and equipment which by any other means would be difficult to locate. For those concerned largely with operational, rather than technical, history illustrative material, is, of course, much less important while the recollections of 'those who were there' is relatively more so. In fact, I have often noticed that many writers of operational histories, while relating brilliantly the events and personalities of a period, have a very weak grasp of the technicalities of the warships themselves. For such histories, however, this is not normally of great importance although it does lead occasionally to misleading statements particularly where the events concerned have strong material aspects, Jutland for example.

Whatever the depth of research, or the lack of it, there is a place for most Warship publications; even the ABC of ships – future enthusiasts have to start somewhere and I have never understood why some books are criticised for being superficial and adding nothing to what has already seen published. Easy to say if you have a large library containing, among others, books long since out of print! One type of publication I do dislike however, is that in which original research has been distorted by the author's lack of ability to interpret correctly the information he has gathered – that after all is the true function of an historian. Such works usually contain sufficient material to give an air of authority which tends to obscure those items out of context and misquoted and adds strength to comments often unsupported by any factual evidence.

WARSHIP PHOTOGRAPHIC SERVICE

Having expounded the advantages of research I have, unfortunately, to inform readers that one section of research material, albeit a small section, will no longer be available. Owing to the inevitable rising costs and a general shortage of customers the *Warship Photographic Service,* from which those *Conway Picture Library* photographs used in this journal could be obtained, will cease to operate after this issue.

John Roberts

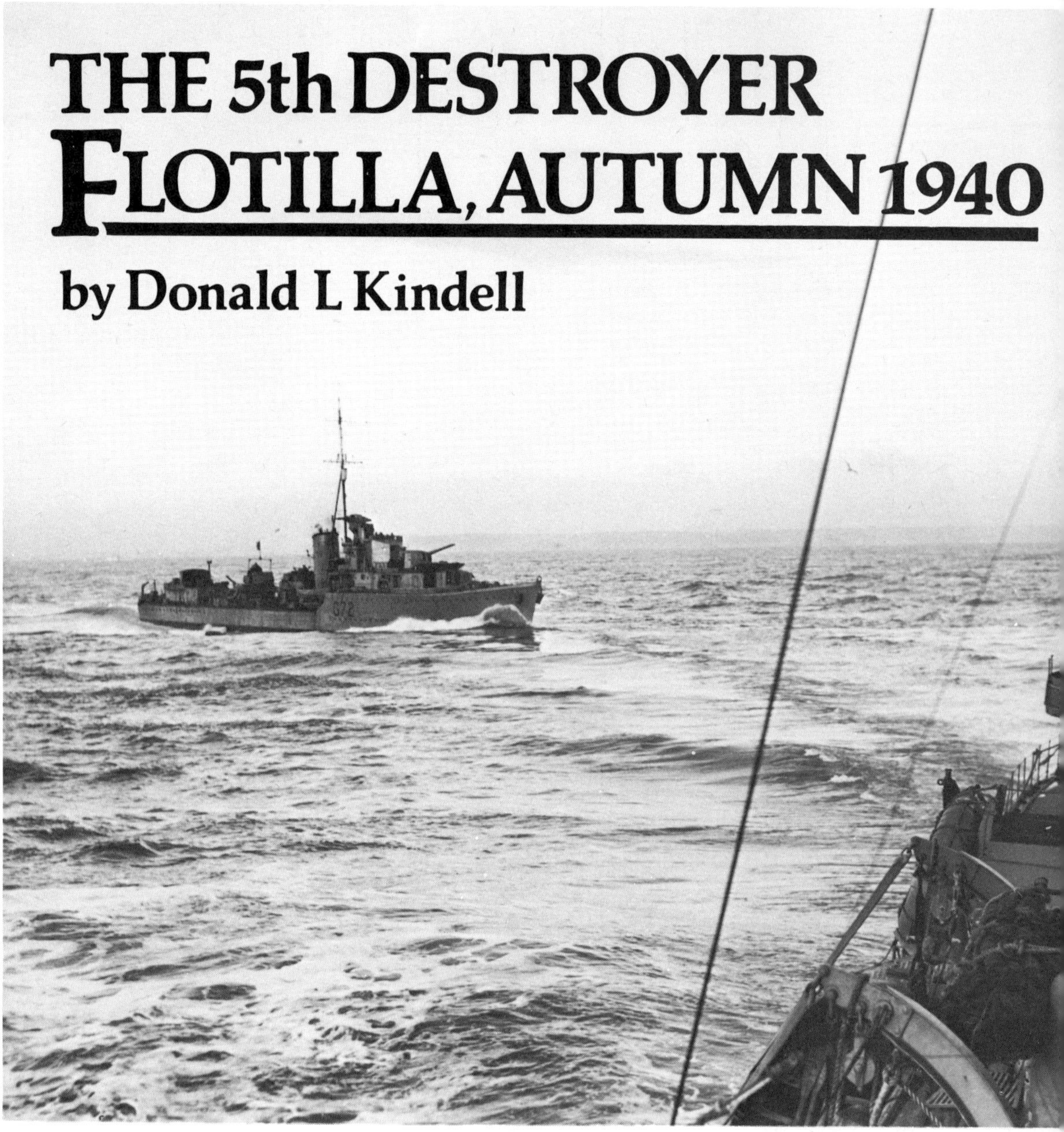

THE 5th DESTROYER FLOTILLA, AUTUMN 1940

by Donald L Kindell

The *Jersey,* seen from the *Kelly,* in January 1940 during gunnery exercises shortly after the latter ship had completed damage repairs *IWM*

Arriving at Harwich, from Scapa Flow, at the end of July 1940 for anti invasion duties, the British 5th Destroyer Flotilla served on unending patrol and escort missions, first at Harwich and Immingham, then later at Plymouth. Composed of destroyers *Javelin, Jupiter, Jackal, Kelvin, Kashmir* and *Kipling,* soon to be joined by *Jaguar* which was completing repairs of aerial bomb damage sustained during the evacuation of Dunkirk on 29 May, the Flotilla was under the comman of Captain Lord Louis Mountbatten.

At Harwich they joined destroyers *Malcolm* and *Ambuscade* (plus *Antelope* which was repairing collision damage received on 13 June 1940) of the 16th Flotilla, *Wivern, Whitshed, Worcester, Verity, Veteran, Wild, Swan* and *Venomous* (plus *Montrose* which was repairing bom damage of 27 July; in addition, on 31 July, *Whitshed* struck a mine off

Harwich and was towed to port by *Wild Swan)* of the 18th Flotilla and the minelaying *Esk, Express, Icarus, Ivanhoe, Intrepid* and *Impulsive* of the 20th Flotilla. In addition the 21st Destroyer Flotilla, consisting of *Campbell, Vivacious, Windsor, Vimy, Vesper, Versatile* and *Venetia* (plus *Vanessa* repairing defects and *Walpole* repairing bomb damage received on 27 July), was nearby at Sheerness.

Only two other units of the J/K class destroyers were in Home Waters; the *Jersey* at Hull, repairing damage she had received when the German destroyer *Erich Giese* torpedoed her off Cromer on 7 December 1939 and *Kelly,* at Hebburn, repairing damage received when torpedoed by the German MTB *S31* in the North Sea on 9 May 1940. *Jersey* completed repairs on 23 September 1940 and departed the Humber on 10 October to rejoin the Flotilla but as she was leaving the Humber Estuary a mine exploded close aboard and she had to proceed to London for further repairs, arriving 13 October. Damage was not severe and she completed in time to arrive at Plymouth on 29 October. *Kelly* left Hebburn for Scapa Flow on 18 December 1940 and, after working-up she departed Scapa Flow on 7 January 1941 to rejoin Mountbatten at Plymouth. Six destroyers of this type were overseas; three Js were serving with the Mediterranean Fleet and three Ks in the Red Sea. Finally, the last unit, *Khartoum,* had already been lost on 23 June 1940 off Perim in the Red Sea.

MINELAYING MISSION

One of the most memorable, albeit disastrous, activities of the 5th Flotilla in the first two months of this assignment was the escort of the 20th Flotilla on a minelaying mission in late August. The *Express, Esk, Icarus, Intrepid* and *Ivanhoe* departed the Humber on 31 August to lay mines off the Dutch coast under the escort of destroyers *Kelvin, Jupiter* and *Vortigern.* When British air reconnaissance sighted a German naval force steering west off the Dutch coast, the minelaying force was ordered to intercept the German ships, thought to be part of the invasion forces. Hastening to intercept, *Express* struck a mine 40 miles northwest of Texel; she was badly damaged and the flotilla commander was fatally wounded. The *Esk* and *Ivanhoe,* going to her aid, also struck mines, the former sinking outright and the latter being badly damaged. Too seriously hurt to be towed back to England, *Ivanhoe* was scuttled by *Kelvin* shortly after midnight. The *Express* was towed back to Hull, but required 13 months to repair. Another casualty of this sortie was the light cruiser *Galatea* of the 2nd Squadron, which, while returning with sister ship *Aurora* (after meeting *Jupiter, Kelvin* and *Vortigern* off Sheringham Light Float) struck a mine at 07.36 on 1 September off Cleaner Shoal Buoy, near the Humber Light Vessel, which exploded on the port side abreast B turret causing light damage. The German force they were to have intercepted was a small German minelaying force en route from Cuxhaven to Rotterdam.

ANTI INVASION PATROLS

British Home Forces issued an invasion warning at 20.07 on 7 September 1940. Increased marshalling of German invasion craft from Den Helder to Le Havre, coupled with an extremely heavy air bombing of London during the afternoon and evening of that day made the invasion of England seem absolutely imminent. RAF heavy bombers began nightly bombing raids on the Channel ports that night and by 21 September, 21 transports, 214 barges, and 5 tugs had been sunk or damaged, some 12 per cent of the entire shipping allocated by the German Command for operation Sealion, the invasion of England.

The Royal Navy was active every night with partrols and sorties into the Channel. Encounters with German ships were rare and the outcomes indecisive but the Navy's presence brought German coastal shipping by night to a virtual standstill. Although usually limited to destroyers and patrol craft, the patrols were reinforced on the night of 8/9 September by the light cruisers *Aurora* and *Galatea* of the 2nd Squadron from Sheerness.

During the night of 13/14 September, the destroyers , *Highlander, Harvester, Bulldog* and *Beagle* conducted a brief bombardment of Cherbourg, while sweeping from Cherbourg to Le Havre, and the *Malcolm, Wild Swan* and *Venemous* bombarded Boulogne, while sweeping from Boulogne to Cap Gris Nez. The destroyers *Campbell, Garth* and

Vesper were to have bombarded Ostend but poor visibility cancelled the mission. While these sweeps were underway, the destroyers *Broke, Witherington, Vansittart, Burza* and *Blyskawica* were sweeping west from Roches Douvres, southwest of Jersey.

Operation Sealion was postponed on 17 September as a result of losses inflicted by the RAF, both to the Luftwaffe and invasion shipping, and the Royal Navy's domination of the Channel. That night, the RAF conducted another heavy bombing mission on the invasion ports in which, at Dunkirk alone, 84 barges were sunk or damaged. On 19 September, the work of dispersing the German landing craft to protect them from further bombing began, but in a manner to allow them to be readily marshalled when the invasion was rescheduled.

GERMAN OPERATIONS

Depleted in the Norwegian campaign, the German Navy was, nevertheless, active during this period. German torpedo-boats had begun escorting minelayers engaged in laying an offensive minefield in the northwest North Sea on 17 July and these operations continued – in the North Sea, Dover Strait and English Channel – as the year progressed. German destroyers began similar operations with the torpedo-boats on 31 August in the English Channel. The operational German destroyers were transferred to Brest in September and, in their first Channel operation, laid mines in Falmouth Bay on the night of 28/29 September. (See Warship 14, p115)

On 8 October, German torpedo-boats *Greif* (flagship of Korvetten Kapitan Wolf Henne, 5th Torpedo-Boat Flotilla), *Seeadler, Kondor, Falke, Wolf* and *Jaguar* departed Cherbourg to raid British shipping off the Isle of Wight. They returned without incident on 9 October.

1 The destroyer *Express,* with her bows blown off by a mine, seen from the forecastle of *Kelvin* on 1 September 1940. The party on *Kelvin's* upperdeck are hoisting in the seaboat.

IWM

2 Lord Louis Mountbatten, commander of the 5th Destroyer Flotilla, on the bridge of the *Kelvin* in September 1940. On his left is Commander J Allison, DSO, CO of the *Kelvin,* and on his right Commander H M Garrett. At this time Mountbatten's own ship, the *Kelly,* was under repair at Hebburn

IWM

3 HM King George VI leaving the *Kelly* during a visit to Plymouth on 20 March 1941. The vessel alongside *Kelly* is one of the ex American 4 stack lend-lease destroyers

IWM

G37

he *Kelvin* in late 1940. Note the AA gun ounted in place of the after torpedo tube ounting – a modification carried out in the ajority of British destroyers during 1940 ollowing the realisation of their great ulnerability to aircraft attack. In this case, as ith all the J and K class ships, the gun is a in but old 12pdr guns were also used xtensively for this purpose

WM

our of the ships of the 5th Flotilla steaming t high speed in line abreast on 28 October 940. The three nearest are *Kashmir, Jaguar* nd *Javelin.* Although not visible in this lustration the original photograph shows the erials of surface warning radar type 286 on he masthead of the *Javelin*

WM

IMS *Kipling* lays a smoke screen while unning at full speed

WM

OMBARDMENT OF HERBOURG

t 20.00, 10 October, the attleship *Revenge* (Captain E R rcher) sailed from Plymouth scorted by destroyers *Javelin* (Cdr F Pugsley), *Jupiter* (Cdr D B Wyburd), *Jackal* (Cdr C L Firth, DSO), *Jaguar* (LCdr J F W Hine), *Kashmir* (Cdr H A King) and *Kipling* (Cdr A St Clair-Ford) of the 5th Flotilla (Mountbatten in *Javelin*) for the bombardment of Cherbourg. Screening forces also put to sea from Plymouth in the form of the cruisers *Newcastle* (Captain E A Aylmer DSC) and *Emerald* (Captain A W S Agar, VC, DSO), destroyers *Broke* (Cdr D B Scurfield) and *Wanderer* (Cdr J H Ruck-Keene) of the 17th Flotilla, and Polish destroyers *Burza* and *Garland.* A flotilla of MLs from Portsmouth acted as an anti E-boat screen for the *Revenge* force before and during the bombardment. The RAF also cooperated by dropping flares over Cherbourg.

The *Newcastle* force, 40 miles off Cherbourg, observed the start of the RAF raid at 02.30, 11 October and 10 minutes later *Newcastle* fired 9 starshell bouquets at one minute intervals, followed by *Emerald* firing 6 starshell bouquets at the same interval. The cruiser force stationed itself to the west of Cherbourg to screen *Revenge* from the Brest based German destroyers.

At 03.32, *Revenge* and the 5th Flotilla opened fire on the harbour works, oil installations and shipping in Cherbourg harbour at a range of 16,000yds and 18 minutes later, the bombardment ended. German shore batteries, surprised by the bombardment, did not open fire until 03.52. At 04.01, when the salvoes began falling close, *Revenge* and her destroyers made smoke and started to snake course away. By 04.30, the British force were out of range of enemy shore guns and at 04.59 the fires caused by the bombardment had passed from view. Although German light naval forces were reported at sea, no contact was made with them.

The *Newcastle* force arrived safely back at Plymouth at 08.00, 11 October while *Revenge* and her destroyers went to Portsmouth. The 5th Flotilla destroyers soon put to sea again on other duties and, on 15/16 October, *Revenge* returned to Plymouth, escorted by destroyers *Wanderer, Witch* and *Burza.*

3

CHANNEL RAIDING

The German 5th Torpedo-Boat Flotilla, less *Jaguar,* departed Cherbourg to raid again off the Isle of Wight on 11 October and, off Portsmouth, *Greif* sighted an English frieghter which was able to escape and make a contact report. Early on the 12th, the German Flotilla engaged a force of Allied light naval forces and sank the Polish manned French submarine chasers *CH6* and *CH7.* They returned to Cherbourg 12 October. On the same day operation Sealion was indefinitely postponed and the landing craft were further dispersed and reassigned.

When loud explosions were heard south of the Lizard at 18.40, 16 October, the destroyers *Broke* and Polish *Burza* were ordered, at 20.03, to investigate. No contact was made with what was determined to be E-boats operating in the area and the two destroyers returned to Plymouth at 09.01 on 17 October.

German destroyers *Hans Lody* (KKpt von Wagenheim; also flagship of Bey), *Karl Galster* (KKpt von Bechstolsheim), *Fredrich Ihn* (KKpt Wachsmuth), and *Erich Steinbrinck* (KKpt Johannesson) departed Brest on 17 October to raid Allied shipping in the western exit of the Bristol Channel while the 5th Torpedo-Boat Flotilla, *Greif* (flagship of Henne), *Seeadler, Kondor, Falke, Wolf* and *Jaguar,* sailed from Cherbourg to act as a support force. The German destroyers were sighted by a British aircraft at 07.19 early on the outward journey in position 48.24N, 05.33W not far from Brest and steering west. At 08.51 the *Newcastle* was ordered to sail from Plymouth with the *Emerald, Jackal, Jupiter, Kashmir, Kipling* and *Kelvin* to intercept the German force; commanding officers were as on the 10th October sortie with Mountbatten in *Jackal,* except that the *Kelvin,* which did not sail on 10 October, was commanded by Cdr J H Allison, DSO.

Three British convoys were in particular peril. At 08.59, convoys OG44, escorted by the sloop *Wellington,* SL50, escorted by the armed merchant cruiser *Pretoria Castle,* and HG45, escorted by the minesweeper *Gleaner,* were ordered to steer west until further notice to take them away from the danger area.

At Plymouth, the British ships raised steam and at 10.21 *Newcastle* weighed anchor and got under way passing the gate at 10.35. *Emerald* weighed anchor at 10.26 , getting underway at 10.40 and passing the gate 6 minutes later. By 11.00, all ships were at sea proceeding at 26 knots; *Newcastle* leading, *Emerald* on station four cables astern and the destroyers in screening disposition. A light breeze from the northwest did nothing to impair the British ships as they proceeded on course 180 to intercept Bey. At 12.30 they altered course to 260 and commenced zig-zagging.

At 16.00, *Newcastle* sighted one destroyer of the *Maas* class, 12 miles distant in position 49.25N, 06.44W, 30 miles southwest of the Scillies. The ships turned to course 200 and at 16.04 *Emerald* had all four of Bey's destroyers in sight. All ships came up to full speed for pursuit and by 16.11, the elderly *Emerald* had worked up to her top speed of 33kts. *Newcastle* opened fire from A and B turrets at 16.07 and 8 minutes later *Emerald* too opened up, firing spasmodically. The German destroyers immediately began to veer towards Brest and started to make smoke. The British force altered course to 160 and *Newcastle* launched her Walrus amphibian at 16.51 for gunnery spotting.

The *Jupiter,* who had been unable to keep up since 16.21 owing to mechanical problems, was ordered at 17.37 to investigate two sailing ships sighted on bearings 213 and $122\frac{1}{2}$.

The German destroyers launched torpedoes at the British ships but at 17.41 the Walrus reported the oncoming torpedoes and *Newcastle* and *Emerald* were able to successfully comb the tracks by swinging to course 196. *Erich Steinbrinck* claimed torpedoing one of the British cruisers, but both excaped harm. None of the British ships indicate in their logs any explosions that could have been mistaken by the Germans for a hit upon one of the British cruisers. The torpedoes, however, were no[t] the only danger. German Do17 aircraft appeared on the scene to shadow the British force and at 18.01 a stick of bombs exploded astern of *Newcastle.**

Due to the lack of adequate air protection, the *Newcastle* force wa[s] advised, at 17.56, to break off the action at their discretion. At 18.0[0] the German destroyers were out o[f] sight in the smoke and haze and b[y] 18.10 the engagement had been broken off. Two minutes later *Newcastle* altered course to 110 as two bombs exploded in the sea ahead of her. *Emerald* reported th[at] from 18.01 to 18.25, course was frequently altered to avoid bombi[ng] attacks. At 18.21 a defect in *Newcastle's* number 3 boiler room reduced the cruiser's speed to 26k[ts;] she altered course to 070 and, at 18.25, turned again to 310.

With the air attacks ended, *Newcastle* stopped at 18.45 to recover her Walrus amphibian, whose pilot had been wounded in an engagement with a Do17. She had worked back up to 12kts by 19.00 when course was altered to 000 to secure the aircraft and, at 19.10, with this accomplished, she again altered course to 310 and increased speed to 16kts with *Emerald* in station astern.

The voyage back to Plymouth was uneventful. At 19.30, the forc[e] began zig-zagging at 20kts and continued until 22.30 when they turned to course 350. The zig-zag was resumed at 23.00 with course again altered to 080, and, at 08.27 18 October, they arrived back at Plymouth. The German force had also arrived at Brest; neither force had inflicted damage upon the other.

The *Jaguar* and *Kelvin* were detached in mid November to escort the cruisers *Manchester* and

* Pierre Hervieux reports that the Germa[n] destroyers mistook these aircraft bombs fo[r] torpedo explosion(s) which, although occurring over 20 minutes after the torped[oes] were launched, would be a reasonable assumption if the torpedoes were fired at extreme range – resulting in a lengthy runn[ing] time.
Editor.

The *Jersey* rounds the Humber lightship during the period in which the 5th Flotilla was operating out of Immingham
IWM

Southampton, which left Scapa Flow on 15 November to carry troops to Alexandria. All four ships arrived without event at Gibraltar on 21 November and took part in Operation Collar and the Battle off Spartivento on 27 November.

Mechanical problems plagued the German destroyers. When, on 24 November, Bey attempted another sortie against British shipping only *Karl Galster, Hans Lody* (now commanded by KKpt Pfeiffer) and *Richard Beitzen* (KKpt Davidson) were operational. These ships operated off Plymouth and sank two allied merchantships of 2156 tons before returning to Brest safely on 25 November.

Three days later on 28 November, the three German destroyers again sortied from Brest to operate off Plymouth. They were sighted en route by British reconnaissance aircraft and the *Javelin* (Mountbatten on board), *Jackal, Jupiter* and *Jersey* (LCdr A F Burnell-Nugent) and *Kashmir* were ordered to sortie from Plymouth and patrol between Lands End and Start Point to intercept. Meanwhile, Bey's force encountered the British tug *Aid* (134 tons) and Barge *BHC 10* (290 tons) 8 miles off Start Point and sank them both early on the 29th.

The *Javelin,* on patrol nearby, sighted the gun flashes of the engagement at 05.40. Closing the area, she sighted the German destroyers, fired a starshell and swung parallel to the German ships to fire torpedoes. She was, however, torpedoed herself in both bow and stern; the bow being blown off to beyond A mounting while half an hour later, her stern also dropped off leaving only 155ft of the ship's original 365½ft overall length.

The *Jackal* reported, at 05.53, that she had sighted the three German destroyers bearing 270 degrees in position 50.37N, 04.00W. She added a minute later that *Javelin* had been torpedoed. At 06.09, she briefly engaged one of the German destroyers with shellfire but did no damage. *Jackal* then returned to *Javelin* and stood by her until dawn when she took off *Javelin's* wounded and carried them to Plymouth.

The *Kashmir,* with *Jersey* and *Jupiter,* had in the meantime reported, at 06.34, that they had lost touch with the German destroyers after a short action. They swept on bearing 240 at 28 knots but were not able to recontact the German force which had escaped behind a smoke screen and later returned unharmed to Brest. *Kashmir* however, radioed at 07.00 that she believed that she had damaged one of the German destroyers.

Later, a tug arrived from Falmouth, passed a wire around *Javelin's* after torpedo tubes and began towing her stern first at 2kts; she was joined en route by a second tug. Enemy air attacks on the damaged destroyer were beaten off by covering Spitfires and three German bombers were shot down. *Javelin* arrived at Devonport on 30 November and remained under repair for 13 months.

The *Kashmir, Jersey* and *Jupiter* returned to Plymouth but soon departed again on the final noteworthy activity of the Flotilla in the autumn of 1940. On 30 November, these three destroyers sailed from Plymouth and arrived at Gibraltar on 2 December. Six days later, together with *Kelvin,* they left Gibraltar escorting the cruiser *Manchester* and arrived back at Plymouth on 11 December, *Manchester* went on to Scapa Flow arriving 13 December.

SELECTED BIBLIOGRAPHY

PUBLISHED SOURCES

Kuhn, Volkmar:	Torpedoboote Und Zerstorer Im Einsatz 1939-1945, Motorbuch Verlag 1977; Stuttgart
March, Edgar J:	British Destroyers, Billings and Son, Ltd., 1966; London
Rohwer, Jurgen:	Chronology of the War at Sea 1939-1945, Arco 1972; New York

OFFICIAL DOCUMENTS

ADM 53/112149	Emerald log
ADM 53/112285	Galatea log
ADM 53/112842	Newcastle log: Cherbourg operation
ADM 53/112849	Newcastle log: Destroyer action
ADM 53/113124	Revenge log
ADM 189/various	Pink List
ADM 199/372	Western Approaches Diary of Events
ADM 199/2063	Damage reports: Whitshed, Esk, Express, etc.
ADM 199/2212	War Diary Note

USS Galveston
The First Talos Guided Missile Cruiser

by Commander T Blades USN (Ret)

A broadside view of *Galveston* on 11 January 1959 showing her appearance as converted. The missile and search radar aerials, from forward, are the SPS10 (surface search) and SPS17 (air search) on the forward lattice, the SPS42 (air search) on the main lattice, the SPS8B (air search) on the platform above the after superstructure and two SPG49 (target tracking) and two SPW 2 (guidance transmitters) on the after superstructure. The SPW2 are the small aerials forward and abaft of their associate SPG 49 aerials

USN

The first Japanese Kamikaze attacks on US ships occurred during the sea battles surrounding the return of US forces to the Philippines in October 1944. In due haste the Bureau of Ordnance stepped up the urgency of a study being carried out by scientists and engineers at the Johns Hopkins University Applied Physics Laboratory (APL) and originally initiated in July of that year. They were looking for shipboard guided-weapons to attack long-range high-altitude bombers to which they could now add the Kamikaze threat. The team, led by Dr Wilburn H Goss, who had earlier been involved in perfecting proximity fuzes and adapting them to Navy AA gun projectiles, recommended the development of a supersonic, ramjet propelled, guided-missile for both jobs. The Bureau gave the go-ahead in December 1944, and so began the programme given the code name 'Bumblebee.'

MISSILE DEVELOPMENT

The concept of the ramjet had been around since the early years of the century but it was known it would not operate efficiently, if at all, unless it could first be boosted to a velocity near the speed of sound. The APL team believed that solid propellant rockets developed during the war could do that. On 15 June 1945 a 6in diameter ramjet was successfully flown over the sand dunes of the New Jersey coast, reaching a velocity of 1200 mph. In October a similar size ramjet engine demonstrated positive thrust over aerodynamic drag. A replica of this 'flying stovepipe', credited to co-designers Drs W H Goss and D H Sloan, is in the custody of the US Smithsonian Institution. It is believed to have been the world's first supersonic ramjet engine.

By October 1945, the Second World War was over but the threat of the Kamikaze had been replaced by that of high-speed jet aircraft while that from high altitude bombers, now carrying nuclear bombs, remained, so the need for shipboard AA guided missiles remained and development

continued. In addition to perfecting a workable propulsion system; efficient flight control, in flight guidance, launching and fire control systems had to be designed and tested; effective warheads and fuzing techniques had to be devised, etc; and the Navy was anxious to get a workable weapon to sea as soon as possible. There was also the parallel problem of what ship or ships would be the first to have AA missiles. The ideal then desired was to fit them in destroyer size ships but it was not known yet what the size and weight of the system would be. In the process of the ongoing 'Bumblebee' programme some options began to appear. The basic guidance scheme for the missile was passive beam riding, where the missile flies slaved to a radar beam into intercept with a target. Around 1948 a test vehicle using a solid rocket motor, rather than the ramjet engine, was developed to prove-in the guidance idea. The vehicle proved so successful that it was recognized that it could be the basis of an interim, shorter range, missile that could be introduced into the fleet fairly quickly. Therein lay the kernel of the idea that missiles would be fitted first in existing ships, converted for the purpose rather than waiting to introduce them in new construction. The 'Bumblebee' test vehicle became the prototype of the Terrier missile and, in 1952, the Navy ordered the conversion of the *Baltimore* class cruisers *Boston* (CA 69) and *Canberra* (CA 70) to carry them. The modification involved the replacement of their after 8in gun turret and aftermost 5in gun mount with two twin Terrier launchers, and the addition of two prototype missile fire-control systems. Recommissioned as CAG 1 and CAG 2 in November 1955 and June 1956 respectively, they were the world's first guided missile cruisers.[1]

The decision to develop Terrier took some of the time-critical pressure off the work on the ramjet missile system, now named Talos.

The programme was realigned to aim at providing the longest range, most advanced missile and fire control system that current technology could then make possible. To attain the desired long range, with an adequate warhead, using an airbreathing vehicle the missile was going to have to be a large one (the operational missile is about 21ft long with a 30in diameter) so a large ship was called for. There were discussions in 1952 about converting the *Alaska* class battlecruiser *Hawaii* (CB 3), which had never been commissioned, into the first Talos ship but the idea was abandoned in 1954. Soon thereafter it was decided to make use of another ship that had been completed but never put into service, the USS *Galveston* (CL 93). Her conversion to the world's prototype Talos guided missile cruiser was authorised by Congress in the Fiscal Year 1956 programme.

THE GALVESTON

CL 93 was one of the *Cleveland* class of 6in gun light cruisers, the largest class of cruisers ever built for the US Navy. Early in 1939 the US Navy General Board approved the design of two new 6in gun cruisers, CL 55 and CL 56, of 8000 tons, the maximum displacement allowed under the 1936 London Naval Treaty, but in September, after the outbreak of the Second World War and the announcement by Britain of her suspension of all treaty restrictions, the US followed suit and the CNO ordered the reconsideration of the design of those two ships, scheduled for inclusion in the 1940 building programme. Events moved rapidly, and by the time of the fall of France in 1940, two huge expansion bills had been passed by Congress aiming at a US 'two ocean' navy. As part of this expansion, the *Cleveland* (CL 55) class became a 10,000 ton standard displacement ship, armed with 12-6in/47 low angle guns (4 x 3), and 12–5in/38 dual-purpose guns (6 x 2). Most also mounted 28–40mm Bofors (4 x 4, 6 x 2) and 10–20mm Oerlikons (10 x 1), and had an actual full load displacement of 13,870 tons. Thirty two were authorised before the US entered the war and 5 more later; 29 were built and the remainder cancelled on 13 August 1945.

Galveston was completed by her builders, W Cramp Shipbuilding Company of Philadelphia, Pennsylvania, on 26 May 1946, but a month later it was decided to put her in mothballs rather than fit-out and commission her. The ship was moved from her berth among Reserve Fleet vessels at the Philadelphia Naval Shipyard to the production facilities there on 15 August 1956 to begin conversion. The initial work involved the removal of the two 6in turrets and three 5in mounts, aft, along with their associated fire control systems, as well as the removal of the 40mm and 20mm systems altogether. In addition, most of the superstructure abaft the second funnel was removed to make room for a redesigned one. The removal of the two turrets, with their barbettes and magazines, did not provide sufficient space below the main deck for the missile magazines, so these had to be constructed as the first level of the new after superstructure. This consisted of three tandem compartments, each separated by watertight and flametight bulkheads. From forward to aft they were: a storage magazine with capacity for 30 separated missiles and boosters or combinations thereof, a ready-use magazine holding 16 mated missiles

1. Although a Terrier system was installed in a *Gearing* class destroyer, the *Gyatt* (DD 712, redesignated DDG 1), it proved impractical because Terrier, having a separate solid rocket booster as well as its solid rocket sustainer motor, took up too much space. The US Navy would have to await the development of a dual-thrust rocket motor in the Tartar missile–another 'Bumblebee' offspring–before it would get guided missiles to sea in a destroyer size ship. That would be in the new construction *Charles Adams* (DDG 2) class.

1 The *Galveston* running trials on 16 June 1962 at the end of her first major overhaul at the Philadelphia Navy Yard. A new main lattice mast had been fitted and the aerials of SPS43 and SPS30 have replaced the SPS17 and SPS8B respectively

USN, courtesy A D Baker III

2 The 6in guns of *Galveston* bombarding the coast of South Vietnam on 7 August 1965

USN

and boosters, and a wing and fin assembly area. Within these spaces were contained almost all the machinery – automatic, semi-automatic, and manual–for receiving, handling, and testing Talos missiles, and for automatically loading them onto the twin rail launcher on the main deck aft of the assembly area. With one exception, these compartments totally housed the ship's supply of missile ammunition within their boundaries. The exception was a separate warhead magazine below the main deck. Talos was designed to be either a conventional or a nuclear weapon and when *Galveston* first went to sea she actually carried the two separate versions. The below deck magazine permitted the stowage of a certain number of spare warheads of either kind.

Considering that the equipments involved were all newly conceived and designed, it is remarkable that most of them worked well. Throughout the life of the ship, and her sisters, the only missile equipments that gave a lot of trouble and had to be modified many times were those for receiving and striking down missiles and spare warheads while underway.

FIRE CONTROL

Turning to the subject of fire control, it will be necessary first to briefly describe the design of the overall Talos system. As said earlier, the basic guidance concept was radar frequency beam-riding. In the shorter range rocket-propelled missiles, such as the original Terrier, the missile simply rode the narrow beam of an automatic target tracking radar by means of coding and attitude sensing schemes. In order to achieve maximum range, without the missile becoming gigantic, the Talos airbreathing ramjet needed high fuel efficiency–low fuel consumption per pound of thrust. That required that the missile fly most of its trajectory enroute to a target at high altitude. Therefore it would have to fly the electronic beam of a separate guidance transmitter, rather than the low slant-angle beam of a radar locked onto, and tracking, a long-range air-target. The coded beam of the computer controlled guidance transmitter carried the missile rapidly to its optimum altitude, leveled it off and, in the terminal phase, the beam, with the missile still following it, was programmed to come into collimation with the track radar beam. To aid actual target intercept, for the last few seconds of flight a receiver in the missile began scanning to try to pick up radar pulses reflected from the target and if so, through the principle of interferometry, use this intelligence to steer itself into a collision course. If it could not pick up adequate signals, it continued to beam ride. Warhead detonation was by proximity fuze.

Two complete Talos missile fire control systems were installed in the *Galveston* class. The after superstructure above the missile deck house was designed essentially to optimally position two AN/SPG 49 monopulse tracking radars and two AN/SPW 2 guidance transmitters. This superstructure also housed the SPW 2 control room and the Weapons Control Station. Several decks below the missile house were the SPG 49 control rooms and the MK III missile fire-control computer room, which equipment controlled the entire system. The ship's Combat Information Center (CIC), unfortunately, was separate and far forward well below the main deck; this arrangement was too complicated and hindered the ship's operations throughout her life.

SEARCH RADAR

Many other, then current, systems were installed during the conversion to make her a versatile warship but I will detail only her search radar suite as particularly pertinent to her prime mission as a guided-missile cruiser. For long-range early-warning and detection she was fitted with the two-dimenstional AN/SPS 17 radar and, for designating targets to the Talos fire control radars, as well as for air control and general precise air-search, the three-dimensional AN/SPS 42. For best performance it was necessary that the aerials of these long range radars be mounted as high above the sea surface as practicable. As a result, the two aerials were mounted atop new lattice masts, just forward of the first funnel and just abaft the second respectively, the height of which almost totally dominated the ship's silhouette which was originally comparatively low. For stability, this arrangement also required the addition of much ballast along the ship's keel.[2] Lastly, as regards radars, the foremast also mounted an AN/SPS 10 aerial for surface search and navigation, while an additional medium-range three-dimensional air-search set, the AN/SPS 8B, was fitted with its aerial on a platform atop the after superstructure forward and above the Talos equipments.

Finally, the whole gunnery fit forward was retained, less the 40mm and 20mm systems noted earlier. The conversion of her two sisters in the class, *Little Rock* (CLG 4) and *Oklahoma City* (CLG 5), also former *Clevelands,* was authorised in the Fiscal Year 1957 programme, but these were made fleet flagships before recommissioning and retained only one 6in turret and one 5in twin mount forward.

TRIALS

Upon completion of the conversion *Galveston* had increased in full load displacement to 15,142 tons at 25ft draught and had a new complement of 125 officers and 1270 enlisted men. She was commissioned at the Philadelphia Naval Shipyard on 28 May 1958 and was sponsored for the ceremony by Mrs Clark Wallace Thompson, wife of a Texas Congressman of the ship's namesake city, who had also been its sponsor when it had been launched on 22 April 1945; her first commanding officer was Captain J

---

2. From some miscalculation, the original distribution of this ballast caused *Galveston* to be down by the bow slightly at full load and to hog uncomfortably in a seaway. It was redistributed properly during an overhaul in 1965.

An aerial view of *Galveston* on 16 June 1962
USN, courtesy of A D Baker III

3 Colwell, USN. The first task at hand, after shaking down as any new warship, was to prove-in the Talos shipboard system and, by 24 February 1959, a target drone was downed in the first successful Talos firing at sea. All-in-all, however, the first three years of *Galveston*'s operations were to prove very sobering. It was an agonizing period of trying to make the system consistently functioning and reliable and many changes had to be made. When not at sea, she returned most often to the Philadelphia yard where modification work could better be performed rather than to her Norfolk, Virginia homeport. Thus, the families of most officers and men lived in the vicinity of Philadelphia.

The experience gained was invaluable. Many, many missiles were fired in test, including an extreme range target hit, which record stood for a number of years, and a successful two missile salvo. Many design changes were worked out and tested; those that could be incorporated immediately were, more extensive ones had to await the first overhaul. Perhaps the most significant contribution of those first years was the nurturing of a cadre of devoted Talos sailors, both officers and enlisted. Although they did not all serve in ship's company, those that did not had sailed in her as technical observers during one or another of her tests or were aboard as nucleii missile complements for follow-on ships; all were grounded in CLG 3's experiences. Many of them would have a profound influence upon the entire field of US Navy surface ship missilery in the years ahead.

SYSTEM MODIFICATIONS

By 1961, when it was time for *Galveston*'s first overhaul, the other Talos ships were at sea or were being readied. CLGs 4 and 5, already mentioned, had been recommissioned. In addition the Fiscal Year 1957 programme provided for a new construction nuclear powered cruiser, that was to be armed with Talos and Terrier, which ultimately became the USS *Long Beach* (CGN 9) and, the Fiscal Year 1958 programme, authorised the conversion of 3 *Baltimore* class heavy cruisers to Talos and Tartar armed missile cruisers, to be designated CGs. These 4 ships were in the shipyards when the time came to put in some of the major system redesigns that had been developed during the previous 3 years. Some of these were not to go into the *Galveston* class, only into CGN 9 and CGs 10, 11 and 12, but all were derived from 'lessons learned' in *Galveston*. In that category were a completely redesigned fire-control computer, the still electronic analog rather than digital MK III Mod 1, a more compact and efficient missile launching system and more sophisticated digital computerised Weapons Direction systems to work in conjunction with the Navy Tactical Data System in the centralised CICs.

Within the Talos weapons system itself the major changes were also applicable to the *Galveston* class, and these were installed during her overhaul of 20 August 1961 – 23 July 1962. The most significant change was a major overhaul of the tracking radar and guidance transmitter designs. *Galveston*'s experience with mechanical and electronic alignment of these equipments, ie ensuring that the two came into near-perfect collimation on a target at the extreme Talos ranges, was a prime mover in that modification. Among other things the change incorporated the addition of a continuous wave (CW) transmitter to the tracking radar for the terminal phase of the missile intercept. This provided a much stronger signal for the missile to home on during its semi-active interferometry period, and eased the stringent collimation problem. Also, there was now a new single missile design, with interchangeable high explosive or nuclear warheads and integrated CW homing.

'TALOS GET WELL PROGRAMME'

When the overhaul was nearing completion conditions were such that, although the Bureau of Weapons would have preferred to test these, and other design changes, on one of its more modern Talos ships, it was not possible. The completion of *Long Beach* had been delayed by a labour strike in the building yard, conversion of the CGs had not progressed far enough and the Navy did not wish to give up the use of the other Talos CLGs as fleet flagships for any extended period. Although, as it turned out, *Long Beach* was completed and at sea before *Galveston* finished her overhaul, she had not yet shaken down and, as the world's first nuclear powered surface warship,

The *Galveston* on 24 August 1964 with the aerial of SPS39A on her main lattice in plac of the SPS42

USN, courtesy A D Baker III

the Navy had other plans for that ship too. Thus, just as *Galveston* had been the pioneer test ship for Talos, she was to pioneer the prove-in of its significant redesign.

After completion of the overhaul, and a brief operability checkout and transit to the US west coast to join the Pacific Fleet, the ship began an extensive test programme, under the auspices of the Bureau of Weapons and the US Navy's Operational Test and Evaluation Force, that was not completed until the end of June the following year. This programme was not only the prove-in of the new missile and the SPG 49 and SPW 2 changes, but also of many other proposed modifications and corrections. In the aggregate, these constituted what came to be known unofficially as the 'Talos get well programme'. Owing to the timing of this testing, while the CGs were still being converted and the other two CLGs were to start overhaul, most of the changes that were successful were able to be worked into all Talos ships quickly.

One of the most significant improvements was the introduction of a digital range-tracking transmitter between the track radar and the system computer. This replacement of electronic analog with coded bit signals vastly improved reliability and track smoothing, especially needed with the airbreathing missile to conserve fuel; it resulted in consistency in achieving Talos extreme range hits for the first time. Another important result of the programme was the introduction and perfecting of a new scheme of system diagnostic and corrective maintenance. The key tool was the Daily System Operability Test (DSOT) which greatly increased confidence in the system's readiness without the need to expend expensive missiles. Talos system designers and operators pioneered this concept. Test firings confirmed that the DSOT was indeed an accurate gauge of reliability. The scheme has proved its worth throughout the years since, not onl with Talos but with all the other surface ship missile systems. Durin this time it was demonstrated that Talos AA missiles also had an antiship potential, with devastating results on a target vessel, even without a warhead. Only minor modification to the missile was required but it was a potential whic was not further developed. The USN had only recently (around 1958) abandoned cruise missiles or its surface ships and submarines by retiring the Regulus I and II, either installed or planned, and in 1963 there were still 4 years to go before the Israeli destroyer *Elath* would be sunk by Styx antiship missiles fired by Egyptian missile boats.

OPERATIONAL SERVICE

The test programme just described ended in June 1963 and in the following month, after 5 years as essentially a test ship, *Galveston* became a regular fleet unit. As a key contributor to the overall Talos programme, the ship went into eclipse. The attention of the material establishment understandably shifted to *Long Beach,* and to *Albany, Chicago,* and *Columbus* (CGs 10, 11 & 12 respectively). There was a larger investment there and, more important, all those ships had better long-range search radars and centralised command and control systems. While it would still be years before efficient, minimum-time, target handling from these types of systems, and their offspring fitted in even newer ships, was to be achieved, this justifiably became the major technological focus. The decision not to modernise *Galveston*'s search radar/command and control suites doomed her to slip into backwaters.[3] She received only one more assignment as a test ship: in the fall of 1966 she demonstrated that a Talos anti-radiation version missile could be successfully used against land radar sites at maximum missile range without having to equip the firing ship with highly accurate navigation devices.

During the remainder of her years in commission, however, she distinguished herself in other ways as a fleet unit. She served in five deployments, three in the western Pacific and two in the Mediterranean. It is ironic that as a deployed warship, in actual combat operations, her guns were used extensively and her missiles not at all. Throughout the late summer and autumn of 1965, and again in early 1969, *Galveston* fired thousands of rounds of 6in and 5in ammunition in support of ground forces in Vietnam. That became a general rule with all the older missile cruisers, the CAGs and CLGs; in the latter case, particularly those that had not lost most of their front-end gun systems while being fitted as fleet flagships. This was, of course, because the USN had so few heavy-gun ships left in commission by the late sixties and early seventies.[4] In fact, instead of taking her regular turn in the western Pacific in the winter and summer of 1967, *Galveston* exchanged deployments with the Atlantic Fleet's *Boston* (CAG 1), and thus was supporting US interests in the Mediterranean during the 6 Day War in the Middle East

A similar swapping of deployments brought her back to that ocean during the spring to autumn of 1969, for the Navy naturally preferred 8in guns to 6in guns so long as there were real gunfire support missions to be fulfilled. As a result, during *Galveston*'s relatively short active service, she passed through the Panama Canal four more times.

It was indeed a short period of active service for a warship of any kind. *Galveston* went out of commission on 25 May 1970, just 3 days short of 12 years since her active life had begun. She had performed the role of being the lead Talos ship admirably, and that system itself remains active in the US fleet better than a decade later. However, like the prototype, the old converted cruisers became more expensive to maintain and to man. *Little Rock* and *Columbus* have also since been decommissioned; *Oklahoma City, Albany* and *Chicago* are to follow during Fiscal Year 1980. Thus, 10 years after *Galveston* was retired only the new construction, nuclear powered, *Long Beach* will remain of the seven Talos cruisers.

And what of Bumblebee? It lives on in the form of the Standard missile, medium and extended range versions.

3. The ship's search radars were exchanged for newer equipments during overhauls in 1961/62 and in 1965. The AN/SPS 43 replaced the SPS 17, the AN/SPS 39A replaced the SPS 42 and the AN/SPS 30 was substituted for the SPS 8B but nothing was done to modernise command and control systems.

4. The more modern Talos cruisers, CGN and CGs, with only minimal self defence guns fitted, were useful during the Vietnam conflict mainly because their high-capacity aircraft tracking, and their command and control systems were well suited for marshalling sorties from and to American aircraft carriers and early warning of enemy aircraft activity. These ships, however, did occasionally fire Talos missiles in combat. *Long Beach* downed MIG fighters over North Vietnam in May and September 1968, each at ranges greater than 100km, and *Chicago* shot down another in May 1972 at about 75km.

NOTES ON SOURCE MATERIAL

The information on the early history of the Bumblebee programme is from *Countdown 60,* the first of USS *Galveston*'s self-published cruise books, and from personal interviews with a few participants. Sources concerning dates and decisions with regard to the conversions of guided missile cruisers are from various entries in *The Illustrated Encyclopedia of 20th Century Weapons and Warfare,* Ed Bernard Fitzsimons (Purnell & Sons Ltd, 1967, 1969; Phoebus Publishing Company/BPC Publishing Ltd, 1971/77/78, distributed by Columbia House, New York), and from various editions of *Jane's Fighting Ships.* The vital statistics on USS *Galveston* are also from *Jane's* and from John S Rowe and Samuel L Morison, *Ships and Aircraft of the US Fleet,* 9th edition (Annapolis: Naval Institute Press, 1965, 1972). The reason more primary sources were not used is that they are not much available. Most of the record of deliberations and decisions of the US Navy Ship Characteristics Board, which began in 1945 to replace the long standing General Board as the deciding body, was allowed to be destroyed in 1975 by the National Records Center (a branch of the US National Archives).

The information on the *Cleveland* class is from the official Navy design history of that class (National Records Center, accession no 344-74-564, box 2); the details of CLG 3's conversion and subsequent history has been derived from many sources including *Jane's* the *Dictionary of American Naval Fighting Ships,* Vol III (Navy Dept, Office of the CNO, Naval History Division, 1968), ship-published cruise books, *Thunder in the West, WESTPAC 1965, Med, 1967,* and *USS Galveston (CLG 3) 68-69,* and from personal records and experiences of various people directly involved with the ship, including my own. For example, the author was the Bureau of Weapons project officer during the 'Talos get well program,' and was the ship's Weapons Officer from 1965 to 1967. Another source used was the data held about the ship by the US Naval History Center's Navy Department Library and its Operational Archives. The details on how the Talos system functions are readily available in many unclassified publications and documents; the information on its combat hits is from *Jane's Weapons Systems, 1977,* p 115.

AMPHIBIOUS FIRE SUPPORT

by Norman Friedman

Post War Development

In 1945 the US Navy posessed an immense force of specialised amphibious support-ships, both improvised – LCI(G), LCI(M), the early LSM(R) – and more-or-less specially designed – the improved LSM(R), LCS(L). The small, davit-launched, LCS (S) were quickly discarded but the improved LSM(R) and the small LCS(L), derived from the LCI(L) infantry landing ship hull, were retained for the postwar fleet. Probably one of the greatest surprises of postwar US naval development was the near total failure to develop a successful replacement for either class, despite several attempts to do so. Thus the only new amphibious fire-support ship, the *Carronade,* was designed for mass production but never repeated. An ambitious programme for a new class of fire-support ships, the LFS, began in 1965 but produced no final design, let alone a new ship. Nor were a series of programmes for new fire-support weapons entirely successful. Thus, for example, naval short-range fire support in Vietnam was still the province of the LSM(R)'s, which had been adjudged unsatisfactory even well before the Korean War.

WAR EXPERIENCE

Late in 1945 the Pacific Fleet convened a series of Boards to collect the materiel lessons of the war; the Comstock Board examined issues of ship and aircraft Characteristics. It suggested that both the LSM(R) – in its advanced form – and the LCS(L) be retained in the postwar fleet as a basis for further development. Neither was considered entirely satisfactory, however. The LSM(R) had inadequate living quarters, a limited fresh water supply and suffered from considerable interference between her single 5in/38 gun and her four 4.2in mortars. The Board suggested the removal of 2 of the 10 rocket launchers (Nos 1 and 10) and the substitution of a 3in/50 for the 5in/38; elimination of No 1 rocket projector would provide additional living space. As for the LCS(L), it had insufficient rocket fire but 'the present size of the LCS(L) does not well lend itself to additional armament features. However, it is considered worthy of consideration to provide a slightly larger unit fitted with the (trainable) Mark 1, Mod 0, 4.2in mortar. The automatic weapons armament should be retained in its present status, since it provides excellent

1 The 'ultimate' LSM(R)s survived long enough to fight in Vietnam. Here USS *St Francis River,* then designated *LFR-525,* steams to pick up ammunition at Cam Ranh Bay, July 1969. She is little changed from her wartime appearance, a notable exception being her semi-enclosed bridge. All the LSM(R)s were named on 1 October 1965.

USN

2 Rocket firing by *LSM(R)536, White River,* off Vietnam in July 1966. The launcher in the foreground is in the vertical (reload) position, aligned with ammunition hoists below it, while the other two twin mounts are elevated for firing. This mark 102 rocket launcher, based on a twin 40mm power-operated gun mount, was controlled by a Mark 51 director on the bridge of the LSM(R). It was credited with a rate of fire of 30 rounds per minute, firing 6in spin-stabilised rockets.

USN

1

Two photographs of the *Carronade* the ultimate US fire-support ship, taken nearly a decade apart. Although a great deal was hoped for, she never really received much more than a standard LSM(R) weapon and sensor fit; for example, a sophisticated mortar-locating radar planned for her was never fitted. The greatest single difference between these two views appears to be the relocation of the rangefinder to a position above and abaft the Mk 52 director with its Mk 26 radar.

USN

features for patrol and picket missions. Combined with the LSM(R) (2), the LCS(L) can effectively perform all landing craft fire-support missions and should therefore be retained.' Admiral Nimitz's staff added that, 'the LCS(L) has an important place in amphibious operations, in that it is small and of light draught, and can take its relatively large armament in close to the beach to deliver accurate close-in fire. We must guard against increasing the size of the craft and thus jeopardising its characteristics of being able to get in close to the beach.'

Both types had one very great flaw – speed. In 1947 an Amphibious Type Conference recommended that future attack transports and LSDs be designed for sustained speeds of 20kts, and indeed the 20kt amphibious force became an important US naval objective, albeit one not to be realised until the late 'sixties. A variety of new classes were designed and attempts were even made to develop a 20kt LST which, in 1947, appeared to require screws forward and a ramp at the stern, the LST turning and running aft onto a beach. The LCS(L) had always been envisaged as having a secondary escort capability, at least for anti-aircraft fire, and in any case a very slow support craft would negate the goal of the 20kt amphibious convoy. Thus in 1948 the search began for replacements for the two wartime designs.

One 20kt hull available in considerable numbers was the DE. Even as early as 1943 the Bureau of Ships had studied a specialised shore-bombardment version, with three 5in/38; indeed, at one point Admiral King had authorised the conversion of 20 units, upon the end of the war in the Atlantic. In March, 1948 the Bureau briefly studied a DE with LSM(R) armament: one 5in/38, 10 Mk 105 rocket launchers, 4 stabilised 4.2in mortars, 2 twin (or one quad) 40mm, 4 to 6 twin 20mm. However, neither deck space nor stability would support this battery. Even with 4 rocket launchers and 2 mortars the ship would gain 90 tons, which was 25 over the limit set by stability. The project had to be abandoned in favour of a specialised design.

REQUIREMENTS

Thus the Ship Characteristics Board allocated two Project numbers, 36 and 37, to the fire-support programme. SCB 36 was described as a Fire-Support Ship Close-In, and SCB 37 as a Rocket-Support Ship Close-In. In fact the issue of rocket fire predominated, no doubt because at this time there were many guns in the Fleet which would be available to support an amphibious operation; the two requirements were combined as SCB 37, which became the *Carronade*. In view of the failure of the DE conversion study, BuShips was asked, as a matter of some urgency, to develop a new design for construction under the FY 50 programme. Requirements, as understood in April 1948, included sufficient speed to accompany a 20kt assault convoy, as well as size small enough 'to operate in shallow water and since such operation may result in accidental grounding, that it be so compartmented and of such strength that such grounding will not render it entirely unfit for further operations.' The Marines wanted it to be able to deliver stabilised rocket fire at 300 rounds per minute, as well as stabilised mortar and 5in fire, 'the latter for destructive effect against pillboxes enfilading the beach. All these fires should be delivered through as wide an unobstructed arc as possible which indicates that superstructures should be well aft in the ship. (One 5in gun should be sufficient)'. SCB 37 would also incorporate AA weapons for self defence. In effect she would be a redesigned, faster equivalent of the LSM(R).

SCB 37 was not in fact authorised until FY 52, despite the urgency of the internal memoranda of 1948. Preliminary Characteristics of February 1951 called for operation in water as shallow as 10ft; rocket fire was the primary function of the ship. The battery was to follow that of the existing LSM(R): one 5in/38, 10 Mk 105 rocket launchers, 2 twin 40mm. One unusual feature was to be a specialised mortar locating radar, to make counterbattery fire more effective. Mass production was contemplated. Since the new ship would be, at best, very specialised, she would have to be very inexpensive. For example, although the President of the Board of Inspection and Survey thought that her rockets would prove effective against junks such as were being encountered in Korea, this was not true; it was argued by Op-343 (March, 1951) that 'the limited range (10,000yds maximum) and relatively poor accuracy (plus or minus 500yds at medium and low elevations) imposes severe restrictions on the employment of rocket launchers and projectiles. There has been considerable correspondence and experiment to determine the most useful weapons and method of fire against junks, none of which have included use of rockets . . .'

Design proved straightforward, but BuShips proposed several modifications to the original Characteristics. For example, the Mk 105 rocket launcher could fire 48 rounds per minute, so that 8 would more than meet the original Marine requirements; indeed, to provide two more would cost an additional 18ft of hull length. To reduce damage from inadvertent grounding, the ship was to be provided with a double bottom over about 80% of her length. Moreover, diesel fuel tanks would be fitted for ballasting, so that in the event of grounding water might be pumped over the side to reduce her draught. It appeared that if a speed of 15kts were acceptable the ship might be built on a waterline length of 230ft and a displacement of 1340 tons fully loaded, or 1080 tons light; overall length would be 238ft. By way of comparison, a '20kt' design would be 273ft long (281ft overall) and would displace 1595 tons fully loaded. However its speed would be limited by a requirement that existing diesels be used. By increasing power from the 3100bhp of the original design to 6000bhp the '20kt' ship would increase its trial speed from only 15 to 18.6kts, and its sustained speed from 14 to 17.5, which results must have been disappointing. An alternative '13.5kt' study (1570bhp, 13.2kts on trial, 12.2 sustained) came to 1300 tons fully loaded, and 223ft on the waterline (231 overall). By way of contrast, the existing LSM(R) was 205ft long (waterline; 206ft overall), displaced 1220 tons fully loaded (790 in light condition) and might make 13kts on 2880bhp. One peculiarity of these designs was that since 'no protection for the magazines is required . . . they are unprotected except by the shell of the ship which is only about $\frac{1}{4}$in thick. The rocket magazines are located above the waterline.' Later Characteristics filled this gap, requiring 30lb ($\frac{3}{4}$in) STS around the rocket magazines, extending down to the waterline. In addition, the

BuShips suggestion that the number of rocket launchers be reduced to eight was accepted. As for the speed, 20kts proved far too expensive. In any case, in the interest of economy, the new fire-support ship was to use existing (ie Second World War) engines in stock in large numbers.

THE CARRONADE

The final design showed some growth, to a length of 237ft (waterline; 245ft overall) and a displacement of 1475 tons fully loaded (1200 tons light); it was expected that 3100bhp would drive her at 15.4kts on trials, or 15.1 sustained, and her endurance was to be 5,000nm at 12kts. At this time the goal of the 20kt amphibious force must have seemed far away; the new LSTs, for example were capable of no more than 15kts. It was even argued that fire-support ships might approach an objective independently, since they would have little ability to cover an amphibious force at sea.

The shallow-draught hull form was developed from an existing shallow-draught form, that of the 220ft aluminium minesweeper; a full midship section was selected to keep draught small, while the forebody was made fuller than resistance considerations would normally have dictated in order to provide space for large rocket-magazines to accommodate 6060 rounds – which, however, would only provide about 20 minutes of fire. At the same time a narrow stern was adopted to avoid yawing and broaching when approaching a beach in a following sea. Mass production considerations showed in a decision to forego deck camber, achieving somewhat the same effect by employing a flat deck along the centerline, with portions sloping to the outer edge. Each of the twin shafts was powered by a single 1550bhp diesel and the ship was to be fitted with variable-pitch propellers which 'improve manoeuverability, eliminate the necessity for reverse gears or reversible engines . . . and comply with the characteristics requirement that detachable propeller blades shall be provided to facilitate repairs.' The latter was part of the shallow-water operating requirement; blades might easily be damaged. As for arrangement, the entire battery, except for one twin 40mm mount, was located forward of the bridge. The rockets were all located on the first platform deck, ie generally above water. In the course of the design, BuOrd and BuShips suggested that the rocket launchers be redesigned with longer hoists, so that the magazines could be relocated deeper in the ship; one incidental saving was 28 tons of STS, no longer required as magazine protection, as the rocke would be largely below the water line and thus protected from shellfire. Crewmen originally to have been quartered in the hold were moved up to the second deck which was a less dangerous locatic 'from which it is easier to escape in the event of serious magazine damage; habitability in general wi be greatly improved by placing the crew's berthing and living spaces above the waterline much closer to messing and sanitary facilities.' Th increased depth of the rocket magazines would even permit som saving in their length.

This design emerged in 1955 as the *Carronade,* IFS-1 (later LSR-1 in a new Inshore Fire-Support series. Although she had been designed for mass production, she was not repeated; after all, she was little more than a redesigned LSM(R) with a new rapid-fire rocket system: all 48 of the revised LSM(R) survived. Thus, although *Carronade* received considerable publicity at the time, she served only 5 years in the active fleet, and was decommissioned in 1960.

Design drawing for LSM(R) Reconstruction – Scheme F

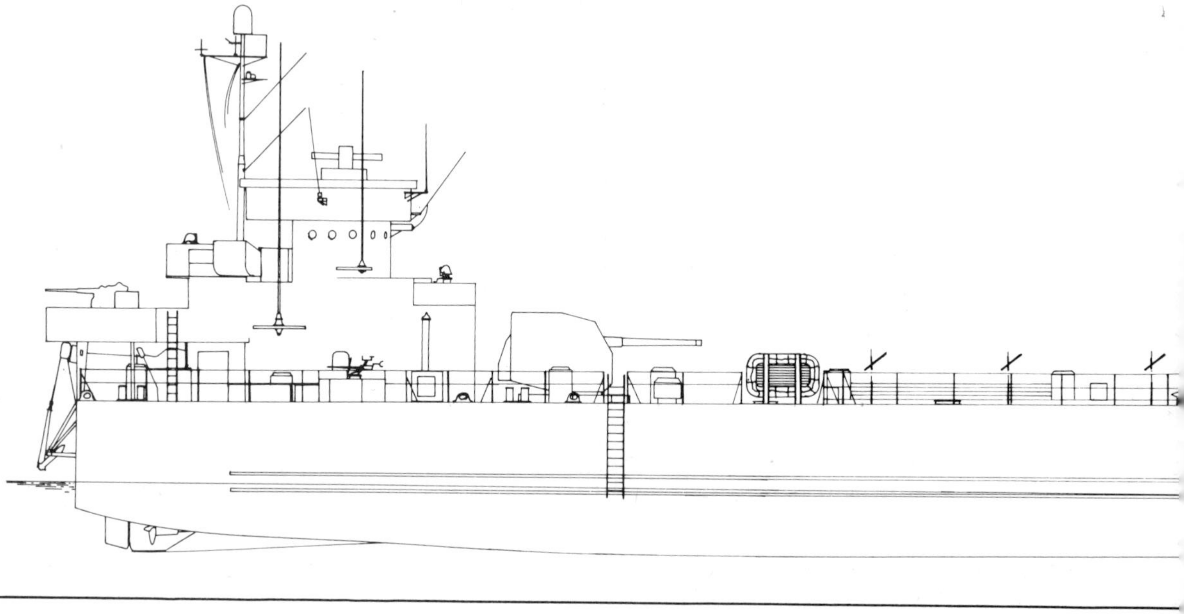

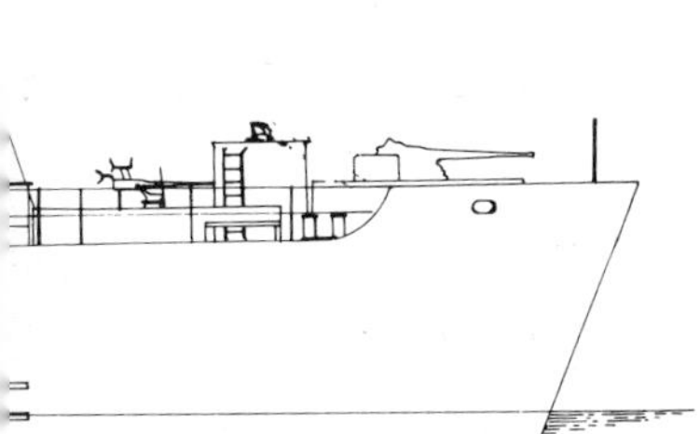

ASSAULT FORCE REAPPRAISAL

Of the large wartime programmes, all of the LCS(S), which had been redesignated LSS(L), were discarded by 1957; it does not appear that any were recommissioned for Korea. Most of the 106 in existence in 1950 were transferred to friendly navies: 6 to Thailand in 1950, 6 to Italy in 1951, 6 to France in 1951 for service in Indo-China, 5 to Korea, 50 to Japan in 1953, 3 to Taiwan in 1954, and 3 more to France the same year. The LSM(R) were more highly valued, 17 being recommissioned in 1951. Even so, 35 out of the original 48

USS *Blackstone River,* LSM(R)403, in 1959. Note the differences in bow contour between this 401-series unit and the 501-series. She also retains her wartime SU surface-search radar, as well as the standard wartime IFF aerials at the ends of her yardarm, while the two 40mm barrels have been removed from her bow mount. Note, too, the non-standard lettering. She was stricken in 1959.

USN

LCS(L)(3)(50), as completed, September, 1944. Note the 3in/50 gun in No 1 position and the details of the Mk 51 director just forward of the after twin 40mm mounting

Courtesy James C Fahey Collection, US Naval Institute

were discarded in 1959, leaving only 11 serving as LSM(R), plus the *Carronade*. At the time, all of these ships were described as at best obsolescent, symbols of an obsolete form of warfare.

One reason for this was a much altered Marine Corps philosophy of amphibious assault. With the advent of nuclear weapons, the Marines felt that the classical mass assault over the beach was no longer practical. Massed amphibious shipping in itself would constitute too inviting a nuclear target. The alternative was to disperse the amphibious transports, using some type of high-speed craft to move troops to the objective area. Although experiments with hydrofoil landing craft were pursued, the Marines favoured the technique of 'vertical envelopment' using helicopters landing well inland. Only heavy equipment would be moved across the beach, and beach defences would be neutralised by naval gun-fire support. There would never be a sufficient density of traffic towards the beach to justify a rocket-firing assault ship. Indeed, the landing craft would be directed to a relatively undefended area, which would be isolated from enemy attack by carrier air strikes.

In a dispersed assault short-range smothering fire was not nearly as valuable as was long-range precise shellfire. The Marines began to think in terms of the number of 5in and 8in guns needed per assault. Unfortunately, the number of such weapons in the fleet as a whole was declining very rapidly. The general trend towards a missile fleet, designed above all for anti-air and anti-submarine operations, quickly led to a mixture of conversions of existing cruisers (in which guns were exchanged for missiles) and new ships (in which there had never been heavy guns). In the face of limited funds, the all-gun cruisers, which required large crews, were gradually withdrawn from service. By the early 'sixties it was clear that quite soon there would only be 5in guns left in the Fleet, and that even they would be in short supply as the Second World War destroyers wer[e] retired. Moreover, the existing 5in gun was no match for some standard Soviet artillery pieces, suc[h] as the 130mm, which would almos[t] certainly be encountered in any shore bombardment mission.

Thus studies of the long-term character and composition of the US fleet led, about 1963, to a requirement for a new large-calibr[e] naval gun, which became the Majo[r] Calibre Lightweight Gun, the Mar[k] 71 8in/55. There was, however, no hope of funding for a new cruiser-size warship; the new weapon would have to be accommodated aboard a destroyer-size ship. Even so, the rapidly increasing cost of the missil[e] and ASW ships, and indeed of the new helicopter assault ships, made funding of a new fire-support type difficult at best, and indeed ultimately made its procurement impossible, given the additional fiscal strains of operations in Vietnam under a 'guns and butter' budget. It did not help that the new ship ran foul of the variety of new defence management concepts introduced by Secretary of Defenc[e] McNamara.

Taken shortly after completion at the Albina Engine & Machine Works, Portland, Oregon September 1944 – *LCS(L)(3)50.* Note the winch and anchor aft, for pulling herself off a beach.

Courtesy James C Fahey Collection, US Naval Institute

THE LFS

In 1965 *Carronade* and three LSM(R) were reactivated for service in Vietnam and, at about the same time, Navy advanced planning documents began to show a requirement for 4 Landing Fire-Support ships (LFS) to replace them in the amphibious force of the future. This was almost a replay of the original IFS story, except that with the decline of cruiser and destroyer gun batteries the new ship had to emphasise both gun *and* rocket fire, and some of her gun fire might well have to be delivered at long-range. One problem was that gun and rocket fire-support tactics were quite different. A rocket ship would have to move in close to the beach, would fire several salvos and then withdraw to reload, whereas the gun ship would be able to remain on station far longer and fire from a greater range. Shallow draught conflicted with any requirement for good seakeeping qualities.

At this time the major new fire-support weapon in prospect was the Major Calibre Lightweight Gun, or MCLWG. In 1965 it was envisaged as a Navy adaptation of the Army's 175/60 and, indeed, this weapon even received a Navy Mark number (Mk 69, or Mk 70 when it was adapted to take longer projectiles). There was also a project for a Landing Force Support Weapon, a long-range guided missile which would be compatible with the Terrier launchers of the two elderly missile cruisers *Boston* and *Canberra*; one alternative was an adaptation of the Army's Lance. Ultimately the gun was preferred in view of its greater flexibility and, even more, in view of the great potential of two new gun technologies, rocket assistance and laser guidance. Other possible LFS weapons were the Mk 105 5in spin-stabilized rocket launcher, already mounted aboard the

Caronnade, and a variety of 5in lightweight guns – the Mk 45, which ultimately armed the *Spruance* class, the proposed Mk 65 rapid-fire version and the proposed Mk 66 lightweight twin mount. Mk 66 was sometimes described as an area neutralisation weapon, in effect a competitor to the rockets. Throughout the LFS programme there were two complementary missions – destruction of point targets, which was to be accomplished by heavy shellfire, and area neutralisation by rockets ('keeping heads down as the assault occurs').

BuShips made its first studies in October 1964 and, early in 1965, the Ship Characteristics Board selected the LFS as the prototype for application of Secretary McNamara's new CF/CD (Concept Formulation/Contract Definition) management approach. In principle the Navy would no longer design its ships. Rather, it would carefully formulate a basic concept on which alternative contractors would bid with their own designs. The *Spruance* and the *Tarawa* class LHA were the primary products of this technique; the LFS never went beyond the CF phase. Indeed, in the absence of firm guidelines LFS funding decreased year after year from 1965 onwards and a planned CD phase (FY 69) was never reached.

The objective of Concept Formulation was to examine as many alternatives as possible, in case some relatively unconventional design might prove far more efficient. Thus in 1966, beside a base case ship with four 175mm or three 175mm and two 5in, there were (i) a reactivated battleship; (ii) a reactivated heavy cruiser; (iii) a new 16in gun ship (although the Navy had no facilities to produce new 16in guns) and, although there were few targets specifically requiring such weapons; (iv) a combined gun and rocket ship; (v) a converted LST (which, however, would be too slow); (vi) separate gun and rocket ships; (vii) an air-cushion (or surface-effect ship) LFS (which would have the advantage of extremely shallow draught); or (viii) a modular gun ship, for maximum flexibility.

No choice was made at the time, and the Navy design staff at the Naval Ship Engineering Center (NAVSEC) had other, more urgent, projects. Some cost and feasibility studies were conducted during 1967-8 showing a considerable increase in projected cost ($80 million in FY 68 rather than $37.4 million in the 1964 studies) but details have not survived. However, the final cycle of studies began in February 1968. At this time the new fast (20kt) LST seemed a good basis for a shore bombardment ship, at least on paper. As yet there were no firm mission guidelines. It appeared that the LFS would have to replace not merely the *Carronade* and the LSM(R) (which were, after all, very similar in battery and function) but also the ageing light fast transport, the APD. By the mid 'sixties the 20kt amphibious force first demanded just after the Second World War was a reality and all of the 1968 designs were expected to make 20kts.

As an APD replacement the LFS would have to land raiding and reconnaissance parties and she would support Underwater Demolition Teams (UDTs), which in turn meant that she would have to be able to launch and recover their boats. NAVSEC proposed to adapt the LST hull by eliminating its bow ramp, retaining the much smaller stern ramp, and in effect replacing most of her cargo capacity with 8 to 10 rocket launchers and 10,000 rockets. There would also be a third twin 3in/50 and the Basic Point Defense Missile System (Sea Sparrow: BPDMS). This ship was described as an LFS(R), for LFS (Rocket), and, by March 1968, sketches showed a 551ft hull displacing 4855 tons (light; 6835 tons fully loaded).

An SCB working group meeting in August noted that the MCLWG was needed, although perhaps not in the small hull; meanwhile other studies suggested that a combination of 175mm and 5in guns would be most cost-effective. Thus a revised LFS(R) of September 1968 showed two 175mm guns as well as rockets and BPDMS on similar dimensions and displacements (550 × 55.5 × 15.9ft, 4970 tons light, 6933 fully loaded). There were also exotic proposals. Sketches (which were nc backed by detailed calculations) showed either a triple 16in battery (unarmored) or a twin 12in/70 forward, the latter presumably a rebuilt 16in/50, together with a 5in gun in the bows and a 175mm gun abaft the superstructure, all on 900 tons fully loaded and dimensions of 480 × 72ft.

By this time NAVSEC had much the present type of computerised ship design system and a great variety of LFS models could be tested. the LFS studies of the Fall o 1968 were armed with two 175mm and four 5in guns, as well as 8, 10 o 12 rocket launchers; BPDMS was an option. Hull size was to be no larger than that of the new LST anc the LFS would have a helicopter platform but no hangar. A sketch shows such a ship with her rocket launchers on the forecastle, dimensions of 550 × 57.8 × 17.2ft displacement 5487 tons light and 7825 fully loaded, powered by two diesels (15,600 bhp for 20kts) and with a complement of 309. BPDM was included. The profile, with three levels of guns fore and aft anc two macks, is as close as the postwa Navy ever got to the traditional image of the destroyer or light cruiser, with a heavy gun battery or a relatively compact hull.

That is, the LFS was the only postwar US warship design in whic raw fire-power was more importan than electronics or command and control facilities. She was about the size of a large destroyer but, unlike the destroyer, she was not associated with a carrier task force. Thus she was a weapon platform, not a sensor platform. Nor was much of internal space consumed b machinery, since she was not intended to move very fast. The irony is that, had the LFS been built, her extremely formidable appearance might well have made for gross mis-use. In fact, as in the case of the *Carronade*, the LFS was extremely limited in capability. An attempt to go beyond her natural limits would have made for costs well beyond those the Navy could bear for so specialised a unit, as the NAVSEC analysis showed again and again.

CONCLUSIONS

The final series of cost studies began in November, 1968. It then appeared that locating guns and rockets in the same hull would not be the most cost-effective. Thus one option for the pure gun LFS showed two MCLWG (which were now 8in guns) and 2 twin Mk 66 5in, with twin diesel power for 20kts. A second alternative added the Improved Point Defense Missile System and a third called for 3-8in and 3-twin 5in; a fourth option again added a defensive missile. None was pursued and, indeed, systematic LFS development ended late in 1968.

What killed the LFS was yet another alternative. In 1968 a new destroyer, DX, was in the CF stage. It was quite large enough to accommodate the heavy gun and the DX hull was suggested as an LFS option, armed with 2 175mm guns but no rockets. The DX hull would have the advantage of providing some ASW capability, so that the fire-support ship might help protect the amphibious convoy en route to its objective. Once an LFS based on the DX had been accepted, however, it was a short step to dispensing with the LFS entirely and providing an option in the DX design for the replacement of one or both of its 5in weapons with heavier guns. That was actually done in the *Spruance* design, which was the culmination of the DX project. Only one gun could be provided, because an 8in gun mount aft would have interfered with the propeller shafts, given the limited hull depth available. Thus the heavy-gun option in the *Spruance* is the sole surviving relic of the entire LFS Programme. Now, even it appears to be gone, since the 8in programme itself was a victim of higher priorities in a period of badly constricted budgets.

The beach neutralisation function fared even worse. In 1970 the *Carronade*, and the 3, more primitive, LSM(R) were laid up and within a few years all had been scrapped. A decade later there is no rocket bombardment capability left in the US Navy, even as the United States finds itself increasingly faced with the possibility that it will need some means of assaulting defended beaches in the near future. Beyond the 5in guns of the destroyers, there are only 6 heavy-gun ships retained in reserve, the 4 battleships and 2 rapid-fire 8in cruisers, which have been inactive for 20 years. In theory, naval aircraft should be able to take over many of the functions of the specialised assault craft but in fact the inventory of such aircraft – and of their carriers – is quite limited.

In 1980 only one Navy in the world retains anything like an extensive rocket assault capability – the Soviet Navy. In its modern landing ships (*Ropucha* and *Polnocny* classes) it has a limited shore-bombardment battery (generally two 18-tube launchers); the advantage of mounting such weapons directly aboard the assault craft is that rockets are always available at the point of landing even in a dispersed operation. On the other hand, magazine capacity cannot be extensive, nor is fire control likely to be very elaborate. Even so, the Soviet example may be worthy of emulation, particularly as the US amphibious force is blessed with far more capacious units.

LCS(L)24, later redesignated LSS-24. Although it is dated 1959. this must be a much earlier photograph as all of the LSS were gone by then. Note the single 40mm gun in No 1 position characteristic of the earlier units.

Courtesy James C Fahey Collection, US Naval Institute

WHERE THE POWER GOES

by D K Brown

In the late 20th Century people have become obsessed with the need to save fuel and hence to find out where losses of precious energy occur. This was not always so; in 1840 the cost of coal was almost negligible and interest in fuel saving was small; hence detailed records do not exist.

In nearly all surface warships power is derived from burning a fossil fuel, either coal or oil. In both cases much of the available energy goes straight up the funnel. Various methods have been used to recover this loss; in early steam ships water was warmed in a tank round the funnel (which led to Brunel's injury in an explosion in such a device on the Great Eastern) and, more recently, domestic water has been heated by the exhaust from diesels or gas turbines.

The thermal efficiency of various modern engines is compared in Table 1.

The useful power developed has generally been measured in Horse Power – ie, the rate of doing work. James Watt defined one horse power as equivalent to raising 33,000lbs through one foot in one minute. He was a shrewd businessman and used a very puny

The British destroyer leader *Swift* in dry dock showing 4 three-bladed propellers. This photograph was taken during her trials period in 1909 when a large number of propellers, of varying form, were fitted in a vain attempt to improve efficiency sufficiently to push her trials speed up to the designed 36kts. She was finally accepted at a little over 35kts; it is not known if the propellers shown here were those finally fitted

NMM

TABLE 1: Thermal efficiency %

	Full-power	25% power
Diesel	44	40
Early gas turbine	28	16
Current gas turbine	35	22
Steam turbine	19	13

It should be noted that gas turbines, particularly the early ones fall off rapidly in efficiency as the power reduces.

Table 2: COMPARISON OF NHP AND IHP

Ship	Engine Builder	nhp	ihp	ihp/nhp
Dee	Maudsley	200	272	1.36
Rhadamanthus	Maudsley	220	400	1.82
Locust	Maudsley	100	157	1.57
Porcupine	Maudsley	132	285	2.16
Jackal	Napier	150	455	3.03
Harpy	Penn	200	520	2.6
Spitfire	Butterby	140	380	2.7

Note: Arranged in chronological order, hence the apparent inferiority of Maudsley engines is probably not genuine. In fact, it probably reflects Maudslay's early use of indicators! However, since nhp was a measure of size it proved fairly reliable to quote cost of machinery in terms of nhp.

horse, so that, in comparison, his new steam engines would show well in the sales literature. It was difficult to measure the real power developed by an engine in the early days so 'nominal horse power' (nhp) was derived based on the geometry of the engine and some assumed design parameters (for older, British readers it was somewhat akin to the old RAC rating for cars).

$$\text{nhp} = 7 \times \text{Area of Piston} \times \frac{\text{equivalent piston speed}}{33000}$$

For paddle steamers the Admiralty assumed piston speed =

$$(\text{stroke})^{\frac{1}{3.38}} \times 129.7$$

or

Stroke (ft)	Piston speed (ft/min)
3	180
6	220
9	248

The formula gave 'real' horse power if the piston speed was as shown and the mean pressure was 7lbs/sq in. In practice, this was rarely so and Table 2 shows the comparison between nhp and indicated horse power(ihp) – the power available in the steam which can be recovered by expansion in ideal circumstances. This can be measured by a mechanical 'indicator' which came into use about 1830.

The ihp is given by:

$$\frac{PLAN}{33000}$$

where P is the mean pressure of the steam (lbs/sq in)
L is the stroke (ft)
A is the area of the piston (ft)
N is the number of revolutions per minute.

The output of the engine, ie the power transmitted to the shaft (shp) falls well short of ihp. There will always be frictional losses inside the engine while other losses result from the need to drive auxiliaries to the engine (air pumps, feed pumps, etc). For small craft, with high efficiency engines, up to 90% of the ihp reached the shaft but 80% was probably more usual and it could be much lower.

Quoted values of ihp are usually the maximum value which could be reached over a 4 hour trial. Over a longer period the normal maximum

The light cruiser *Enterprise* entering Portsmouth harbour in July 1938. She and her sister ship *Emerald* were designed for a maximum speed of 33kts with 80,000shp but, at almost half this power, they could still match the 29kts of the earlier *C* and *D* class ships

Wright and Logan

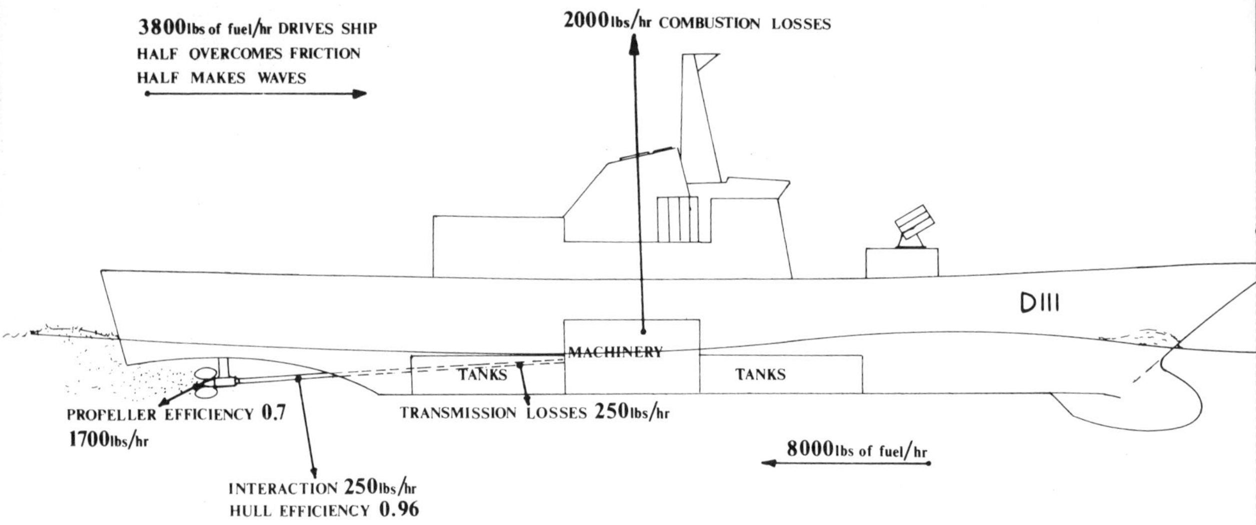

Fig 1: Where the fuel goes

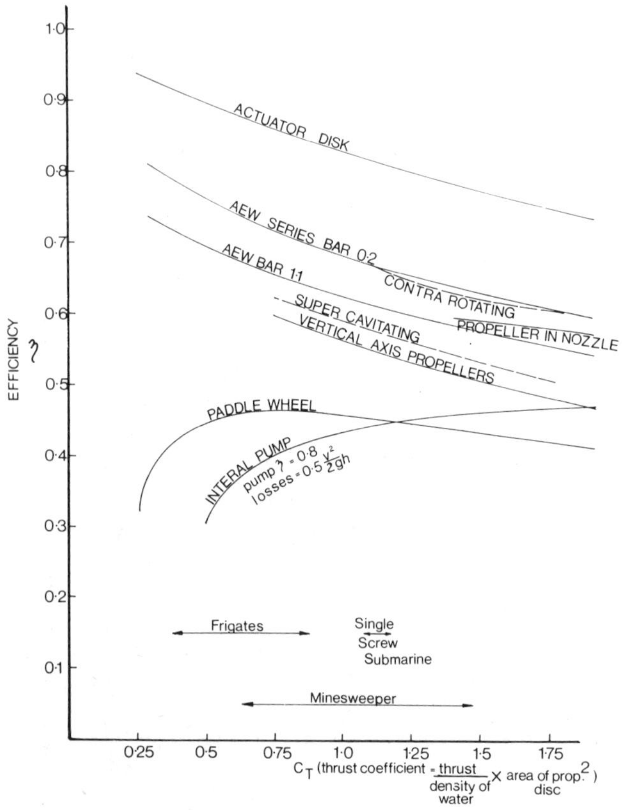

Fig 2: Comparison of Propulsive Devices

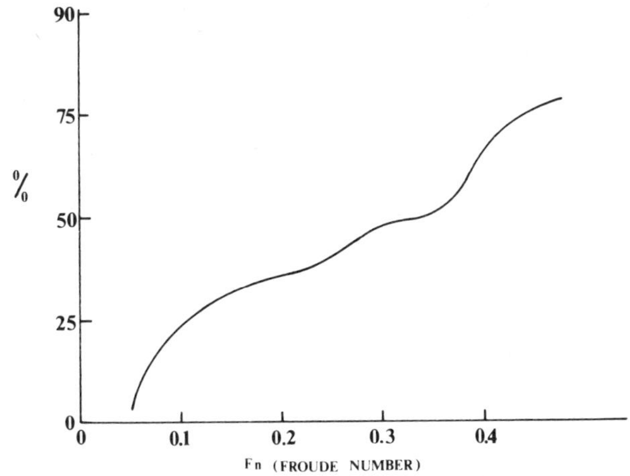

Fig 3: Wave making as a percentage of total resistance

TABLE 3: Comparison of Leonidas (geared) and direct drive L class Turbines

Ship		rpm (machinery)	rpm (propeller)
Leonidas (geared drive)	HP	3000	380
	LP	1800	
L class (direct drive)		650	650

HP = high pressure turbine LP = low pressure turbine

would be about 60 to 70% of the trials figure (RN 'with all despatch'). For ships using forced draught the normal power might only be some 40% of the trial figure. Modern gas turbines are particularly sensitive to both pressure and temperature changes and a 1% pressure loss in the intake will lead to a 2% loss of power while a similar pressure loss in the exhaust will involve a further 1% loss of power. These engines are even more particular about air temperature and a rise in inlet air temperature from 15° causes about 20% loss of power.

The power leaving the engine will be diminished by a further 2-5% before it is delivered to the propeller (delivered horse power dhp), mainly by friction in the gear box and bearing. In passing, one must note that the development of a satisfactory stern gland bearing was one of the most difficult problems in introducing the screw ship, only solved by Penn with the use of lignum vitae. A leaking stern gland nearly caused the loss of HMS *Royal Albert* in December 1855 at the Isle of Zea.

PROPELLERS

In early steam ships the usual device for converting torque (rotational power) in the shaft to forward thrust was the paddle wheel. RN ships usually used a wheel with fixed blades, thought to be more robust and less liable to be put out of action in battle than the slightly more efficient 'feathering' wheel introduced by Galloway in 1829 and developed by Morgan later. Sometimes, each blade was broken into sections, staggered along the circumference of the wheel. This 'cycloidal' wheel, due to John Field, reduced the loss of energy as a blade hit the water and was slightly more efficient than the simple wheel. However, the wheel was cumbersome and inefficient, at best 45% (see Fig 2) and was superseded during the 1840's by the screw propeller (see *Warship 1*) which had a much higher efficiency. High results were obtained very early in the history of the screw propeller and SS *Great Britain*'s first had a peak efficiency of 70% and HMS *Rattler's* (now at the RN Museum Portsmouth) was of the same order.

More modern propellers have not improved significantly in efficiency over these early designs but they are quieter and cause less vibration. More important, the matching of the propeller characteristics to those of the machinery is better understood so that the combined efficiency is raised even if one component may suffer at the expense of the other. The classic example of this is in the early turbine ships where the rpm used was too low for the turbine and much too high for the propeller. In this case the solution was obvious – geared drive – though it required much hard work from the engineers before it became both reliable and quiet.

The first geared turbines went into the British destroyer *Leonidas* particulars of which are given in Table 3. The gearing weighed 14 tons per shaft but gave a 10% improvement in overall efficiency (and hence fuel consumption and endurance) at full power and 26% at low power so that the power could be reduced from 24,500 to 22,500 for the same speed (30kts). The use of gearing was only made possible by Mitchell's invention of the thrustblock which took the thrust loading off the gears.

The screw propeller acts by increasing the speed of the water passing through it; a change of momentum which is reflected in the generation of forward thrust. There is a very convenient mathematical fiction known as the 'Rankine actuator disk' from which many of the laws of real propellers can be deduced. In particular, efficiency only reaches 100% at zero load! The efficiency of an 'actuator disk' is plotted in Fig 2 and no real propeller can equal or even approach such efficiencies. The frequent claims to have done better are, if honest, mistaken.

The difference between the 'ideal' or actuator disk efficiency and that of a real one is made up as follows (typical figures).

Actuator disk efficiency	.88
losses due to finite number of blade	.02
losses due to friction	.11
losses due to rotation in the slipstream	.05
Total	.18
Real propeller efficiency	.70

In general, a small blade area is most efficient at low speeds and a high blade area at high speeds when cavitation losses become significant. In one example, a destroyer model was tested with propeller blade equal to half the area of the disk (Blade Area Ratio - BAR 0.5) The propeller efficiency was 0.68 at 15kts and dropped to 0.57 at 31½kts while a bigger propeller with overlapping blades (BAR 1.1) had an efficiency of 0.64 at all speeds.

Big, slow running propellers are the most efficient but they require big gearboxes and both the gears and the propellers are difficult to fit into the shapes of a conventional ship and a normal dry dock. Some idea of what can be done is shown in Table 4 though, it must be admitted, some of these values are unrealistic.

There are many other tricks which can be played by the propeller designer such as altering the number of blades, skewing, altering the blade outlines etc, but these are intended to reduce vibration or noise and have little effect on efficiency. Controllable pitch propellers (CPP) are very often used on modern ships, gas turbine or diesel, to avoid the cost and weight of a reversing gearbox. However, a CPP will reduce the overall efficiency of a twin screw warship by about 6% (3% on the propeller – due to the thick root sections – and 3% due to the added drag of the bigger shafts and shaft brackets).

The flow around the after end of a ship is extremely complicated and the effects of the hull and the propeller mutually interact. The propeller accelerates the water ahead of it and reduces its pressure and these tend to add to the resistance of the ship usually referred to as the thrust deduction fraction 't'. Water passing along the hull is slowed by friction and

The British Polaris submarine *Resolution*. This and other nuclear boats are propelled by a single screw which gives a combination of high wake and large diameter for high overall efficiency

MoD

affected by the waves produced by the hull. This change of velocity into the propeller, due to the hull, is defined by the wake fraction 'w'. In general, propeller performance is improved if the water velocity into it is reduced and hence a high value of

TABLE 4: PROPELLER EFFICIENCY IN A DESTROYER

Propeller diameter (m)	10	8	6	5	4.5	4	3.4	2.5
rpm	63	104	157	224	300	361	446	2000
efficiency	.86	.84	.82	.78	.74	.72	.67	.35

wake will usually mean a good propeller efficiency. This is the main reason for going to a single shaft for fast submarines such as the British *R* class of 1916 and modern nuclear boats. The combination of high wake and large diameter gives very high overall efficiency and so higher speed.

HULL EFFICIENCY

The effects of wake, thrust deduction and another small term allowing for the effects of irregular flow are lumped into the parameter – hull efficiency – which gives an overall figure to hull/propeller interaction. Typical values for hull efficiency are:

Battleship	0.96
Cruiser	0.95
Destroyer	0.94
Single Screw Submarine	1.30
Merchant ship (single screw)	1.12
Merchant ship (twin screw)	1.00

The power actually put into the water is called the effective horse power (ehp) and it has to overcome the resistance of the hull and its fittings to motion through the water. It is usual to follow William Froude's great approximation and divide the total resistance of the hull into two components, frictional and residuary mainly wave making (Fig. 3). The frictional resistance depends on the area of the hull surface below water and on how smooth it is. Hence low speed ships are usually short and fat (*see Inflexible, Warship 15*) to reduce the area. The effect of roughness and fouling has been dealt with before (*see Warship 12*) and must not be forgotten.

Reduction of wave making resistance is much more complicated and will be dealt with in a later article. In general, the

The 'Battle' class destroyer *Aisne,* after conversion to a radar picket. A large proportion of the power developed by ships is absorbed in overcoming hull and wave making resistance

MoD

TABLE 5: COMPARISON OF BRITISH LIGHT CRUISERS

	Columbo	**Dunedin**	**Emerald**
Length (ft)	445	465	560
Displacement (tons)	4270	4530	8160
Installed shp	40,000	40,000	80,000
Power for 29kts on trial	37,125	41,100	41,000
shp/Dispt x speed	0.3	0.26	0.17
(Displacement/L_3) × speed	1.7	1.6	1.6

high pressure area near the bow of a ship generates a train of waves beginning with a crest, while the suction area under the ship makes another set of waves beginning with a trough. By choosing a length appropriate to the speed, the crests of one set may cancel out the troughs of the other producing the least surface disturbance and hence least loss of energy.

In general, a fast warship should be as long a possible, a point clearly demonstrated in Britain's First World War cruisers, *Columbo, Dunedin* and *Emerald. Columbo* was built with 5-6in guns, all on the centre line, and *Dunedin* was lengthened 20ft to get in another 6in. *Emerald* was a much bigger ship to get in twice the power and a 7th 6in gun. Table 5 compares these three ships and their power requirements at 29kts. As can be seen at 29kts *Emerald* required little more power than *Columbo,* of half her displacement. The remaining 39,000 shp available to *Emerald* produced a mere 3.9kts extra speed although this did make her the fastest RN cruiser of the Second World War. The ratio of displacement to the cube of the length is rarely quoted, except in text books of Naval Architecture, but it (or its inverse, length/cube root of volume) is the most signifiant measure of 'fineness'.

The appendages to the hull also add to the resistance and this can be considerable, in a frigate for example figures (in percentage of total resistance) would be:

Rudders	0.8
Bilge keels	3.3
Stabiliser fin	2.0
Shaft and shaft brackets	1.3
Sonar dome	5.0

The percentages for a submarine are much higher and, in particular, the flooding holes in the casing can account for 10% of the total resistance.

The wind resistance of modern ships is quite small and can often be neglected but the drag of a full set of masts, yards, rigging and furled sails in the early steamships seems to have been about 10-15%of the total in still air.

The Captain of *Gorgon* in 1849 said that his ship was 2-3kts faster, as a steamer, with top masts housed than under full rig. This was heading into a 'strong double reefed topsail breeze' which would correspond to wind force 5 on the Beaufort scale or about 14kts. Since such winds would generate waves of about 8ft in height; this would also account for some of the speed loss.

It is the art of the naval architect to balance thermal efficiency, shaft losses, propeller efficiency and drag of the hull together with many other conflicting requirements to produce the complete warship.

The frigate *Leopard,* steaming at high speed on a smooth sea, creates only a moderate hull wave

MoD

The S Class Destroyers 1918–1945

by Commander H St A Malleson RN (Ret)

The S class destroyer *Stalwart* steaming at full speed, probably during her trials

Coming into service late in the First World War, the handsome ships of the *S* class never found a long-term niche in post war fleet distribution (unlike the V/W's), despite being suited to certain peacetime requirements. As examples, individual units, rather than squadrons or flotillas, were required to be spread around the Eastern Mediterranean and the Black Sea during 1919–1922. Similarly, a flotilla of S boats was employed on police duties in Southern Ireland during the 'troubles' of 1920–1921.

Thus, in 1919 some 22 S's were allocated to the Mediterranean Fleet and, at any one time, most of them would be found by themselves at one of the many ports or harbours stretching from the Black Sea to the Aegean and Levant. They assisted in rescue work from South Russian ports during the evacuation by the White Russians under Denikin and Wrangel and, after that, helped to keep the peace between Greek and Turk, all round the Asia Minor coast, during the turmoil of the Greek invasion. Among other tasks, one destroyer was always on guard duty over the Turkish battlecruiser *Yavuz Sultan Selim II* (ex *Goeben)* at Ismid.

After the Chanak crisis in 1922, the S class vessels were brought home, being replaced by V/W's in 1923 and 1924. Thus ended the most active part of the life of the class as a whole and some of the most envied commands of all stations at that time. Subsequent dispositions and details of transfers are given in Table 2, which also gives final disposal dates.

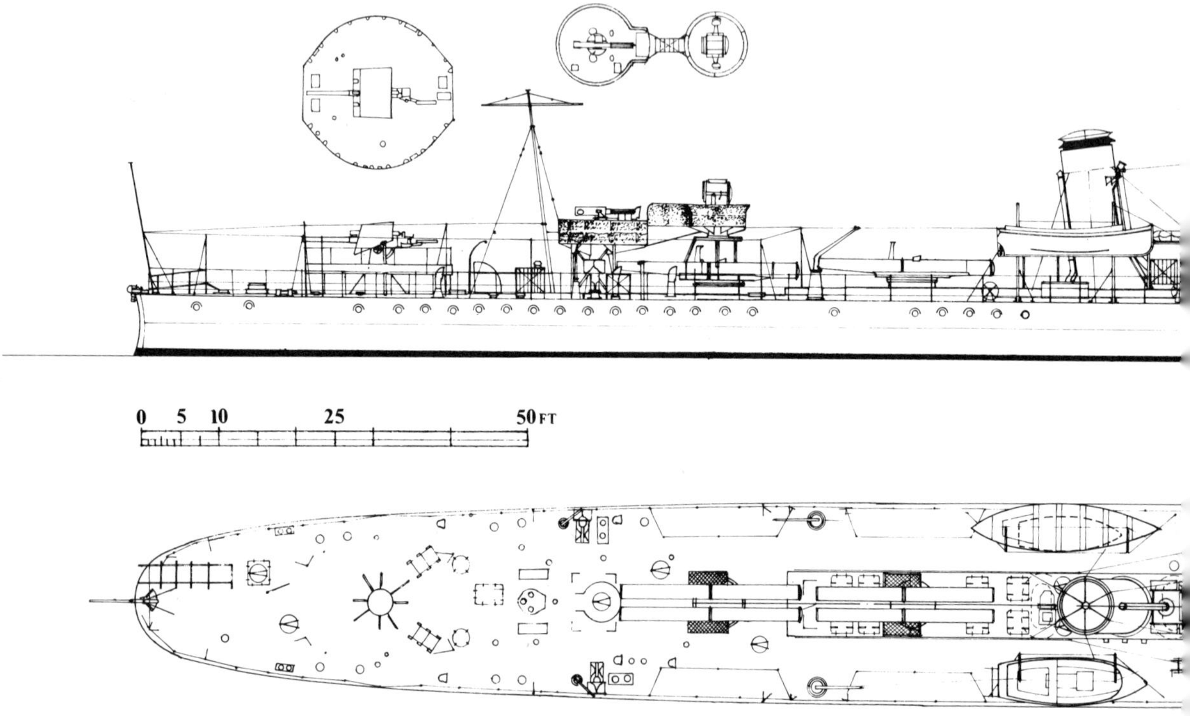

ADMIRALTY S CLASS DESTROYER SARDONYX AS COMPLETED, 1919
As designed the S class were to carry a single 18in torpedo tube on each side of the upper deck, at the break of the forecastle, for use in close-range action with enemy destroyers and submarines. These were, however, only mounted in the early units and were removed from these vessels soon after the war.
Drawing based on official 'As Fitted' plans held by the National Maritime Museum

HANDLING

The combination of small size and powerful engines made the S class highly manoeuverable – the diameter of the propellers in the geared turbine boats was some 10ft. The view astern from the bridge was not as clear as in the V/W's and the bridge itself was only marginally drier than those of the R's.

APPEARANCE

(a) Admiralty type; as shown in the photographs, the principal features were the long sheered forecastle, with slight turtle back, and the distinctive funnels – the rather high and thick fore funnel, just abaft the bridge, with a shorter and thinner after funnel spaced well aft, between the boats' davits. Their appearance was, in fact, very similar to that of the modified R class, which had the same dimensions and the same long forecastle but less sheer and no turtle back.

The 5 built by J Samuel White had distinctive funnel fittings, otherwise there were only minor differences between ships in this category.

(b) Yarrow type; the main outward difference between this and the standard type was the characteristic sloping stern, introduced in the 7 Yarrow later M class. A minor feature was the close fitting of the funnel cowls.

(c) Thornycroft type; there were two main features which distinguished this type from the other two – the high flat funnels and the raised platform for 'A' gun on the forecastle.

GENERAL ARRANGEMENT OF HMS TOURMALINE – THORNYCROFT S CLASS

Key:

1 Oil fuel; **2** Magazine; **3** Boiler room; **4** Engine room;**5** 4in Mk IV gun on CPIII mounting; **6** 2pdr pompon on HA MkII mounting; **7** Director; **8** 24in searchlight; **9** 25ft whaler (13½ dinghy under); **10** 20ft motor boat; **11** Twin 21in Mk IV torpedo tubes

TABLE 1: DETAILS OF S CLASS DESTROYERS AND OTHER CONTEMPORARY BRITISH DESIGNS 1915–1917

Class	Number built	Length oa (ft–in)	Length pp (ft–in)	Beam (ft–in)	Full-load displacement (tons)	Designed shp	Design speed (kts)	Trials and Remarks
ADMIRALTY S	55*	275	265	26–8	1180	27,000	36	Deep load – 32kts Light load – 35kts
THORNYCROFT S	5	276	267	27–4	1200	29,000	36	Trial speeds between 36 and 38 kts
YARROW S	7	269–6	260–4	25–9	1045	23,000	36	Trial speeds between 35 and 39kts
ADMIRALTY MOD R	10	275	265	26–9	1190	27,000	36	
ADMIRALTY R	39	275	265	26–9	1160	27,000	36	
THORNYCROFT R	5	274	265	27	1150	29,000	35	Trial speeds between 37 and 40kts
ADMIRALTY V/W	48**	312	300	29–6	1470	27,000	34	Speed at deep load – 32kts
ADMIRALTY MOD V/W	16	312	300	29–6	1500	27,000	34	Speed at deep load – 32kts
THORNYCROFT V/W	6***	312	300	30–7	1520	30,000	35	Trial speeds between 36 and 37kts
TOTAL	191							

* Includes 5 transferred to Royal Australian Navy.
** Includes *V Leaders.*
*** Includes 2 Modified type (*Wishart* and *Witch*).
Note: The average draught of all these classes was about 9 to 10ft, but the 'sweep' of the propellers extended about 2 or 3ft below this.

ORIGIN

Although built more or less in parallel with the larger V/W's, their smaller size, armament, and endurance, while being adequate for the North Sea, made the S's less suitable for ocean warfare. They were, however, slightly cheaper[1], needed smaller complements and required less extensive armament and stores support. From the Admiralty's point of view, there were also savings in the cost of Government-supplied items, such as guns, ammunition, torpedoes, etc.

Whatever their relative merits, no less than 67 S class boats were completed, compared with 70 of the combined V/W classes (a much more successful design). However, only 7 survived for the 1938 Munich Crisis mobilisation at home, although there were another 4 in the Far East. In contrast, 57 V/W's were available; and there would have been a few more if they had not been tragically scrapped in the *Majestic* deal.

The paradox here is that, when the majority of the S's were disposed of, their hulls were in good shape and, in general, the ships were good for several years service, having been fairly lightly used. The V/W's, on the other hand, were showing their age in the 1920s, especially the earlier built ones; strains, leaks and loosened rivets in the way of the forward magazine were not uncommon. With hindsight, it would perhaps have been wiser to concentrate new destroyer building at that time (1917/18) on V/W's but in many cases the building berths were already occupied by S class boats.

TRANSFERS

In 1919, 5 S class vessels were transferred to Australia, the *Success, Swordsman, Tasmania, Stalwart* and *Tattoo,* and, in 1928, *Torbay* and *Toreador* were transferred to Canada and renamed *Champlain* and *Vancouver* respectively.

GENERAL FEATURES

The S class carried the same weight of gun and torpedo armament as the R class – their predecessors, although the 4in guns, torpedoes and tubes were of a later pattern. They also incorporated the successful propulsion machinery of the R class (three 3-drum Yarrow type boilers, coupled with two sets of single reduction geared turbines. Incidentally, this same arrangement of main machinery was continued from the S's to the V/W's and their modified successors.

Exceptionally, the 7 Yarrow boats had direct drive turbines, also driving twin screws. With less shp, smaller dimensions and lower displacement than the Admiralty design, they could claim higher trial speeds (trial speeds are commented on later). Their endurance was smaller, however, on account of reduced fuel stowage and higher specific consumption. J Samuel White's boats employed White-Forster boilers – similar to Yarrow boilers but with curved water tubes (to ease the tube replacement problem – but making cleaning difficult). Thornycroft-built boats were slightly longer and beamier than the standard type and again, high trial speeds were claimed and these were supported

Tobago of the Thornycroft S class in 1919; she was mined and badly damaged in the following year and declared a constructive total loss, the wreck being sold at Malta in 1922. The main appearance variations of the Thornycroft type from the Admiralty boats, the funnels of equal height and the raised platforms for the forecastle guns, are clearly visible in this photograph

IWM

by an additional 2000 shp.

Both Thornycroft and Yarrow enjoyed a 'special relationship' with the Admiralty, due primarily to their historical specialisation in high speed small craft and destroyers. Both firms could submit modified designs to a basic Admiralty specification and there was usually a good chance of approval, provided the general design concept was complied with. Thornycroft were given the same degree of design freedom in the contemporary V/W class and their later modified W sisters; they had also produced their version of the R class in 1917. The main point of difference in the Yarrow boats was in the main engines, while in the Thornycroft versions the forecastle gun was mounted on a raised platform – a positive improvement.

All the S class shared the distinguishing feature of a semi-turtle deck stretching from the high and prominent stem to the break of the forecastle. This was a partial reversion to the design of the old torpedo boats of 1908 but a better parallel, perhaps, is the turtle deck of the P-boats of 1916, although in this case it extended over the full length of the ship.

ARMAMENT

The three 4in semi-automatic guns (fixed ammunition) fitted were Mark IV elevating to 30°, 'A' gun on the forecastle, 'B' between the funnels and 'Y' on the quarter deck, both the latter on 'bandstands' – a configuration started in the L class in 1913. One disadvantage of this arrangement was 'B' gun's ammunition supply (over a long route from the forward magazine) and its restricted arcs of fire. 'A' gun was not very effective in much of a sea, although Thornycroft's modification was an improvement, in this respect. Simple director firing gear was fitted.

The two twin 21in torpedo tubes accommodated Mark IV* torpedoes, which could be fired from the bridge as well as the mounting. The remaining armament comprised a 2pdr pom-pom and a 0.303in Vickers machine gun – this latter was for landing purposes, for which some 20 Lee-Enfield rifles were also supplied.[2]

ACCOMMODATION

Officer's cabins were aft and the ratings' mess decks forward – the usual arrangement at that time. Although the Captain used a bunk in the chart house at sea, for easy access to the bridge, other officers, coming from aft to change watches, sometimes had a hazardous journey along the upper deck, when in a seaway. At such times, too, the mess decks, always somewhat crowded and uncomfortable, were often leaky and wet. There were cabins for 5 officers – Captain, 1st Lieutenant, Engineer Officer, Sub Lieutenant and Gunner (T). The remainder of the crew totalled about 85 in peacetime.

COMMUNICATIONS

The W/T apparatus originally fitted was Type 4/15, meaning a low-power spark transmitter, supplemented by a higher power arc set, which could transmit C/W (continous wave). Receivers used bright emitter valves.

TRIAL AND SERVICE SPEEDS

Builders' records of acceptance trials are not easily available these days – many sources have dried up or have been destroyed – but what is fairly certain is that trial figures between different yards for similar ships are not necessarily comparable. Much has been written on the subject, which covers such aspects as draught and trim of the ship, depth of water, weather conditions, accuracy of distance and time measurement, helm and handling during the trials, etc. This point can also apply to similar ships from the same yard.

Judging, however, from the figures in the *Denny List Part 3,* a realistic maximum speed under full load and when brand new was 32kts; 36kts was seldom achieved, even when lightly loaded, and this applied not only to the S class but to the same builders' production of *R* class destroyers.[3].

Under practical conditions at sea, and when destroyers were in a formation, 28kts was the normal maximum for station-keeping purposes in fine weather. In such circumstances some of the Thornycroft and Yarrow-built boats might have had a knot or two in hand over the others.

1

2

1 The *Taurus* and *Teazer* (background) of the Thornycroft R class in the 1920s

2 A close-up of the Admiralty S class destroyer *Steadfast* as completed in 1919

3

HMS *Thanet* of the Admiralty S class at Devonport in 1934. She was sunk in action with Japanese destroyers off Malaya on 27 January 1942

NMM

TABLE 2: S CLASS DESTROYERS – SUMMARY OF SERVICE

Name	Builder	Service
Sabre	Stephen 1918, note – completed by Fairfield	1919 Nore; 1920 RX; 1928 R; 1931 Portsmouth; 1937 AT; 1939 C (war service); 1945 S
Saladin	Stephen 1919	1920 Portsmouth; 1920 RX; 1928 R; 1939 C (war service); 1947 S
Sardonyx	Stephen 1919	1920 Devonport; 1927 R; 1931 T (signal school); 1939 C (war service); 1945 S
Scimitar	Clydebank 1918	1919 Devonport; 1920 RX; 1927 R; 1931 Nore; 1938 Portsmouth; 1939 C (War Service); 1947 S
Scotsman	Clydebank 1918	1919 Devonport, B; 1927 R; 1937 S
Scout	Clydebank 1918	1919 Devonport, B; 1920 RX; 1927 R; 1931 Nore; 1937 C (Far East)
Scythe	Clydebank 1918	1919 Devonport; 1931 S
Seabear	Clydebank 1918	1919 Devonport; 1920 RX; 1927 R; 1930 Nore; 1931 S
Seafire	Clydebank 1918	1919 Devonport, B; 1920 RX; 1927 R; 1936 S
Searcher	Clydebank 1918	1919 Devonport; 1920 RX; 1927 R; 1931 Gibraltar; 1933 Med (detached); 1936 Portsmouth; 1937 S
Seawolf	Clydebank 1919	1919 AF (4DF), B; 1921 R; 1931 S
Senator	Denny 1918	1919 Med (6DF); 1921 Med (7DF)) 1923 AF (7DF); 1925 AF (9DF); 1936 S
Sepoy	Denny 1918	1919 Med (6DF), B; 1921 Med (7DF); 1923 AF (7DF); 1925 AF (8DF); 1927 CH (8DF); 1932 S
Seraph	Denny 1919	1919 Med (6DF); 1921 Med (7DF); 1923 AF (8DF); 1927 CH (8DF); 1932 Portsmouth; 1932 S
Serapis	Denny 1919	1920 Med (7DF); 1921 Med (7DF); 1923 AF (8DF); 1927 CH (8DF); 1932 Portsmouth; 1934 S
Serene	Denny 1919	1919 AF (4DF), B; 1920 Devonport; 1927 R; 1937 S
Sesame	Denny 1919	1919 AF (4DF), B; 1920 Devonport; 1929 Af (detached); 1932 Portsmouth; 1934 S
Shamrock	Doxford 1919	1919 Af (4DF), B; 1921 Nore; 1937 R; 1930 Gibraltar; 1936 Portsmouth; 1937 S
Shikari	Doxford 1919 (completed by Chatham Dky)	1924 AF target duties (radio control of *Agamemnon* – later *Centurion*); 1932 with HF; C for war service 1938/39
Sikh	Fairfield 1918	1919 Med (6DF); 1921 Med (7DF); 1923 AF (7DF); 1925 AF (9DF); 1926 S
Simoon	Clydebank 1918	1919 Devonport; 1928 R; 1930 S
Sirdar	Fairfield 1918	1919 Med; 1920 (7DF); 1923 AF (8DF); 1927 CH (8DF); 1932 Portsmouth; 1934 S
Somme	Fairfield 1918	1919 Med; 1920 (6DF); 1921 (7DF); 1923 AF (8DF); 1927 CH (8DF); 1932 S
Sparrowhawk	Swan Hunter 1918	1919 Portsmouth; 1920 Med (6DF); 1921 (7DF); 1923 AF (7DF); 1925 AF (9DF); 1926 Devonport; 1927 R; 1931 S
Spear	Fairfield 1918	1919 Med (6DF); 1921 (7DF); 1923 AF (7DF); 1925 AF (9DF); 1926 S
Speedy	Thornycroft 1918	1919 Med (6DF); 1921 (7DF); 1922 – sunk by collision in sea of Marmara
Spindrift	Fairfield 1919	1919 Devonport; 1920 RX; 1927 R; 1937 S
Splendid	Swan Hunter 1918	1919 Nore; 1920 Med (6DF); 1921 (7DF); 1923 Gibraltar; 1930 Portsmouth; 1931 S
Sportive	Swan Hunter 1918	1919 Med (6DF); 1921 (7DF); 1924 Nore; 1928 R; 1937 S
Stalwart	Swan Hunter 1919	1919 RAN; 1936 S
Steadfast	Palmer 1919	1919 Med (6DF); 1920 Nore; 1927 R; 1929 Devonport; 1934 S
Sterling	Palmer 1919	1919 Nore; 1921 Queenstown; 1920 RX; 1922 Devonport; 1925 AF (8DF); 1927 CH (8DF); 1932 S
Stonehenge	Palmer 1919	1919 AF, B; 1921 Wrecked
Stormcloud	Palmer 1919	1919 AF (4DF), B; 1921 Queenstown; 1922 Devonport; 1925 AF (8DF); 1927 CH (8DF); 1932 Devonport; 1934 S

Strenuous	Scott 1919	1919 AF (4DF), B; 1920 Devonport; 1921 Queenstown; 1922 Devonport; 1928 R; 1932 S
Stronghold	Scott 1919	1919 Portsmouth; 1921 RX; 1928 R; 1931 Tender to *Vernon*; 1934 HF (air target); 1942 Sunk S of Java
Sturdy	Scott 1919	1919 AF (4DF); 1920 Portsmouth & RX; 1927 R; 1932 Nore; 1934 HF (air target); 1936 Portsmouth; 1939 C war service; 1940 Wrecked
Success	Doxford 1919	1919 RAN; 1926 S
Swallow	Scott 1918	1919 Med (6DF); 1921 (7DF); 1923 AF (8DF); 1925 AF (9DF); 1926 Chatham; 1928 R; 1937 S
Swordsman	Scott 1919	1919 RAN; 1926 S
Tactician	Beardmore 1918	1919 Devonport (tender to *Defiance*); 1927 R; 1930 S
Tara	Beardmore 1918	1919 Portsmouth (tender to *Vernon*); 1931 S
Tasmania	Beardmore 1919	1919 RAN; 1936 S
Tattoo	Beardmore 1919	1919 RAN; 1936 S
Tenedos	Hawthorn Leslie 1919	1919 AF (4DF), B; 1921 Nore; 1927 R; 1930 Devonport; 1932 Portsmouth; 1934 Devonport; 1937 FE; 1942 Sunk (bombs) at Colombo
Thanet	Hawthorn Leslie 1919	1919 AF (4DF), B; 1920 Nore; 1928 Devonport; 1937 FE; 1942 Sunk off Malaya
Thracian	Hawthorn Leslie 1922 (completed by Sheerness dockyard	1922 Nore; 1925 AF (8DF); 1927 CH (8DF); 1932 Nore; 1934 R; 1936 FE; 1941 Sunk and salved by Japanese; 1946 S
Tilbury	Swan Hunter 1918	1919 Med (6DF); 1920 Nore; 1920 Queenstown; 1922 Portsmouth; 1928R; 1931 S
Tintagel	Swan Hunter 1918	1919 Nore; 1920 R; 1929 Devonport; 1932 S
Tobago	Thornycroft 1918	1919 Med (6DF); 1920 Damaged by mine in Black Sea; 1922 S
Tomahawk	Yarrow 1918	1919 Med (6DF); 1921 (7DF); 1923 Devonport; 1928 S
Torbay	Thornycroft 1919	1919 AF (4DF), B; 1920 Portsmouth and RX; 1928 Transferred to RCN; 1937 S
Torch	Yarrow 1918	1919 Med (6DF); 1921 (7DF); 1923 Devonport; 1924 AF (8DF); 1925 Nore; 1929 S
Toreador	Thornycroft 1919	1919 Portsmouth; 1920 RX; 1928 Transferred to RCN; 1937 S
Tourmaline	Thornycroft 1919	1919 AF (4DF), B; 1920 Detaced with *Turquoise* to Med; 1921 Portsmouth; 1923 Gibraltar; 1931 S
Tribune	White 1918	1919 Med (6DF); 1921 (7DF); 1928 AF (7DF); 1925 AF (9DF); 1926 Portsmouth; 1927 R; 1931 S
Trinidad	White 1918	1919 Med (6DF); 1921 (7DF); 1925 AF (7DF); 1926 R; 1929 Chatham; 1932 S
Trojan	White 1918	1919 Portsmouth; 1920 RX; 1927 R; 1937 S
Truant	White 1919	1919 Portsmouth (tender to Signal School); 1931 S
Trusty	White 1919	1919 Portsmouth; 1920 RX; 1927 R; 1937 S
Tryphon	Yarrow 1918	1919 Med (6DF); 1920 Wrecked
Turbulent	Hawthorn Leslie 1919	1919 AF (4DF), B; 1920 Nore and RX; 1927 R; 1937 S
Tumult	Yarrow 1918	1919 Med (6DF); 1921 (7DF); 1923 Nore; 1928 S
Turquoise	Yarrow 1919	1919 AF (4DF), B; 1920 Detached with *Tourmaline* to Med; 1921 Portsmouth; 1927 R; 1931 S
Tuscan	Yarrow 1919	1919 Portsmouth; 1920 RX; 1921 Med(7DF); 1923 Devonport; 1924 AF(8DF); 1925 Devonport; 1927 R; 1922 S
Tyrian	Yarrow 1919	1919 AF(4DF),B; 1921 Portsmouth; 1927 AF (air target); 1930 S

Key: C – Commissioned, R – Reserve, T – Tender, AF – Atlantic Fleet, MED – Mediterranean Fleet, CH – China, FE – Far East, RX – Reserve Fleet exercises 1920, AT – Air target, S – scrapped, B – Baltic

The Yarrow S class destroyer *Turquoise*
IWM

CASUALTIES

There were no losses among the class during the 1914–1918 War but two were wrecked (*Stonehenge* and *Tryphon*) in 1919/1920. *Tobago* was mined in the Black Sea in 1920 and sustained serious damage to her engine room but was patched up enough to float (in the Constantinople Arsenal) and towed by the battleship *Centurion* the 1000 miles to Malta. In the end, she was scrapped, being uneconomic to repair. *Speedy* sank after collision with a tug in the sea of Marmara in September 1922.

When the *Tomahawk's* port turbines stripped in 1921 and were damaged beyond repair (luckily at Malta), it was decided to salve and fit the duplicate engine from the *Tryphon*, still lying half submerged at Tenedos. Happily, this gamble came off – in a subsequent trial, 31kts was obtained without undue effort and that with half the power coming from a turbine which had been under water for two years!

In the 1939–1945 War, nearly half the surviving *S* class destroyers were lost, as follows:–

Stronghold Sunk 2/3/42 south of Java.

Sturdy Wrecked 30/10/40 on Tiree Island.

Tenedos Bombed 5/4/42 at

\ submariners nightmare on the quarterdeck
›f *Scimitar* 12 September 1940

Iod

2 The *Scimitar* at Portsmouth on 12 September 1940 as modified for escort work. The foremast and after funnel have been shortened, the midships 4in gun replaced by two quadruple 0.5in AA mountings, the forward torpedo tube mounting has been replaced by a 12pdr AA gun and the after torpedo tubes, 4in gun and mainmast, by an extensive depth-charge armament (8 throwers, two stern chutes and substantial deck stowage for depth-charges) while, to improve seaworthiness, the forecastle side plating has been extended aft abreast the bridge

Mod

2

The *Thracian* in 1936 with mine rails, extending from the searchlight platform to the stern, fitted. She was captured by the Japanese at Hong Kong in December 1941 but survived the war, to be returned in October 1945 and scrapped in 1946

The *Sabre* was modified for escort work along the same lines as *Scimitar* except that the after funnel was not shortened. She is seen here in 1943 with later modifications (also applied to *Scimitar*) including the addition of 4-20mm AA (2 in place of the 0.5in mountings and 2 new platforms abreast the after superstructure) and radar Type 291 (aerial at head of foremast)

NMM

Colombo.
Thanet Sunk 27/1/42 off Malaya.
Thracian Bombed and beached 24/12/41 at Hong Kong. Salved by Japanese and returned 1945.

CONCLUSIONS

Outclassed between the wars by the V/W types and overtaken by foreign designs too, it was clearly a problem what to do with the recently built S class once the Mediterranean political situation (particularly at the Eastern end) was more or less stabilised.

After sending a flotilla to China and allocating several training and experimental jobs to individual ships, there were still too many with no role in sight. Periods with reduced complements and little steaming were followed by maintenance reserve and, in the apparently peaceful days of the mid 1930s, scrapping for the majority was inevitable.

It is open to argument that, if more of them had been retained, the necessary modifications for modern war would have interfered with new construction and with vital improvements to the V/W class; in my view it would not have been cost-effective, to use a modern term.

Within their limitations, they performed well and reliably, due in large part to their excellent boilers and engines. These units, designed first for the R class and subsequently employed in the Modified R class, the S class (except in Yarrow boats) and on to the V/W's and Modified W's, were effectively standard production items for some 168 ships (not counting the 16 contributed by Thornycroft to these classes, which had slightly more powerful machinery). The exceptional reliability of these installations, which mainly used Brown Curtis turbines, was demonstrated by the 'continuous performance' of the V/W's throughout the 1939–45 war, particularly in the Western Approaches and with the Arctic convoys[4].

Finally, as this article is about the S class, 4 of the remaining 7 at home played a famous part in the evacuation of the BEF from Dunkirk in 1940, the *Sabre, Saladin, Scimitar* and *Shikari*. They all survived to serve through the remaining years of the War.

The Far East Detachment (*Stronghold, Tenedos, Thanet* and *Thracian*) were all sunk or destroyed by the Japanese Navy.

ACKNOWLEDGEMENTS

National Maritime Museum (Perkins Collection) for invaluable career details and photographs. Also for profile drawings (Admiralty and Thornycroft). *Janes Fighting Ships 1919* for building dates, yards, etc, *British Warships of World War Two* (Lenton & Colledge) for 1939/1945 data. Private contemporary Journal for some Mediterranean details.

1 S class average £185,000, V/W class average £200,000 (*The Denny List, Part 3*; by D Lyon; NMM)

2 This applies to the pre-1939 era. In the few that survived at home – employed as local escorts, aircraft targets, submarine escorts etc – short range AA guns replaced most of the original armament, leaving only 'A' gun.

3 The writer was serving in the *Vivacious* (Yarrow) in 1924. A full power trial was held after the Flotilla (the 5th) had left Port Edgar, fully fuelled, etc. The ships started level and, after the 4 hours, *Wallace* (our Leader) was about 2 miles ahead of us, while the rest were spread out astern, *Vampire* being almost out of sight. Our speed over the 4 hours was 31.6 knots (by observation of shore points). The ship was completed in 1917.

4 One of the R class – the *Skate* – escaped the scrapping programmes somehow and served through 1939–1945 in training and other roles: another tribute to her machinery – the same as in the S class.

The Original Whalecatcher Escorts

by Antony Preston

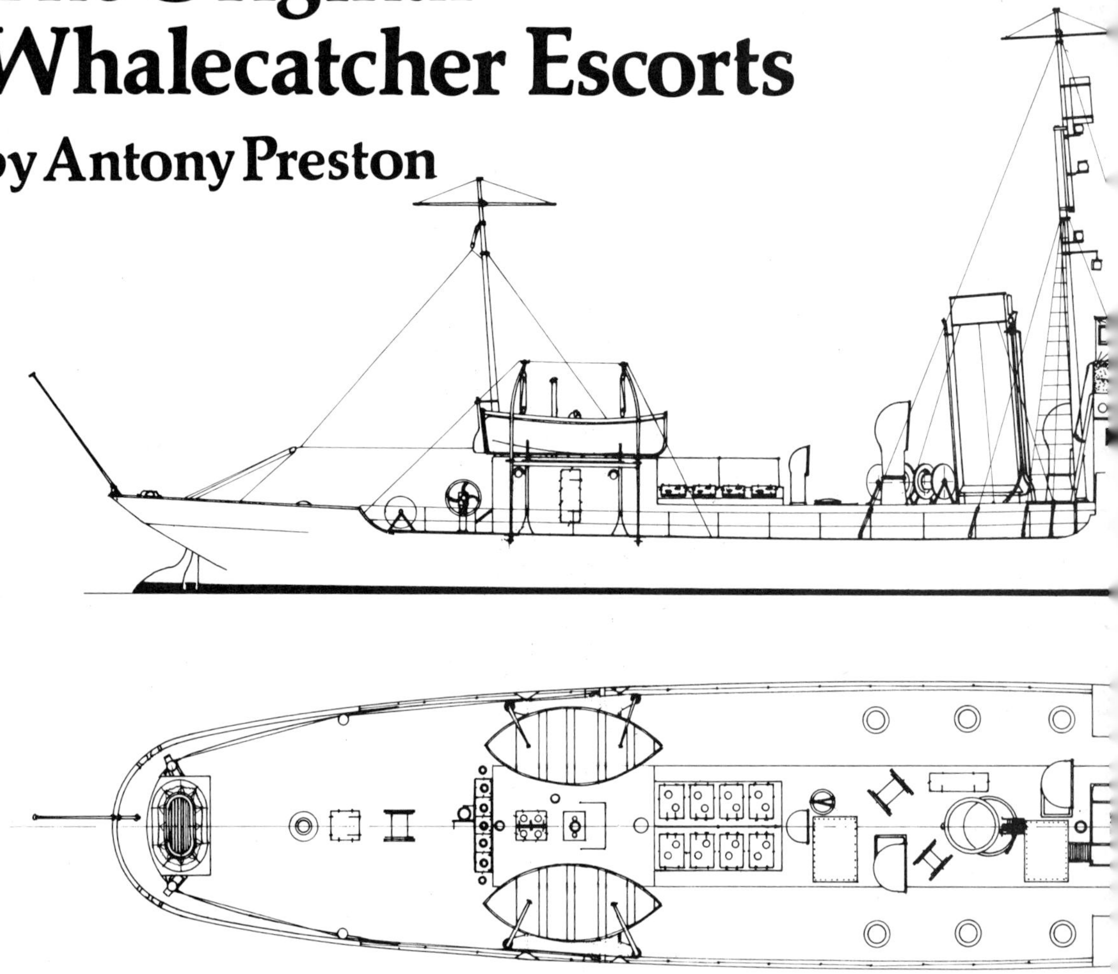

The story of the First World War 'Flower' class sloops is very well known as is the tale of their illustrious successors, the 'Flower' class corvettes of a generation later but it is not widely known that another escort class was based on a whalecatcher design.

Early in 1915 the threat from U-Boats was driving the British Admiralty to ask for patrol craft to relieve destroyers from escort duties. This led to the officially designed 'P-Boat', essentially a utility warship with steam turbines and a good turn of speed, but there was also a need for a slower craft for escort and patrol work. Opinion favoured the whalecatcher, renowned for its manoeuvrability rather than its speed, and so on 15 March 1915 the Admiralty ordered Smith's Dock Company of Middlesbrough to stop all work in hand to leave themselves free to build 15 new escorts as fast as possible. The plans were to be provided by the departments of the Director of Naval Construction (DNC) and Engineer-in-Chief (E-in-C) but they were in fact adapted from a design ordered by the Imperial Russian Government.

DESIGN AND CONSTRUCTION

Delivery of the first vessel was planned to be three months after approval of the plans, followed by the remaining 14 at 10-day intervals, subject to the delivery of boilers and ancillary equipment. The principal features of the design were:

1. rapid manoeuvring;
2. minimum draught, to reduce risk from mines and torpedoes;
3. maximum speed possible on the displacement;
4. a stiffened stem to allow for ramming U-Boats;
5. a raised forecastle, stiffened to carry a 12pdr (3in) gun;
6. subdivision into six watertight compartments and a watertight lower deck abaft the machinery, for maximum safety on the displacement.

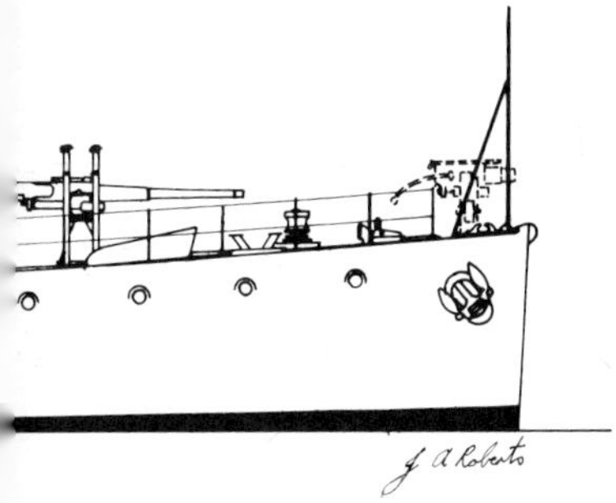

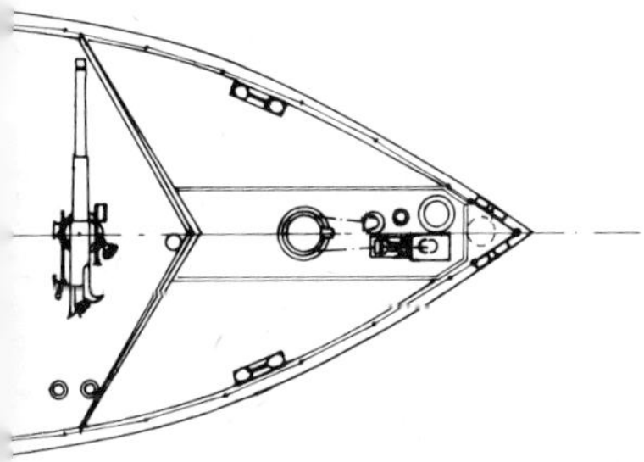

The hull was constructed of mild steel and propulsion comprised a single-shaft triple-expansion surface-condensing engine (cylinders 14.5in, 23.5in and 39in dia × 24in stroke), working at 200 rpm on full power. The propeller was 8ft in diameter and a pair of coal-fired White-Forster water-tube boilers provided steam at a working pressure of 200lbs per sq in. A total of 60 tons of coal was carried in bunkers abreast of the boiler room.

The auxiliary machinery comprised a steam steering engine and a steam-driven generator supplying current for lighting and the wireless. For the record, electrical power was provided by an inverted single-cylinder engine coupled directly to a multi-polar compound-wound dynamo giving 152 amps at 100V. It provided current for 70 lights, a 20in searchlight and a 15amp Type IV W/T set similar to that provided for contemporary destroyers.

The cost of the first eight craft (designated *Z1—8)* averaged £27,195 each but this was exceeded in the last seven as the machinery had to be sub-contracted to avoid delays.

ARMAMENT

The 12pdr 18cwt gun was fitted on the forecastle deck immediately over No2 bulkhead and stowage was provided in the magazine and shellroom below for 300 rounds. The Q-Ships were already in being and it was felt that these armed whalecatchers should also try to look as innocuous as possible. The 12pdr was arranged to stow athwartships under a canvas cover and a dummy harpoon gun was provided to complete the disguise. For the same reason the searchlight on the navigating bridge was fitted with a canvas screen painted to resemble a wheelhouse.

Apart from *Zedwhale (Z1)*, which mounted an additional 6pdr (57mm) anti-aircraft gun, Admiralty records show no addition to the listed armament but they were undoubtedly given two depth-charges when these were issued in 1916 and the outfit presumably increased as more DCs became available in 1917–18.

THE CLASS

The first to be completed was *Z1*, launched on 29th May 1915 and completed on 20th July 1915. Her sister *Z2* was launched on the same day and completed on 11th August. *Z3* and *Z4* were launched on 26th June and completed 31 and 9 August respectively. *Z5* and *Z6* were launched on 28 June and completed on 18 August and 11 September. *Z7* and *Z8* were launched on 28 July 1915 and completed on 11 and 21 September. *Z9* and *Z10* were launched on 12 August and completed on 9 and 3 October. *Z11* and *Z12* were launched on 10 September and completed on 16 and 23 October. *Z13* and *Z14* were launched on 24 September and completed on 30 October and 3 November, while the last of the class, *Z15* was launched on 7 October and completed on 9 November, 1915.

Before launching, the following names were given to the class – *Z1* became *Meg*, but in January 1916 she was renamed *Zedwhale. Z2* became *Balena, Z3* became *Beluga, Z4* became *Bowhead, Z5* became *Blackwhale, Z6* became *Bullwhale, Z7* became *Cachalot, Z8* became *Cowwhale, Z9* became *Humpback, Z10* became *Rorqual, Z11* became *Rightwhale, Z12* became *Icewhale, Z13* became *Finwhale, Z14* became *Pilotwhale* and *Z15* became *Arctic Whale.* They were formed into 3 squadrons, based at Stornoway, Shetland and Peterhead, although from time to time some were stationed in the Humber.

Although they lived up to their reputation for manoeuvrability their seaworthiness proved something of a disappointment, and was considered inferior to trawlers of the same tonnage. The decision was therefore made not to continue with the type, although in April 1915 another four whalers had been bought from Norwegian owners, namely – *Eclair* renamed *Charon; Krabben* renamed *Salamander; Splint* renamed *Rattler* (June 1915), later *Prattler; Etoile* renamed *Styx.*

With other whalers, hired earlier, these vessels served in East Africa but after the Armistice they reverted to their original owners.

Only one of the class was lost, the *Blackwhale,* mined off Fife Ness on 3 January 1918. The first to be sold was the *Zedwhale,* on 12 February

1920. Next were the *Bullwhale, Cowwhale, Finwhale, Humpback, Rightwhale* and *Rorqual* a month later, and the *Arctic Whale, Balena, Beluga, Bowhead* and *Pilotwhale* a month after that. Two survived in the post-war fleet as tenders, the *Icewhale* becoming HMS *Osprey* in March 1924. The *Osprey* was sold in October 1928 but the *Cachalot* survived until 1933.

It is ironic that these humble little escorts should have been found wanting in seakeeping for when, in 1938, Smith's Dock Company was again approached by the Admiralty for help in producing a utility escort, the genesis of the 'Flower' Class corvettes, their experience with the 'Z-Whalers' prompted them to put forward their latest whalecatcher, the *Southern Pride*. And to cap that, the senior designer at Smith's Dock was the man who had designed the original whalers. It had taken a long time to reap the full benefit of his experience.

PARTICULARS

Displacement:	336 tons (normal), 346 tons (full load), 238.8 GRT
Length:	125ft (pp), 132ft 9in (oa)
Beam:	25ft
Draught:	8ft 7in (mean)
Machinery:	1-shaft TE, 1,200ihp = 13kts
Complement:	26

The *Cachalot* during 1922/23, minus her 12pdr gun
NMM

The CAVOUR & DUILIO Class Battleships

by Giorgio Giorgerini

The concept of the single calibre battleship was developed in Italy by Captain Eng Vittorio Cuniberti before the battle of Tsushima. In fact, he wrote about this subject in the German naval magazine *Marine Rundschau* and more deeply in the Italian *Rivista Marittima* in 1902, recommending a fast armoured ship of at least 20kts and exclusively armed with 12in (305mm) guns. Cuniberti developed a design too but the Italian Admiralty, after a careful examination, decided to reject the project and authorised Captain Cuniberti to publish his study abroad.

The 1903 edition of the British naval annual *Jane's Fighting Ships* published an article by Cuniberti entitled *An ideal battleship for the British Navy*. The parallel events which led to the design and construction of HMS *Dreadnought* and the subsequent adoption of the single calibre battleship by all the

Duilio completing at Genoa in 1915

Andrea Doria on completion, at Genoa in March 1915. The differences between the *Cavour* and *Duilio* classes are clearly apparent

main fleets of the world is well known. Oddly, Italy, even though the country of Cuniberti, was the last to build dreadnought type ships. Only in 1908 was it decided to build the first Italian dreadnought, the *Dante Alighieri,* designed by Rear Admiral Eng Edoardo Masdea and armed with 12-12in guns. *Dante Alighieri* was the first battleship in the world designed to have her main armament mounted in triple turrets, a suggestion made by Captain Umberto Cagni and Captain Eng Edoardo Ferrati as early as 1905. Unfortunately, the usual delays of that time in the manufacture and supply of weapons by national and foreign industries (Italy has been subject to foreign supplies in this technical field for a long period of its history) did not allow this advantage to be maintained. Consequently, the Austro-Hungarian triple turret dreadnought *Viribus Unitis,* laid down over one year later and launched 10 months later than *Dante Alighieri,* was commissioned sooner – demonstrating the advantages of guns designed and manufactured by a national industry.

TABLE 1: TRIALS FIGURES

	Displacement (tons)	Power (HP)	Speed (knots)
Cavour	21,612	31,278	22.2
Cesare	21,700	30,700	21.56
Leonardo Da Vinci	24,700	32,800	21.6

TABLE 2: PARTICULARS OF GUNS

	12in/46	4.7in/50	3in/50	3in/40	6in/50
Year of Construction	1909	1909	1909	1917	1911
Weight of barrel (tons)	64.1	3.7	1.1	0.7	6.5
Maximum elevation	+ 20°	+ 20°	+ 20°	+ 70°	+ 20°
Maximum depression	− 5°	− 5°	− 10°	− 10°	− 5°
Muzzle velocity (ft/sec)	2757 (840m/sec)	2817 (850m/sec)	2676 (815m/sec)	2493 (760m/sec)	2724 (830m/sec)
Maximum range (yds)	26,240 (24,000m)	13,120 (12,000m)	10,934 (10,000m)	9,840 (9,000m)	18,600 (17,000m)
Weight of projectiles (lbs)	996.5 (452kg) AP	48.7 (22.1kg)	13.2 (6kg) AP	14 (6.3kg)	104 (47kg)
Rate of fire (Rounds per minute)	1	6	10	10	6

The forecastle of *Andrea Doria* seen from the bridge

On the other hand, the Italian Navy in designing its warships had to keep in mind the political and strategic situation of the time. Italy was allied with Germany and the Austro-Hungarian Empire and the main potential enemy fleet was that of France, possibly reinforced by units of the British Navy. Therefore, comparison of ships should be between Italian and French vessels, although later events brought Italy into the First World War on the side of the Allied powers. Nevertheless, *Dante Alighieri* had just been laid down in 1909 when the Italian Admiralty realised that they risked having a ship that would already be obsolete on completion. Soon after, therefore, the Regia Marina projected a new type of battleship developed from the *Dante Alighieri* but incorporating ideas inspired by the most recent foreign ships derived from HMS *Dreadnought.*

THE BIRTH OF CAVOUR CLASS

The new battleship was to be of increased displacement to allow for additional protection and armament. The maximum calibre of the main armament was to remain at 12in (305mm), as in *Dante,* because this type of gun was considered a good compromise of penetrating power, weight and rate of fire, but the number of main guns was to be increased from the 12 of *Dante* to 13. Specified speed was 21 to 22kts.

Rear Admiral Eng Edoardo Masdea, Chief Constructor of the Royal Italian Navy, was instructed to prepare detailed designs for the new battleships. The project study was soon completed and tenders for the construction of 3 new battleships were immediately approved. By the law of 2 December 1909, the new ships were listed among the ships of the Italian Navy and christened *Conte di Cavour, Giulio Cesare* and *Leonardo da Vinci.* They were ordered respectively from the Spezia Navy Yard, Ansaldo SB Co and Odero SB Co at Genoa, and laid down in the summer of 1910.

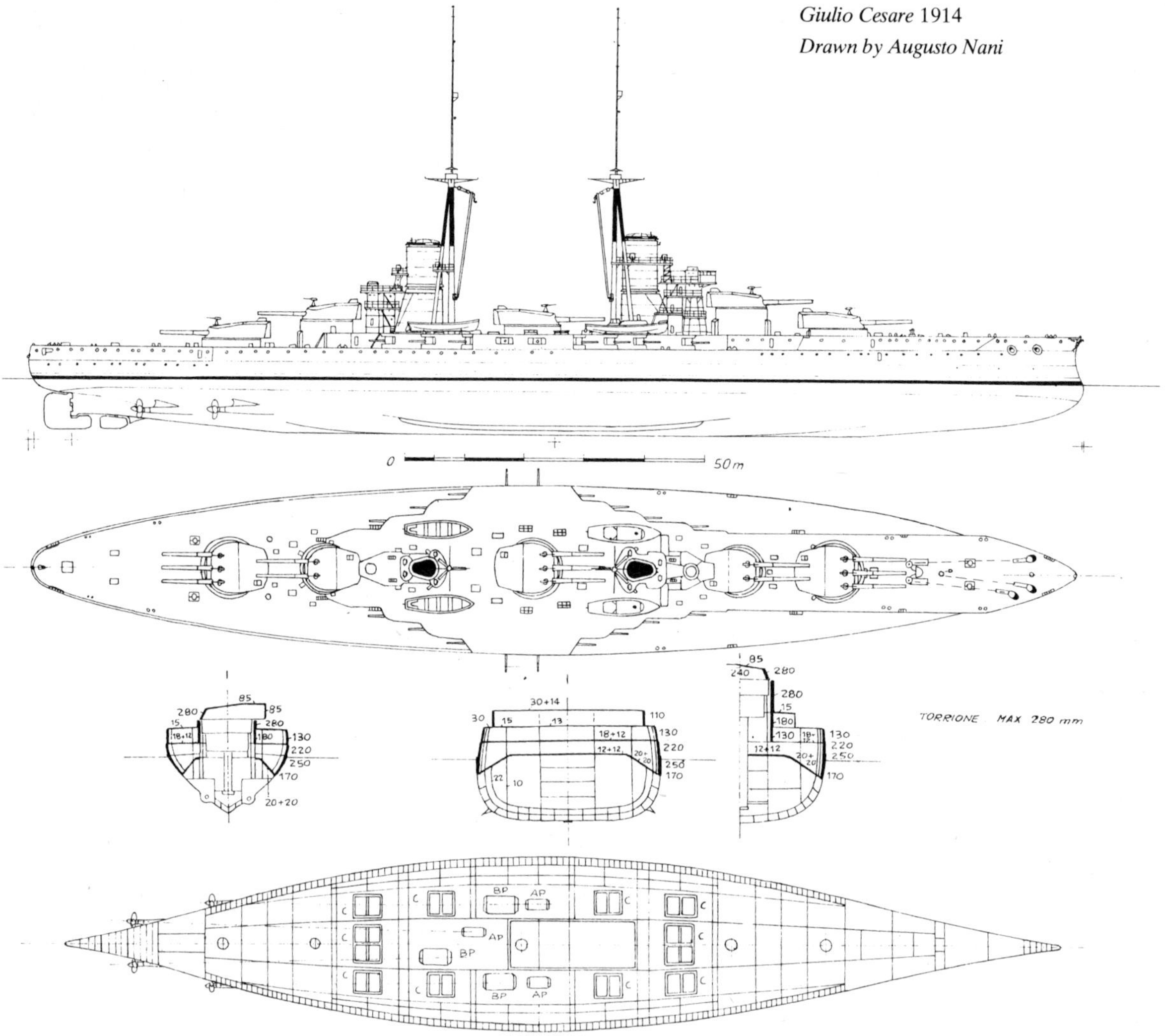

Giulio Cesare 1914
Drawn by Augusto Nani

The *Conte de Cavour* class were to be the most powerful battleships of the Royal Italian Navy, capable of standing with the best of existing foreign designs and in particular were expected to show some superiority in comparison with the contemporary French battleships of the *Courbet* class. However, even as their first steel plates were laid on the stocks they were being surpassed by new British and American dreadnought designs armed with 13.5in (343mm) and 14in (356mm) guns. This was compounded by construction delays due mainly to additional work loads on the yards as a result of extra maintenance and repair work on operational warships during the 1911–12 Italo-Turkish war. Technical difficulties were also encountered in the production of armour plates and the 12in guns. The ships were not, therefore, completed until 1914–15 so, when the class were commissioned, other navies had dreadnoughts of the same age but more heavily armed, better protected and, in some cases, faster.

HULL AND SUPERSTRUCTURE

The hull lines were extensively tested in the experimental tank during the project study and, for the first time in Italy, the ram shaped bow was abandoned in favour of a clipper ram which was curved forward only to increase the length to beam ratio. The stern was of the usual rounded cruiser form with two rudders, of different areas, placed centrally on the longitudinal axis of the hull. The larger had a surface area of 343.58 sq ft (32 sq m) and the smaller 107.37 sq ft (10 sq m).

The hull, for the most part in high-tensile steel, had a complete cellular double bottom divided into many cofferdams by 23 longitudinal and transverse bulkheads. There were three full decks – the armoured deck, main deck and upper deck. The superstructure consisted of a long forecastle containing the armoured battery of 4.7in (120mm) guns and superimposed above this were two

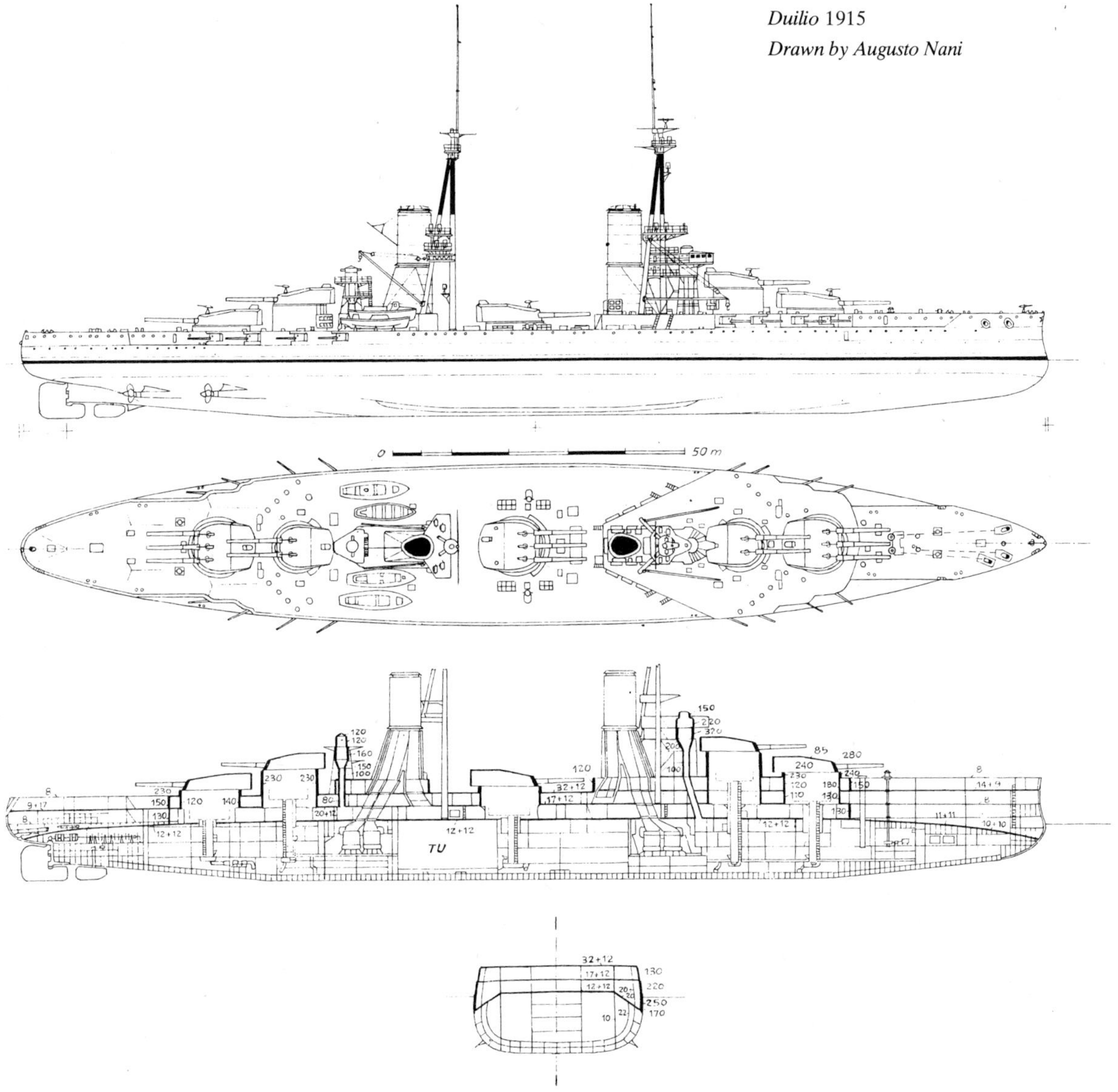

Duilio 1915
Drawn by Augusto Nani

tripod masts and two high funnels, well distanced from each other. The armoured conning tower was placed between B turret and the forward mast and funnel. Q turret, one of the three triples, was installed at the middle of the ship. As a whole the superstructure was kept simple and rational to avoid unnecessary top weight and allow for wider arcs of fire for the main and secondary armament.

A secondary command post was situated between Y turret and the after funnel and mast. The living spaces were well divided but large enough to offer comfortable habitability, by the standards of the time, to both officers and men.

PROTECTION

The ships were protected by a complete water line belt with a thickness of 9.8in (250mm) amidships reducing to 5.1in (130mm) and 3.1in (80mm) towards the stern and bow respectively. The armoured belt extended 3.9ft (1.20m) above and 5.2ft (1.60m) below the water line. From the upper edge of the armoured belt to the lower edge of the main deck there was a complete upper belt of 8.7in (220mm) thickness, 7.5ft (2.30m) high, and above this a third strake of 5.1in (130mm) thickness extending to the upper deck over a length of 452.7ft (138mm), from the bow to X turret, and protecting the 4.7in battery.

The complete sloping armoured deck had a maximum thickness of 1.6in (40mm) on the slopes and 0.94in (24mm) on the flat. Between the main and armoured decks there was another protected deck, running from A to X turrets, with a thickness of 1.2in (30mm). Two transverse armoured bulkheads joined the belt to the decks.

TABLE 3: CONSTRUCTION PARTICULARS OF CAVOUR AND DUILIO CLASS BATTLESHIPS

	Builder	Laid Down	Launched	Completed
Conti Di Cavour	Spezia NY	10.8.1910	10.8.1911	1.4.1915
Guilio Cesare	Ansaldo SB Co	24.6.1910	15.10.1911	14.5.1914
Leonardo Da Vinci	Odero SB Co	18.7.1910	14.10.1911	17.5.1914
Duilio	Castellammare di Stabia DY	24.2.1912	24.3.1913	10.5.1915
Andrea Doria	Spezia NY	24.3.1912	30.3.1913	13.3.1916

TABLE 4: PARTICULARS OF CAVOUR AND DUILIO CLASS BATTLESHIPS

	CAVOUR	DUILIO
Displacement	23,088 tons normal; 25,086 tons full load	22,964 tons normal; 25,200 tons full load
Length	577ft 3in oa; 554ft 8in pp	577ft 2in oa; 554ft 8in pp
Beam	91ft 10in	91ft 2in
Draught	28ft 7in (mean)	29ft 5in mean
Machinery*	20 water-tube Blechynden boilers (8 oil and 12 coal); 3 groups Parsons turbines; 4 shafts (3-bladed propellers); 31,000shp = 21.5kts; fuel – 570 tons coal + 350 tons oil normal, 1450 tons coal + 850 tons oil full load; endurance 4800 miles at 10kts	20 water-tube Yarrow boilers (8 oil and 12 coal); 3 groups Parsons turbines; 4 shafts (3-bladed propellers); 32,000shp = 21.5kts; fuel 606 tons coal + 314 tons oil normal, 1488 tons coal + 886 tons oil full load; endurance 4800 miles at 10kts
Armament	13-12in/46, 18-4.7in/50, 14-3in/50, 3-17.7in torpedo tubes	13-12in/46, 16-6in/45, 19-3in/50, 2-17.7in torpedo tubes
Armour	belt 9.8in max; deck 1.6in to 0.94in; turrets and CT 11in max	belt 9.8in max; deck 1.6in to 0.94in; turrets 9.05in max; CT 11in max
Complement	31 officers, 969 petty officers and ratings	31 officers. 969 petty officers and ratings

**Guilio Cesare* had 24 water-tube Babcock & Wilcox boilers (12 oil and 12 coal)

The front of the main turrets had a thickness of 11in (280mm) and the sides 9.4in (240mm), while the barbettes had 11in (280mm) armour. The sides of the forward conning tower were 11in (280mm) thick and that of the after, secondary, command post had a thickness of 7in (180mm).

The armour plates for the hull were fixed on a wooden backing except abreast the forecastle where they were applied directly to the side plating. The plates were manufactured in Krupp hardened nickel-steel processed by the Terni steel works from raw material (nickel steel) manufactured and supplied by two American and one British firm: the Bethlehem Steel Co, the Carnegee Steel Co of Pittsburgh and the Glasgow Steel Co Ltd.

The weight of the armour steel in each of the *Cavour* class battleships was 5150 tons, that is just over 25 per cent of the designed displacement. The total weight of the armour system as a whole was 6122 tons corresponding to 30.2 per cent of design displacement.

MACHINERY

The Italian Navy has always attached great importance to the speed of its ships and the Regia Marina wanted to give the *Cavour* class 1.5 to 2kts more speed than the standard 20 to 21kts of contemporary dreadnoughts. However, even though the *Cavour* class reached about 22kts on trials, they effectively failed to meet the required speed goal and were outstripped by recent foreign battleships (ie, the British 25kt *Queen Elizabeth* class) even as they commissioned.

The boiler plant consisted of 20 water-tube Blechynden boilers in *Cavour* and *Leonardo* and 24 Babcock & Wilcox boilers in *Cesare*. In each ship 12 of the boilers were combined coal and oil fired and 8 (12 in *Cesare*) exclusively oil fired. Stowage for coal and oil was provided below the armoured deck and in the double bottom, where there were also storage tanks for drinking, washing and feed water. The boilers were separated in two widely spaced groups, one forward and one aft, with the engine rooms positioned between them.

Three groups of Parsons turbines formed the propulsion plant. Each group was contained in a separate engine room; two at the sides and

The launch of a scout aircraft from *Duilio*

one in the centre. The side groups each consisted of a high pressure and a low pressure turbine (the latter incorporating the astern turbine) driving the two outer shafts and their three-bladed propellers. The third group positioned between the first two, consisted of a high pressure and a low pressure turbine each driving one of the two inner shafts. The engine rooms also contained all the auxiliaries, to ensure the independent running of each group. On full power trials the ships attained the results shown in Table 1.

ELECTRIC PLANT

Electric power, at 110 volts direct current, was supplied by 6 turbo-generators of 150kW grouped in 3 power stations placed inside the citadel below the armoured deck. The power stations fed electricity for all board services.

The three sisters of the *Cavour* class; *Giulio Cesare, Conte di Cavour, Leonardo da Vinci*

TABLE 5: LIST OF WEIGHTS, CAVOUR CLASS BATTLESHIPS

Items	Tons	% for unloaded & dry ship	% for full-load ship
Hull	7017	34.6	28.0
Armour	6122	30.2	24.4
Armament	2817	13.9	11.2
Machinery	1732	8.5	6.9
Equipment	1535	7.6	6.1
Miscellaneous	1057	5.2	4.2
WEIGHT OF UNLOADED SHIP	20,280	100.0	80.8
Ammunition	725		2.9
Provisions	175		0.7
Crew and kit	200		0.8
Stores	106		0.4
Drinking, washing and feed water	400		1.6
Auxiliary machines, spares, grease, lub-oil, etc	600		2.4
Coal and fuel oil	2300		9.2
Water for boilers, reserve	300		1.2
FULL LOAD DISPLACEMENT	25,085		100.0

The displacements figures shown in the summaries of technical data for the *Cavour* and *Duilio* classes are an average of values calculated in several technical papers of the engineering services of the Italian Navy and of the building yards. It is a common habit to give detailed displacement figures to the decimal or to the unit. The author has always maintained that this habit is, at least, questionable. Detailed displacements are only theoretical assumptions because ships are continuously subjected to fluctuations of weights. Therefore, in case of *Cavour* and *Duilio* classes (and other ships) it would be sufficient to describe them as ships of 23,000 tons normal displacement, 25,000 full load and 20,000 in designed or light displacement.

ARMAMENT

The 12in/46 cal (305mm/46) guns of the *Cavour* class were of Armstrong design, mod 1909. Those of *Cavour* herself were manufactured in Italy by Vickers-Terni, while those of *Leonardo da Vinci* were built by Armstrongs in the UK and those of *Cesare* by the Italian branch of Armstrongs at Pozzuoli (Naples) with the assistance of the British parent company. The delivery of these weapons was greatly delayed, the average being over one year. For *Cavour* this delay should have been even longer but, by chance, were completed when *Doria* was still far from being completed so it was decided to install them onboard *Cavour*.

1 Three of *Duilio's* 3in/50 guns

2 The forward turrets of *Andrea Doria*

TABLE 6: THE MANOEUVRABILITY OF THE CAVOUR CLASS BATTLESHIPS
The turning circle of the *Cavour* class battleships, when steering with the main rudder only, was:

Angle of helm	**20°**	**25°**	**30°**	**35°**
At a speed of 10kts (yds)	1320	1216	1135	1068
At a speed of 20kts (yds)	1450	1384	1280	1120

The time taken to turn 360° with 35° of helm, was 8min 20sec at 20kts and 14mins 35sec at 10kts (initial speed).

Conte di Cavour at Taranto on 9 June 1917
IWM

The main armament was distributed in turrets placed along the centre line in such a way as to allow the greatest volume of fire possible on all bearings especially forward, aft and abeam where 5, 5 and 13 guns could be brought to bear respectively. A, X and Q turrets were triple mountings and B and Y turrets twin mountings, the latter being superimposed over A and X mountings. The arcs of fire were 300° for A and X turrets, 310° for B and Y and 360° for Q. The normal ammunition allowance for each 12in gun was 40 AP (armour piercing) projectiles and 30 HE (high explosive) fragmentation shells but stowage was provided for a maximum of 100 rounds per gun.

The secondary armament consisted of 18-4.7in/50cal (120/50) Vickers, mod 1909, mounted in a battery located under the forecastle and grouped amidships. Eight could fire forward, 6 aft and 9 on each broadside. Magazine stowage for this battery was 3,600 rounds.

The ships also carried 14-3in/50 cal (76mm/50) Vickers, mod 1909, guns in open mountings distributed on the tops of the turrets and on the forecastle and upper decks. They could be mounted in 30 different positions.

Doria arriving at Taranto on 21 January 1917 after two months on station at Corfu

Giulio Cesare during her trials in 1914

The magazines were in three groups – one forward, one amidships and one aft – all situated below the armoured deck. The ammunition supply for the main armament was hydraulically powered while the main guns were trained and elevated by an electro-hydraulic system. The armament was completed by 3 submerged 17.7in (450mm) torpedo tubes, two forward on the broadside and one aft.

Soon after the First World War the 3in/50 guns were reduced to 13 and concentrated on the tops of the turrets, while 6 new 3in/40 (76mm/40) AA Armstrong, mod 1917, guns were installed abreast the after funnel. The AA armament was then completed with two single 2pdr (40mm/39) Vickers guns installed on the forecastle abreast B turret. The fire control system was also improved after the war, with more modern range finders, target indicators, etc. The armament of the *Cavour* class was not otherwise altered until their reconstructions of the 1930s.

MODIFICATIONS

As built the ships were very similar to each other, only *Leonardo da Vinci* differed a little from her sisters, in the number of derricks she carried and in some details of her bridge structure. The first modification of *Conte di Cavour* and *Giulio Cesare* was in 1925: the forward tripod mast was removed and replaced by a quadruped mast placed in a position between the bridge structure and the funnel. The vertical part of the original tripod mast was cut off just above the upper edge of the funnel and served as derrick post. At the same time both ships were equipped with an aircraft catapult mounted on the port side of the forecastle deck. It was not trainable. The crane for raising and lowering the scout seaplane was fitted on the port battery deck. A total reconstruction of these two ships was begun in the autumn of 1933.

OPERATIONAL LIFE

Wartime service of these battleships was very modest. The Adriatic naval war did not require the employment of the battlefleet, in fact it was exclusively a conflict of light vessels, submarines, patrol craft and fast torpedo boats. Only occasionally were the battleships taken to sea. *Conte di Cavour* carried out only three war missions and *Giulio Cesare* only two for a total of 40 and 31 hours respectively. The *Leonardo da Vinci* suffered a different and unfortunate fate. Based at Taranto, she was moored in the Mar Piccolo when on the night of 2 August 1916, a sudden explosion was followed by a fire which reached one of the magazines causing the ship to blow-up and capsize: 248 lives were lost. The loss of the *Da Vinci,* like that of the pre-dreadnought *Benedetto Brin* at Brindisi on 25 September 1915, was attributed to sabotage organised by Austro-Hungarian Intelligence. This fact was later confirmed on investigation by officers of the Italian navy counter-intelligence service who discovered the details of a complete sabotage ring operating against Italian warships from the centre of Austrian intelligence in Zurich, Switzerland. As a result several people were arrested, court-martialed and shot. *Leonardo da Vinci* was refloated in 1919 and righted in 1921, but plans to rebuilt her were abandoned and she was finally scrapped in 1923.

CONCLUSIONS

From a general point of view the *Cavour* class battleships can be considered good ships for their time in meeting the requirements of the Italian Navy. They were fast enough, of good manoeuvrability, excellently sub-divided, of good habitability and armed fairly well considering the comparatively light calibre main guns. Many people insist on comparing these vessels with the Austrian-Hungarian battleships of the *Viribus Unitis* class, forgetting that the *Cavour* class were designed to face French ships of the line. However, comparing the *Cavour* to the *Viribus Unitis,* it can be seen that the former were superior in main armament and speed, radius and habitability and the latter in protection, secondary armament and, perhaps in fire effectiveness and manoeuvrability. Anyway, the *Cavour* class did not require improvements or extensive modifications during the first period of their lives, except for those minor changes already mentioned.

DUILIO AND ANDREA DORIA

The 3 battleships of the *Cavour* class were considered insufficient in numbers by the Italian Admiralty and in 1911 it was decided to increase the number of modern dreadnoughts by 2 ships, largely because France, which was already building its first 4 dreadnoughts – the *Courbet* class – was designing a new class of 3 battleships – the *Bretagne* class.

Responsibility for the new design was given to Vice Admiral Eng Giuseppe Valsecchi, one of the best naval engineers of the Royal Italian Navy. He examined the designs of the *Cavour* class carefully and concluded that it was sufficient to introduce only a few minor alterations and modifications. This view was accepted by the naval authorities and the subsequent decision to repeat the main essentials of the *Cavour* type saved much time in the design phase, which took less than 6 months, and during construction. The ships, christened *Caio Duilio* and *Andrea Doria,* were ordered from Ansaldo SB Co at Genoa and Spezia NY respectively in September 1911 and laid down in February and March 1912.

The protection, machinery, hull form and main armament was basically as in the *Cavour* class. Changes concerned superstructure, secondary armament and some minor details. The forecastle was shortened and the *Cavour's* armoured battery was abolished permitting a general lowering of all superstructure, including the main armament turrets, giving improved stability. The 12in gun turrets fore and aft were placed closer together and the disposition of the conning tower, secondary command post, funnels and masts were improved slightly. Although basically similar to the *Cavour* class they could be distinguished from them by the placing of the tripod mast before the forefunnel instead of abaft it, by the different distribution of the secondary armament and the lowered amidships turret.

The only modification made to the protection was a slight reduction of turret armour and a slight increase in the thickness of the conning tower.

The *Duilio* class was designed for a maximum speed of 22kts, but this was not attained on trials. Although the ships developed more than the designed horsepower, the highest speed value obtained was 21.3kts.

The major difference between the *Duilio* and the *Cavour* classes was in the secondary armament. The former carried 16-6in/45 (152mm/45) guns, positioned in two groups, one below the forecastle and the other abreast X and Y turrets. This allowed for a better concentration of fire ahead, astern and abeam, but the arrangement did prove a little 'wet' particularly in the case of the after group.

A and X turrets had arcs of fire of 286°, B 306°, Q 360° and Y of 310°. The light armament of 19-3in/50 Armstrong, mod 1909, guns was installed in substitution for the more modern 3in/45 Ansaldo-Schneider, mod 1912, guns which were not ready at that time owing to the usual manufacturing delays. These guns were transferrable and there were 39 positions for them. As in the *Cavour* class, the delivery of the main guns suffered long delays. Magazine stowage was provided for 1144 rounds of 12in, 3440 of 6in and about 3500 of 3in.

From commissioning until 1937, *Duilio* and *Doria* were not subject to modernisation or conversion, receiving only minor modifications. In 1925 their light armament was modified to 13-3in/50, 6-3in/40 AA Armstrong, mod 1917, and 2-2pdr (40mm/39) AA Vickers guns. In 1926 they were fitted with a fixed aircraft catapult on the port side of the forecastle deck for a scout plane. Three, more modern, range finders were mounted, one above the other on the forward conning tower.

The *Duilio* and *Doria* were virtually inactive during the First World War and afterwards were employed on normal fleet training missions in the Mediterranean area.

Even if *Duilio* and *Doria* were far from being revolutionary ships, and retained most of the weak points of their predecessors, they were an improvement on the *Cavour* class; steady, well built and handy. In 1937, after experience gained in the reconstruction of *Cavour* and *Cesare,* it was decided they also should be completely rebuilt on the lines of the modern *Vittorio Veneto* class battleship design, but this will be the subject of a latter article.

1 The launch of *Giulio Cesare* at Genoa, on 15 October 1911

2 *Duilio* during her trials in 1916

ML 2840 & Brave Borderer

The accompanying photographs (excluding the two portrait shots) were taken by the editor at St Peter Port, Guernsey, in 1961. They show two of the Royal Navy's coastal craft; one, the *ML2840* – an old Fairmile B motor launch, nearing the end of its career and the other, the Fast Patrol Boat *Brave Borderer*, a brand new vessel commissioned in the previous year.

The 112ft, Fairmile Bs were 20kt vessels intended forA/S and general patrol work in coastal waters but were also adapted for minelaying, minesweeping, air-sea rescue and many other duties. 656 were build (excluding 16 lost on the stocks) of which 53 were lost during the war. The designed standard armament was 1-3pdr, 2-.303in MG and 12 depth charges but subsequent variations were many. Various mixtures of 3pdr, 20mm and 40mm guns, together with a heavier depth charge armament in some, were ultimately mounted and a few units even carried a pair of torpedo tubes. The majority of the class were sold shortly after the end of the war but a few remained in service with the Royal Navy, until discarded in the late 1950s/early 1960s. The *ML2840* (ex *840*), which served as a minesweeper, until derated to 'tender' in 1960, being one of the last to go.

The 75ton *Brave Borderer,* and her sister *Brave Swordsman,* were designed and built by Vosper at Portsmouth with assistance from the Admiralty's design and experimental departments. Derived from experience with the earlier *Bold* class, they were propelled by three Bristol Marine Proteus gas turbines which provided power for a designed sustained speed of 44kts for 10 hours, although much higher speeds could be attained for short periods. Launched in 1958 and commissioned in1960, they were capable of accepting various combinations of 40mm guns and 21in torpedoes for service as torpedo boats or gunboats and could also undertake a variety of other roles including transport of commando raiders and minelaying. Both vessels were laid-up for disposal in 1975.

ML2840: the bridge viewed from above showing bridge wheel, engine controls and compass with, on the port side, the companion way to the wheel/charthouse

The *Brave Borderer* on trials in 1960 carrying two 40mm Bofors guns on single MkVII mountings and four 21in torpedoes on side launching carriages. The standard armaments quoted in most reference works are one 40mm and four torpedoes as a torpedo boat, two 40mm and two torpedoes as a torpedo-gunboat and two 40mm as a gunboat. None of these match the configuration in the photograph but some experimentation during trials may account for this

Vosper-Thornycroft

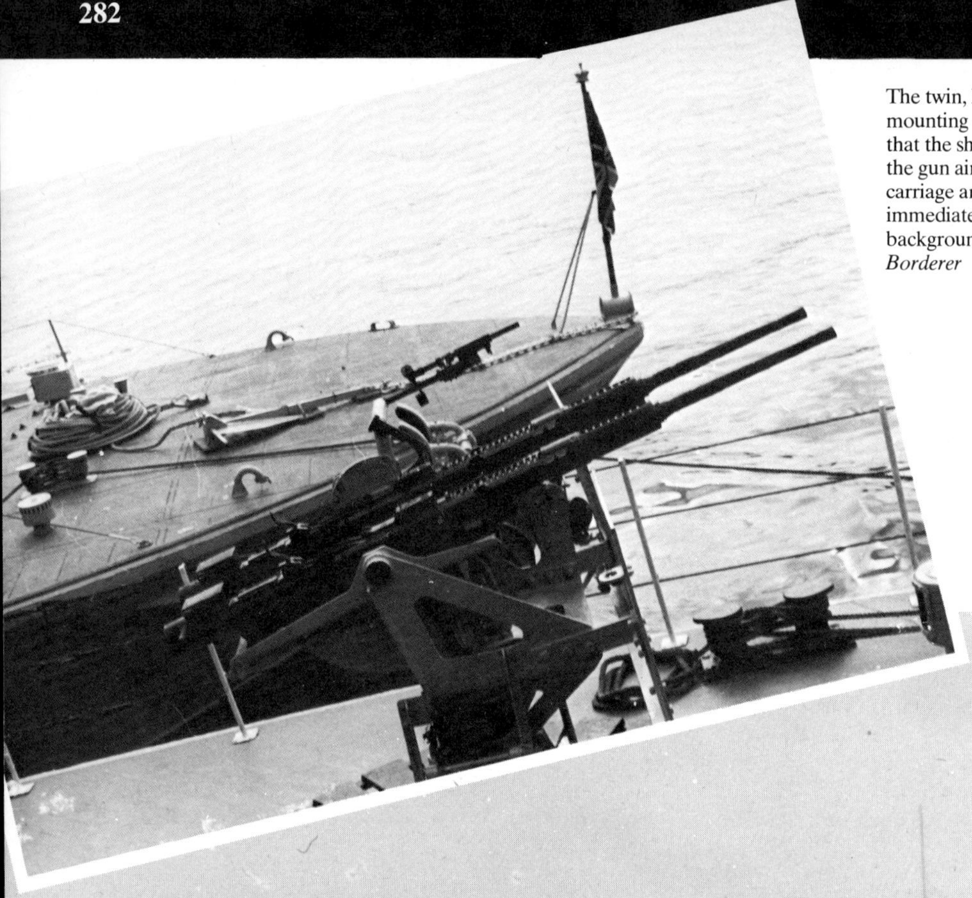

The twin, hand operated, 20mm Oerlikon mounting on the forecastle of *ML2840*. Note that the shoulder rests and sighting gear for the gun aimer are fitted on the left side of the carriage and not, as in the single mounting immediately behind the gun. In the background is the forecastle of *Brave Borderer*

A starboard bow view of *ML2840.* In the background is Castle Cornet which guards the entrance to St Peter Port harbour

ML2840 at Portsmouth in May 1955 while operating as an inshore minesweeper
Wright & Logan

1

2

1 A close up of *Brave Borderer's* after 40mm MkVII mounting

2 *ML2840* with *Brave Borderer* alongside. The apparent difference in size is misleading as displacements were comparatively close, the *Brave Borderer* being a broadbeamed shallow draught vessel, substantially shorter than her companion (98ft 9in oa x 25ft 6in x 6ft compared with 112ft oa x 18ft 3in x 7ft) and only slightly larger

3 The two vessels viewed from astern. At this time *Brave Borderer* carried an armament of 2 single 40mm Bofors, no torpedoes or launching gear being mounted. *ML2840* still carries the two stern davits fitted for service as a minesweeper

3

Warship

BOUND VOLUMES

With the publication in October 1980 of *Warship Volume III* (containing issues 9-12), the first three bound volumes will be available. Volumes I and II, which went out of print very rapidly after their first publication, have been reprinted, enabling new readers to complete their sets. Each volume contains four individual issues in a durable hardback binding with dustjacket.

(Individual issues from Number 2 to date are also available at £2.50 per copy, plus 50p post and packing.)

VOLUME I	VOLUME II	VOLUME III
(issues 1-4)	(5-8)	(9-12)
260 pages	292 pages	292 pages
£12.50	£12.50	£12.50

(Please add £1.50 per volume for post and packing when ordering direct.)

FROM YOUR LOCAL BOOKSELLER

OR CONWAY MARITIME PRESS LTD, 2 NELSON ROAD, LONDON SE10 9JB

Book Reviews

Modern Warship Design and Development *by Norman Friedman, Conway Maritime Press, London 1979* (Price £9.50)

The evaluation of individual classes has always been difficult; one of the primary aims of *Warship* is to provide people with a more firm basis for such comparisons, but it remains an elusive problem on account of the number of variables in the equation. This is more true of modern warships than ever before, partly because of security but mainly because the most important features of a ship are unseen.

This book has been put together to help solve some of the perplexities: it is not a guide to building your own warship, nor is it an abstruse discussion of theoretical naval architecture. It is a concise explanation of the factors which influence warship design and it goes a long way to explain just why modern warships turn out the way they do, and why they cost so much for so little apparent punch.

Dr Friedman takes up his story with the revolution brought about by electronics and electronically controlled weapons. He shows how steel has become a very small component of cost, whereas any improvement in weaponry such as the addition of a second fire-control system sends the cost spiralling upwards. The first case cited is the attempt in 1961 to design an 'austere' escort for the US Navy with only 12 Tartar SAM rounds, on the assumption that cost was proportional to number of missiles carried. The addition of an improved sonar and 1500 miles' endurance was enough to send costs up by more than 50 per cent over the original estimates. What many critics overlook in such cases is that the growth in size to accommodate the equipment has added proportionately little to the total cost so if the chance is taken to improve habitability this has not turned the ship into a 'pleasure yacht'. On the contrary, seaworthiness and endurance can be improved, onboard maintenance is facilitated and, if incidentally recruiting is made a little easier, the customer has gained much for very little extra cost: the expensive items were necessary anyway to allow the ship to carry out her military tasks.

All readers will be interested in the comparisons between Soviet and Western ships. The author does not pour scorn on Soviet designs but he does put them into perspective, pointing out that they differ in appearance and style for the simple reason that they are designed in a totally different way. For a start Russian maintenance relies heavily on shore support, not least because they are chronically short of trained manpower. With fewer maintenance ratings carried on board and rather more primitive standards of habitability crew-spaces can be smaller and the ship can therefore be smaller for a given payload. In many modern classes the risk of a radar set, for example, going 'on the blink' at a crucial moment is insured against by the expedient of putting in a second one. This costs more than mere money: each radar will have significant blind spots and will make heavy demands on space and electrical power.

His main worry in all this is not the underarmed *appearance* of Western ships, for as he shows repeatedly, the efficiency of modern weapon systems cannot be gauged by counting gun-barrels and missile-launchers. After all, major Western escorts are intended to make full use of that excellent but virtually invisible weapon system, air strike. His fear is that the small number of anti-ship missiles carried in Western ships (8 Harpoons in a *Spruance,* 4 Exocets in a *Leander,* for example) could result in a 'High Noon' shoot-out; your opponent holds his fire until you have fired your 8 rounds at decoys and other ships and then closes in for the kill.

The book is well illustrated, both with fully captioned photographs and drawings by John Roberts showing various internal layouts. The various headings cover the major divisions: propulsion, sensors, air defence, A/S defence, guns, protection and future 'unconventional' types of ship. Well worth reading by anyone with an interest in modern navies, and I would go so far as to recommend it as standard reading for naval officers as well. Ignorance of basic concepts of ship-design is common enough, but it is all the more worrying when it is displayed by people in uniform. Warship-design will never be a simple subject, but this book shows that much of it is common sense. All that has really changed since 1945 is the priorities of ship-design, complicated by the fact that nobody has the slightest idea of how modern fleets would fare in battle. We can only echo the author's final wish, that Western navies will not have to endure another Jutland to provide the experience.

Antony Preston

A major new naval book from
Arms and Armour Press

British Cruisers of World War Two

Alan Raven and John Roberts

(authors of British Battleships of World War Two)

The complete development history of the British cruiser from 1912 to the 1950s: *Arethusa* class, early C classes (*Caroline, Calliope, Cambrian* and *Centaur* classes), *Hawkins* class, (*Elizabethans*), later C classes (*Caledon, Ceres* and *Carlisle* classes), D class, E class (*Emerald* and *Enterprise*), County class (*Kent, London* and *Dorsetshire* classes), *York* and *Exeter*, *Leander* class, modified *Leander* class, *Arethusa* class, *Town* class, modified *Town* class, *Dido* class, modified *Dido* class, *Fiji* class, *Uganda* class, *Swiftsure* class and *Tiger* class.

Over 60 sets of line drawings showing designs, as-built and modified configurations (based on official constructors' plans).

Complete data per class – including 20 lists of ship's particulars – and a complete list of modifications to each individual ship.

- 448 large-format 10in x 9¾in pages for maximum size of photograph reproduction.

- £25 net; fully bound with wrap-around jacket painting by John Roberts.

- Over 400 photographs – all carefully selected to show maximum detail and modifications to the 121 cruisers described.

ARMS AND ARMOUR PRESS, Lionel Leventhal Limited, 2-6 Hampstead High Street, London NW3 1QQ. Telephone 01-794 7868 (24 hours).

NAVAL BOOKS

Conway Maritime offer an unrivalled range of authoritative and well-illustrated titles on naval subjects. A free catalogue is available, but some of the leading titles are listed below:

SAILING SHIPS OF WAR 1400-1860
by Dr Frank Howard
A technical history of the wooden warship from the advent of the three-masted ship to the eclipse of naval sail. "... a wealth of information on practically all phases of ship construction ..." — *The Navy*
12"×9½", 256 pages, 150 photos, 32 colour plates, 200 line drawings. ISBN 0 85177 138 6. £12.50 (plus £1.50 p+p)

GREAT SHIPS: THE BATTLEFLEET OF KING CHARLES II
by Frank Fox
The definitive work on Restoration warships, illustrated with a large number of contemporary drawings and paintings. "... packed with detailed fact and equipped with immense amounts of data in several appendices. The superb and numerous colour plates merit particular attention." — *Lloyd's List*
11½"×9¾", 208 pages, 230 photos, 37 colour plates. ISBN 0 85177 166 1. £20.00 (plus £1.50 p+p)

CONWAY'S ALL THE WORLD'S FIGHTING SHIPS 1860-1905
The first complete listing of all warships between the first ironclad and the *Dreadnought*. "... must rank with the all-time great naval reference works ..." — *The Navy*. "... all the thoroughness and attention to detail we have come to expect from Conway Maritime ... excellent value". — *Ships Monthly*
12¼"×8½", 448 pages, 471 photos, 506 line drawings. ISBN 0 85177 133 5. £24.00 (plus £2.00 p+p)

BATTLESHIPS OF THE WORLD 1905-1970
by Siegfried Breyer
A spectacular collection of photographs of all capital ships from the *Dreadnought* onwards. "*Battleships of the World* is a lot of book, and it will certainly enthrall all historians, enthusiasts, modellers and general readers alike." — *Airfix Magazine*
11¾"×8¼", 400 pages, 570 photos, 44 line drawings. ISBN 0 85177 181 5. £25.00 (plus £2.00 p+p)

DESTROYER WEAPONS OF WORLD WAR 2
by Peter Hodges and Norman Friedman
A detailed comparison between British and US destroyer weapons, including mountings, directors and electronics. "... one of the greatest possible additions to the ... range of naval books ..." — *The Navy*
9½"×7¼", 192 pages, 150 photos, 73 line drawings. ISBN 0 85177 137 8. £7.50 (plus £1.25 p+p)

BATTLESHIP DESIGN AND DEVELOPMENT 1905-1945
by Norman Friedman
The first layman's guide to the design process and the factors governing the development of capital ships. "... an eye-opening study of an extremely complex business ..." — *Nautical Magazine*
10"×8", 176 pages, 200 photos, plans and line drawings. ISBN 0 85177 135 1. £8.50 (plus £1.25 p+p)

MODERN WARSHIP DESIGN AND DEVELOPMENT
by Norman Friedman
"... never before have the problems and parameters of modern warship design been set out so comprehensively, informatively and clearly ... the book should be read by everyone with a concern for the modern naval scene, professional or amateur, uniformed or civilian." — *Journal of the Royal United Services Institute*
10"×8", 192 pages, 167 photos, 65 line drawings. ISBN 0 85177 147 5. £9.50 (plus £1.25 p+p)

AIRCRAFT CARRIERS OF THE US NAVY
by Stefan Terzibaschitsch
"... a definitive history of the US carrier fleet from 1920 until the present day ..." — *Journal of the Institute of Marine Engineers*
11¾"×8¼", 320 pages, 322 photos, 94 plans and line drawings. ISBN 0 85177 159 9. £15.00 (plus £1.50 p+p)

THE AMERICAN FLYING BOAT
by Richard C Knott
A comprehensive history of all such aircraft in the US Navy. "... one of the best aviation books to come from the United States." — *Air Pictorial*
11"×8½", 262 pages, 123 photos, 5 colour plates, 31 line drawings. ISBN 0 85177 165 3. £15.00 (plus £1.25 p+p)

CAMERA AT SEA 1939-1945
edited by the staff of *Warship*
"A unique collection of some of the best photographs of World War II at sea" — *Sea Power*
12¼"×8½", 192 pages, 250 photos, 24 colour plates. ISBN 0 85177 124 6. £12.00 (plus £1.50 p+p)

SCALE MODEL WARSHIPS
edited by John Bowen
A complete manual covering every aspect of the hobby. "... a most informative book ..." — *Sea Breezes*
9½"×7¼", 192 pages, 147 photos, 179 plans and line drawings. ISBN 0 85177 070 X. £7.50 (plus £1.25 p+p)

THE BRITISH SUBMARINE
by Commander F W Lipscomb
A classic history of the British Submarine Service, revised and expanded to include the developments of the nuclear age.
9½"×7¼", 298 pages, 66 photos, 12 line drawings. ISBN 0 85177 086 X. £6.50 (plus £1.25 p+p)

FROM YOUR LOCAL BOOKSELLER

OR CONWAY MARITIME PRESS LTD, 2 NELSON ROAD, LONDON SE10 9JB
(when ordering direct please add the post and packing charge noted after the price)